Topics in Current Physics

Topics in Current Physics Founded by Helmut K. V. Lotsch

Modern Three-Hadron Physics

Edited by A. W. Thomas

With Contributions by
R. Aaron I. R. Afnan R. D. Amado
D. D. Brayshaw L. R. Dodd E. F. Redish
A. W. Thomas

With 30 Figures

Springer-Verlag Berlin Heidelberg New York 1977

Dr. Anthony William Thomas

TRIUMF, University of British Columbia
Vancouver, B.C. Canada

ISBN-13: 978-3-642-81072-5 e-ISBN-13: 978-3-642-81070-1
DOI: 10.1007/978-3-642-81070-1

Library of Congress Cataloging in Publication Data. Main entry under title: Modern three-hadron physics. (Topics in current physics; v. 2). Bibliography: p. Includes index. 1. Hadrons-Scattering. 2. Scattering (Physics). 3. Problem of three bodies. I. Thomas, Anthony William, 1949--. II. Aaron, Ronald, 1935--. III. Series. QC793.5.H328M6. 530.1'4. 76-44445.

Preface

Our purpose in this volume is to lead the reader from the fundamentals of three-body scattering theory, to a clear understanding of several important areas of current research in three-body theory and its applications. To make this transition as painless as possible, full algebraic derivations are given wherever possible. Thus, while most of the material presented here, has until now, only appeared in research papers or personal notes, the technical level should be suitable for graduate students with some knowledge of scattering theory.

Let us briefly outline the contents. After a thorough development of the fundamentals of three-body scattering theory, we describe the angular momentum reduction of the Faddeev equations (with separable interactions) in the case of arbitrary spin and angular momentum. There follows (in Chapter 2) a discussion of the analytic properties of on-shell three-body scattering amplitudes; a derivation of the minimal equations guaranteeing unitarity in three-body final states (Chapter 3); a complete explanation of the Boundary Condition Formalism for three-body systems (Chapter 4); and a discussion of the relativistic three-body problem (Chapter 5). All of these developments in three-body theory appear in a text book for the first time. In each case, the very simple example of three identical, spinless particles, interacting only in s-waves with separable potentials, is used as an aid to understanding the essential physics.

With regard to applications, nuclear (and intermediate energy) physicists will find a complete chapter (6) devoted to a review of model three-body calculations of direct reactions, as well as an investigation of how three-body techniques may be used to improve many approximations commonly found in the description of real reactions (e.g. stripping, knock-out ...). As a second example, the discussion of analytic structure (Chapter 2) is basic to the theory of dispersion relations, and experience gained in three-body systems should have considerable impact in many fields. Finally, the formalism developed for three-body final states (in Chapter 3 and part of Chapter 5) is directly relevant in such diverse fields as heavy ion reactions, and the extraction of amplitudes for, e.g., $\pi N \to \Delta\pi$ or $N\rho$, for comparison with quark model predictions.

Of course, the limitation to one volume has meant the omission of many topics. For example, we make only passing reference to the excellent works on numerical methods for solving three-body equations. For this the reader is referred to existing texts by WATSON and NUTTALL *(Topics in Several Particle Dynamics)*, and particularly SCHMID and ZIEGELMANN *(The Quantum Mechanical Three-Body Problem)*. Also, the final chapter deals with applications *only* to many-body systems. The true three-body systems are usually covered adequately by the topical "Few Body" conferences - or in the three nucleon case by some excellent reviews (e.g. LEVINGER in Springer Tracts in Modern Physics Vol.71).

In conclusion, it is a pleasant duty to acknowledge some of the debts I have incurred. I would like to thank my collaborator in three-body investigations I.R. AFNAN for innumerable useful discussions; T.E.O. and M. ERICSON for their support during this work - not to mention the tax payers of the CERN member states, and the CERN typing service. The idea of this book arose at the University of British Columbia, and it is a pleasure to acknowledge helpful comments from D. BEDER, E. VOGT and J. WARREN in the planning stage. The authors who wrote most of this text worked very hard to produce excellent manuscripts in a short time, despite my abuse! Finally, I would like to thank my wife for her help and encouragement during the writing and editing of this manuscript, as well as our families in Australia, from whom we have been separated too long - it is to them that my efforts in this work are dedicated.

Geneva, Switzerland
August 1976

A.W. Thomas

Contents

List of Contributors

AARON, RONALD

Dept. of Physics, Northeastern University, Boston,
Massachusetts 02115, USA

AFNAN, IRAJ R.

School of Physical Sciences, Flinders University,
Bedford Park 5042, South Australia

AMADO, RALPH D.

Dept. of Physics, University of Pennsylvania, Philadelphia,
Pennsylvania 19174, USA

BRAYSHAW, DAVID D.

Stanford Linear Accelerator Center, Stanford University,
Stanford, California 94305, USA

DODD, LINDSAY R.

Dept. of Mathematical Physics, University of Adelaide,
South Australia

REDISH, EDWARD F.

Dept. of Physics and Astronomy, University of Maryland,
College Park, Md.20742, USA

THOMAS, ANTHONY W.

Theory Group, TRIUMF, University of British Columbia,
Vancouver, B.C. Canada/V6T 1W5

1. Fundamentals of Three-Body Scattering Theory

I. R. Afnan and A. W. Thomas

There is a considerable volume of literature on the nonrelativistic three-body problem. FADDEEV has devoted an entire book to a careful treatment of the mathematical properties of his equations [1.1]. Clearly a one-chapter introduction must show a good deal of selectivity, both in material and depth of treatment. Our aim is to present a good coverage of the fundamentals of three-particle scattering theory-naturally with some bias to material needed in later chapters.

To facilitate the understanding of the equations, we will deal first with spinless, distinguishable particles. However, having established this base, we then derive the corresponding results for the case of the most general separable interactions with arbitrary spin and angular momentum. Let us emphasize that while mathematically inclined readers will be provided with appropriate references (and in some cases this may be to a later chapter, such as the discussion of analytic properties in Chap.2), our main concern is not the mathematical soundness of the theory. It is hoped that this heuristic approach will aid the assimilation of the important physical ideas.

Briefly the plan of the chapter is as follows. Assuming the reader has some familiarity with formal scattering theory, we show how an integral equation incorporates both the dynamics of the Schrödinger equation, and the boundary condition. Then there is a detailed discussion of the two-body t-matrix, which is an essential ingredient in the three-body theory. The separable interaction is described as a useful special case.

To introduce our notation, as well as a difficulty in the full three-body problem, we discuss in Section 1.2 the case of two particles interacting in the presence of a third noninteracting particle. In Section 1.3 the problem of the conventional (Lippmann-Schwinger) approach to three-body scattering is explained in essentially the same way as Faddeev - via the coordinate space wave function. By assuming the Faddeev decomposition it is shown that the difficulties of incorporating the boundary conditions evaporate. (The coordinate space picture is emphasized here, because it is needed in Chap.4.)

A rigorous derivation of the Faddeev equations for the case of three free particles initially and finally is given in Section 1.4. The equations are written explicitly in momentum space, with some brief discussion of how to extract bound state scattering from the residue. In Section 1.5, this problem of scattering a projectile from a two-body bound state is examined in detail (including possible rearrangement).

The unitarity property (basic to several later chapters) and the inclusion of identical particles are taken up in Sections 1.6 and 1.7. In Sections 1.8 and 1.9 we treat the problem where all pairwise interactions are separable. This includes a discussion of the three (spinless) boson case, as well as the full angular momentum reduction for the general case of arbitrary spin and angular momentum. (Details of this decomposition are to be found in the Appendix). Section 1.10 serves to tie up loose ends by referring the reader to the best descriptions of topics we have perforce omitted.

1.1 Two-Body Scattering

A sound grasp of the two-body problem is essential to a proper understanding of the three-body problem. Indeed, the input to the three-body equations is the complete solution of the two-body problem. As we shall see, this information is not available from purely two-body experiments, but must come from a dynamical theory such as the nonrelativistic Schrödinger equation.

1.1.1 The Schrödinger Equation

Consider the scattering of two particles (of mass m_1 and m_2, momenta $\underline{p}_1$ and $\underline{p}_2$) via the two-body potential $V_{12} \equiv V_3$.[1] The Hamiltonian is then $(K_1+K_2+V_3)$, with K_i the kinetic energy of particle i ($K_i = p_i^2/2m_i$). Introducing the relative and center-of-mass (CM) momenta, $\underline{q}_3$ (see (1.58)) and $(\underline{p}_1+\underline{p}_2)$ respectively, one can separate the uninteresting CM motion [1.2]. What remains is essentially a one-body problem,

$$(H_0 + V_3)\psi \equiv [(q_3^2/2\mu_{12}) + V_3]\psi = E\psi \qquad (1.1)$$

with $\mu_{12} = m_1m_2/(m_1 + m_2)$ the reduced mass of (1.2). (Translational invariance guarantees that V_3 is a function of only the relative coordinates of 1 and 2). In general V_3 can be nonlocal, in which case (1.1) is an integrodifferential equation, but for the moment we assume it is local (i.e., $V_3 = V_3(\underline{r}_{12})$, where $\underline{r}_{12} = \underline{r}_1-\underline{r}_2$).

[1] In agreement with later conventions, we label properties of a pair by the label of the "spectator" particle, i.e., V_3 for the potential, and $\underline{q}_3$ for the relative momentum.

The solution of (1.1) may be expanded in partial wave form as

$$\psi(\underline{r}_{12}) = (2/\pi)^{1/2} \sum_{\ell m} i^{\ell}\psi_{\ell}(r_{12})Y^{*}_{\ell m}(\hat{\underline{k}})Y_{\ell m}(\hat{\underline{r}}_{12}) \tag{1.2}$$

in analogy with the expansion for a plane wave [Ref.1.3, Eq.(6.58)], where ψ_{ℓ} equals $j_{\ell}(kr)$ - the spherical Bessel function. The functions $Y_{\ell m}$ are the spherical harmonics defined in [1.4]. Using this expansion in (1.1), one obtains the usual second-order differential equation

$$\psi''_{\ell} + (2/r)\psi'_{\ell} + [q_3^2 - U_3(r) - \ell(\ell + 1)/r^2]\psi_{\ell} = 0 \tag{1.3}$$

with $q_3^2 = 2\mu_{12}E$, and $U_3(r) = 2\mu_{12}V_3(r)$.

Outside the range of the potential ψ_{ℓ} is a linear combination of the regular (j_{ℓ}) and irregular (n_{ℓ}) solutions of (1.3) with $U_3 = 0$.

$$\psi_{\ell}(r) \rightarrow \exp\,(i\delta_{\ell})\,[\cos\,\delta_{\ell}j_{\ell}(q_3r) + \sin\,\delta_{\ell}n_{\ell}(q_3r)] \quad . \tag{1.4}$$

The wave function inside the potential is obtained by standard numerical integration of (1.3) to a point where U_3 is negligible. Then the logarithmic derivative is matched to that of (1.4) and $\delta_{\ell}(q_3)$ determined. In the next subsection we show that the two-body data determine δ_{ℓ}, whereas (as we see later) for the three-body problem one needs to know $\psi_{\ell}(r)$ inside the range of the potential.

1.1.2 The Differential Cross Section

The usual laboratory scattering experiment has a wave packet incident on a target, which scatters and is detected at some angle to the incident direction. Any strictly correct treatment must start from this basis. The purpose of "formal scattering theory" is to enable a much easier derivation of the correct scattering amplitude. The price paid for this simplicity is that one must accept rather heuristic arguments at a few crucial points. While adopting the more convenient formal approach here, we refer the "doubters" to the very complete discussion in Chapters 3, 4, and 5 of [1.3].

In this time independent picture (for purely elastic scattering) the total wave function is asymptotically

$$\psi(\underline{r}_{12}) \xrightarrow[r_{12}\to\infty]{} (2\pi)^{-3/2}\,[\exp\,(i\underline{q}_3 \cdot \underline{r}_{12}) + f(q_3,\theta)\,\exp\,(iq_3r_{12})/r_{12}] \quad . \tag{1.5}$$

The first term is the wave function of the incident particle, while the second is the (scattered) outgoing wave. The coefficient of the latter is called the scattering amplitude. Now the result of a scattering experiment is usually given as a differential cross section ($d\sigma/d\Omega$) equal to the flux of particles scattered in a given direction per unit solid angle, divided by the incident flux. Confining the system to a box of unit volume, the incident flux corresponding to (1.5) is $(2\pi)^{-3}v$ (with $v = q_3/\mu_{12}$ the incident velocity). The outgoing flux across a surface dS in the θ-direction ($dS = r^2 d\Omega$) is $(2\pi)^{-3}v(|f(q_3,\theta)|^2/r^2)\, r^2 d\Omega$, and hence

$$d\sigma/d\Omega = |f_3(\underline{q}_3,\theta)|^2 \quad . \tag{1.6}$$

Returning to the expression for $\psi_\ell(r)$ in (1.4), and using the asymptotic forms for $j_\ell(kr)[\to\sin(kr - \ell\pi/2)/kr]$ and $n_\ell(kr)[\to\cos(kr - \ell\pi/2)/kr]$ [1.5], one finds that

$$\psi_\ell(r) \to j_\ell(q_3r) + i^{-\ell}\exp(i\delta_\ell)\sin\delta_\ell \exp(iq_3r)/(q_3r) \quad . \tag{1.7}$$

As usual one expands $f(q_3,\theta)$ in the form

$$f(q_3,\theta) = (4\pi/q_3)\sum_{\ell m} f_\ell(q_3)Y^*_{\ell m}(\hat{\underline{q}}_3)Y_{\ell m}(\hat{\underline{r}}_{12}) \quad . \tag{1.8}$$

Finally, substituting the expansions (1.2) and (1.8) in (1.5), and comparing with (1.7) leads to

$$f_\ell(q_3) = \exp(i\delta_\ell)\sin\delta_\ell = \{\exp[2i\delta_\ell(q_3)] - 1\}/2i \quad , \tag{1.9}$$

which proves our assertion that the complete elastic scattering information is contained in the phase shifts.

1.1.3 The Scattering Integral Equation

Thus far the procedure for treating two-body scattering has two steps. First, one solves the differential equation, and then one uses a knowledge of the correct boundary condition to determine the asymptotic form of the wave function. The phase shift is found by matching logarithmic derivatives. Formally it is much more convenient to incorporate both the Schrödinger equation and the boundary conditions in a single equation - the Lippmann-Schwinger equation.

First one rewrites (1.1) as

$$(E - H_0)\psi = V_3\psi \quad . \tag{1.10}$$

Let us call the (plane wave) solution of the homogeneous version of (1.10) ϕ_0 [$\phi_0 = (2\pi)^{-3/2}\exp(i\underline{q}_3\cdot\underline{r}_{12})$]. The free two-body Green's function $G_0(E)$ is then formally defined as $(E - H_0)^{-1}$. More explicity, in coordinate space it is the solution of

$$\{\underline{\nabla}^2/2\mu_{12} + E\}G_0(\underline{r},\underline{r}';E) = \delta(\underline{r} - \underline{r}') \tag{1.11}$$

where we drop the subscripts from both $\underline{\nabla}_{12}$ and $\underline{r}_{12}$, since this should cause no confusion.

Now the general solution of (1.10) is the sum of the particular solution [$\psi_{part} = G_0(E)V_3\psi$] and ϕ_0. While this presents no formal difficulties, as it stands ψ_{part} is in fact meaningless for real E. To see this, write

$$\begin{aligned}\psi_{part} &= \int d\underline{r}' G_0(\underline{r},\underline{r}';E)V(\underline{r}')\psi(\underline{r}') \\ &= 2\mu_{12}\int d\underline{r}'[(q_3^2 + \underline{\nabla}^2)^{-1}\delta(\underline{r} - \underline{r}')]V(r')\psi(\underline{r}') \quad .\end{aligned} \tag{1.12}$$

Substituting $(2\pi)^{-3}\int d\underline{k} \exp[i\underline{k}\cdot(\underline{r} - \underline{r}')]$ for $\delta(\underline{r} - \underline{r}')$, and performing the angular $\hat{\underline{k}}$ integration, one now finds

$$\psi_{part}(\underline{r}) = \frac{i\mu_{12}}{2\pi^2}\int \frac{d\underline{r}'}{|\underline{r} - \underline{r}'|}\int_{-\infty}^{\infty} dk \frac{k \exp(ik|\underline{r} - \underline{r}'|)}{k^2 - q_3^2} V(\underline{r}')\psi(\underline{r}') \quad . \tag{1.13}$$

Clearly this is not defined because of the singularity at $k = q_3$.

To overcome this difficulty we replace q_3^2 by $q_3^2 + i\varepsilon$, and define the physical solution as $\psi^{(+)} = \lim_{\varepsilon\to 0+} \psi(E + i\varepsilon)$. From a practical viewpoint this solves the problem, since the k-integration can now be done by closing the contour in the upper half k-plane. This leads to the result

$$\psi^{(+)}(\underline{r}) = \phi_0(\underline{r}) - \frac{\mu_{12}}{2\pi}\int d\underline{r}' \frac{\exp(iq_3|\underline{r} - \underline{r}'|)}{|\underline{r} - \underline{r}'|} V(r')\psi^{(+)}(\underline{r}') \quad , \tag{1.14}$$

which asymptotically becomes

$$\psi^{+}(\underline{r}) \underset{r\to\infty}{\longrightarrow} \phi_0(\underline{r}) - \frac{\mu_{12}}{2\pi}\frac{\exp(iq_3 r)}{r}\int d\underline{r}' \exp(-i\underline{q}'\cdot\underline{r}')V(\underline{r}')\psi^{+}(\underline{r}') \quad , \tag{1.15}$$

where we have used $\underline{q}' = q_3\hat{\underline{r}}$, since we only consider elastic scattering. Comparing (1.15) with (1.5) we see that ψ has the correct (outgoing spherical wave) boundary conditions, provided we identify

$$f(q_3,\theta) = -4\pi^2\mu_{12}\int d\underline{r}'\phi_{\underline{q}_3'}(\underline{r}')V_3(\underline{r}')\psi^{(+)}(\underline{r}') \quad . \tag{1.16}$$

For a more rigorous justification of this "$i\varepsilon$-procedure" one must return to the wave packet description. We shall use it without further explanation throughout the book to impose outgoing boundary conditions. At no stage of the preceding argument did we need the condition that V_3 was local. Indeed, using Dirac notation for the eigenstates of momentum (ϕ_0 becomes $|\underline{q}\rangle$), with the normalization

$$\langle\underline{q}|\underline{q}'\rangle = \delta(\underline{q} - \underline{q}') \quad , \tag{1.17}$$

one can rewrite (1.16) as

$$f(q_3,\theta) = -4\pi^2\mu_{12}\langle \underline{q}_3'|V_3|\psi^{(+)}_{\underline{q}_3}\rangle \quad . \tag{1.18}$$

Eq.(1.18) is correct for the most general short-range interaction.

For many purposes it is more convenient to work in the momentum representation. This is easily shown from the operator form of (1.14), namely

$$\psi^{(+)}_{\underline{q}} = \phi_{\underline{q}} + G_0(E)V_3\psi^{(+)}_{\underline{q}} \quad , \tag{1.19}$$

to lead to the equation

$$\psi^{(+)}_{\underline{q}}(\underline{q}') = \delta(\underline{q} - \underline{q}') + (E^+ - q'^2/2\mu_{12})^{-1}\int d\underline{q}'' V_3(\underline{q}',\underline{q}'')\psi^{(+)}_{\underline{q}}(\underline{q}'') \quad , \tag{1.20}$$

with $E^+ = E + i\varepsilon(\varepsilon\to 0+)$, and $q^2 = 2\mu_{12}E$.

Finally, we make an observation crucial to the practical utility of (1.19). That is, in order to be useful, one must be able to solve it numerically. Essentially this means replacing it by a finite matrix equation, which yields a unique solution. The condition for this is that the kernel of the equation ["G_0V" in (1.19)] should be compact. When "$i\varepsilon$" is nonzero, one can show that (for reasonable potentials) G_0V is L^2 (i.e., $Tr[(G_0V)^\dagger(G_0V)]<\infty$), and hence compact in the space of square integrable functions (see [1.1, 1.6]).

Unfortunately the proof fails in the limit $\varepsilon \to 0$. Then, as discussed by LOVELACE [1.6] one can show G_0V is compact in the space of bounded, continuous, differentiable functions, with bounded continuous derivatives on $(0,\infty)$. Thus the equations are completely acceptable for practical applications.

1.1.4 The Two-Body t-Matrix

Let us next relate what we have done to the t-matrix. If one keeps $E + i\varepsilon$ (with ε nonzero) in (1.19) and multiplies to the left by $(E + i\varepsilon - H_0)$, it is one line of algebra to show that

$$\psi^{(+)} = \lim_{\varepsilon\to 0+} i\varepsilon G_3(E + i\varepsilon)\phi \quad , \tag{1.21}$$

where the resolvent (fully interacting Green's function) is

$$G_3(E + i\varepsilon) = [E + i\varepsilon - H_0 - V_3]^{-1} \quad . \tag{1.22}$$

One can easily derive an integral equation for the resolvent by using the operator identity

$$B - A = A(A^{-1} - B^{-1})B \quad , \tag{1.23}$$

with $B = G_3(E^+)$ and $A = G_0(E^+)$ (or vice-versa). This is called the resolvent equation

$$G_3(E^+) = G_0(E^+) + G_0(E^+)V_3G_3(E^+) \tag{1.24a}$$

$$= G_0(E^+) + G_3(E^+)V_3G_0(E^+) \quad . \tag{1.24b}$$

The two-body t-matrix (for V_3) can now be defined in terms of G_3

$$G_3(E) = G_0(E) + G_0(E)t_3(E)G_0(E) \quad . \tag{1.25}$$

Substituting this in (1.21) and using $\lim_{\varepsilon\to 0} i\varepsilon\, G_0(E + i\varepsilon)\phi = \phi$, one finds

$$\psi^{(+)} = \phi + G_0(E^+)t_3(E^+)\phi \quad . \tag{1.26}$$

Formally then, one sees that $V_3\psi^{(+)}$ equals $t_3\phi$, where as for the rest of this chapter, we drop the argument "E^+" of t_3, when there is no possibility of confusion. Multiplying (1.26) to the left by V_3, and substituting $t_3\phi$ on the left-hand side, leads to an operator equation for t_3

$$t_3 = V_3 + V_3G_0(E^+)t_3 \quad . \tag{1.27a}$$

Using this with (1.25) in (1.24b) implies also that

$$t_3 = V_3 + t_3G_0(E^+)V_3 \quad . \tag{1.27b}$$

Eq.(1.27a) (or (1.27b)) is the Lippmann-Schwinger equation.

Let us write (1.27a) in the momentum representation, which is used in almost all practical calculations.

$$\langle \underline{q}'|t_3(E^+)|\underline{q}\rangle = \langle \underline{q}'|V_3|\underline{q}\rangle + \int d\underline{q}''\langle \underline{q}'|V_3|\underline{q}''\rangle \times (E^+ - q''^2/2\mu_{12})^{-1}\langle \underline{q}''|t_3(E^+)|\underline{q}\rangle \quad . \tag{1.28}$$

To begin the discussion of the properties of the t-matrix, defined by (1.28) as a function of three independent variables, we observe that it contains the elastic scattering information. Consider the "fully on-shell" t-matrix, $t_3^{on}(E,\theta)$, which is just $\langle \underline{q}'|t_3(E^+)|\underline{q}\rangle$ with $q^2 = q'^2 = 2\mu_{12}E$. From our earlier relation $t_3\phi = V_3\psi^{(+)}$ (following (1.26)), it is clear that

$$t_3^{on}(E,\theta) = \langle \underline{q}'|V_3|\psi_{\underline{q}}^{(+)}\rangle \quad , \tag{1.29}$$

and comparing with (1.18) one finds

$$f(q,\theta) = -4\pi^2\mu_{12}t_3^{on}(E,\theta). \tag{1.30}$$

(Recall that (1.18) was written for elastic scattering in the two-body CM, and hence q' and q satisfy the on-shell condition). Thus the on-shell t-matrix can be written directly in terms of the phase shifts. Indeed, substituting the partial wave expression

$$\langle \underline{q}'|t_3(E)|\underline{q}\rangle = \sum_{\ell m} Y_{\ell m}(\hat{\underline{q}}')t_\ell(q',q;E)Y^*_{\ell m}(\hat{\underline{q}}) \quad , \tag{1.31}$$

on the right of (1.30) and (1.8) and (1.9) on the left, one finds

$$t_\ell^{on}(E,\theta) = -(\pi\mu_{12}q)^{-1}\exp(i\delta_\ell)\sin\delta_\ell \quad . \tag{1.32}$$

Clearly the t-matrix is more closely related to experiment than the underlying potential. However, this on-shell information occupies only the diagonal part of the (q',q,E) space. Indeed, since the phase shifts determine only the asymptotic scattering wave function (cf.(1.4)) many different short-range potentials can produce the same on-shell t-matrix. (Ref. [1.7] contains an exhaustive discussion of this problem for the nucleon-nucleon system). This extra information, contained in the off-diagonal part of t, is usually referred to as off-shell information. An important area of concern in this book (see Chap.4) is whether three-body processes can yield information about the two-body t-matrix off-energy shell.

The r-matrix: In order to evaluate (1.28) one uses Dirac's result that [1.8]

$$(x + i\varepsilon - x')^{-1} \xrightarrow[\varepsilon\to 0+]{} P(x - x')^{-1} - i\pi\delta(x - x') \quad . \tag{1.33}$$

In fact, making a partial wave expansion of V_3 just like (1.31) for t_3, and using the orthonormality of spherical haromics, (1.28) reduces to

$$\begin{aligned} t_\ell(q',q;E) = V_\ell(q',q) &+ P\int_0^\infty dq''q''^2V_\ell(q',q'')(E - q''^2/2\mu_{12})^{-1}t_\ell(q'',q;E) \\ &- i\pi\mu_{12}q_EV_\ell(q',q_E)t_\ell(q_E,q;E) \quad , \end{aligned} \tag{1.34}$$

where $q_E \equiv (2\mu_{12}E)^{1/2}$. Computationally it is often easier to deal with the real quantity (Hermitian operator) called the r-matrix. This is defined by the operator equation

$$r_3(E) = V_3 + V_3 G_0^P(E) r(E) \tag{1.35a}$$

where $G_0^P(E) \equiv P(E - H_0)^{-1}$ - corresponding to a standing wave boundary condition. After the standard angular momentum expansion, (1.35) becomes (in momentum space)

$$r_\ell(q',q;E) = V_\ell(q',q) + P\int_0^\infty dq'' q''^2 \frac{V_\ell(q',q'') r_\ell(q'',q;E)}{E - q''^2/2\mu_{12}} \quad . \tag{1.35b}$$

Now as discussed on p.227 of [1.3], the r-matrix is related to the t-matrix by the Heitler equation (derived from (1.27), (1.35b) and (1.33))

$$t(E) = r(E) - i\pi r(E)\delta(E - H_0)t(E) \quad . \tag{1.35c}$$

In partial wave form, this becomes an algebraic equation with the solution

$$t_\ell(q',q;E) = r_\ell(q',q;E) - \frac{i\pi\mu_{12} q_E r_\ell(q',q_E;E) r_\ell(q_E,q;E)}{1 + i\pi\mu_{12} q_E r_\ell(q_E,q_E;E)} \quad . \tag{1.35d}$$

From this one readily shows that

$$r_\ell^{on}(E) \equiv r_\ell(q_E,q_E;E) = -(\pi\mu_{12} q_E)^{-1} \tan\delta_\ell \quad . \tag{1.35e}$$

An extremely important constraint on the t-matrix, which is a consequence of it satisfying (1.27), is unitarity. Multiplying (1.27a) to the right by t_3^{-1} and the left by V_3^{-1} implies

$$t_3^{-1}(E^+) = V_3^{-1} - G_0(E^+) \quad . \tag{1.36}$$

Now, as we have mentioned, (1.27) can be used to define t_3 for arbitrary E, and E - io in particular. Subtracting the relation for $t_3^{-1}(E^-)$, analogous to (1.36), from (1.36), one finds

$$t_3^{-1}(E^-) - t_3^{-1}(E^+) = G_0(E^+) - G_0(E^-) = -2\pi i\delta(E - H_0) \quad , \tag{1.37}$$

and thus

$$\text{disc}\{t_3(E)\} \equiv t_3(E^+) - t_3(E^-) = -2\pi i t_3(E^+)\delta(E - H_0)t_3(E^-) \quad . \tag{1.38}$$

This means that $t_3(E)$ has a cut (referred to as the unitarity cut) along the positive real axis. Note that (1.38) is an algebraic constraint in each partial wave

$$\text{disc}\{t_\ell(q',q;E)\} = -2\pi i\mu_{12} q_E t_\ell(q',q_E;E^+) t_\ell^*(q_E,q;E^+) \quad , \tag{1.39}$$

where one has used the fact that $V_\ell(q',q)$ is real and symmetric to show that $t_3(E^-) = t_3^*(E^+)$.

Because unitarity is such an important concept, we shall try to make the physical content of (1.38) a little more transparent. Consider the discontinuity of the forward on-shell scattering amplitude, which from (1.38) is

$$\text{disc}\ \{t_3^{on}(E,\theta = 0)\} = -\ 2\pi\mu_{12}q_E i\int d\hat{\underline{q}}_E|t(\underline{q},\underline{q}_E;E)|^2 \quad . \tag{1.40}$$

Defining the total elastic cross section σ_{tot} as $\int d\Omega(d\sigma/d\Omega)$, and using (1.6) and (1.30) this implies

$$\text{Im}\{f(q_E,\theta = 0)\}[\equiv (\text{disc}\ \{f\})/2i] = [q_E/4\pi]\sigma_{tot} \quad , \tag{1.41}$$

which is known as the optical theorem. While we have only proved it for two-body elastic scattering, (1.41) (with σ_{tot} including all possible inelastic processes) is generally true.

The underlying reason why this holds is conservation of probability - that is, the total flux into the scattering region must leave. This is referred to as unitarity because if the S-matrix (whose elements are equal to the probability for the projectile to scatter or pass unscattered) is unitary, then probability is conserved. On the energy-shell S is defined as

$$\langle\underline{q}'|S(E)|\underline{q}\rangle = \delta(\hat{\underline{q}} - \hat{\underline{q}}') - i2\pi\mu_{12}q_E\langle\underline{q}'|t(E^+)|\underline{q}\rangle \tag{1.42}$$

(we have factored out $\delta(q-q')/q^2$). It is left to the reader to verify that (1.38) guarantees that S is unitary ($S^\dagger S = 1$). Finally we observe that while on-shell unitarity and conservation of probability are equivalent, the extension to off-shell unitarity (e.g., (1.39) with $q \neq q' \neq q_E$) is a much more specific (and therefore more powerful) statement.

To see how powerful this is, and at the same time introduce some of the ideas of Chapter 2, let us assume there is no bound state in the ℓth partial wave - so that the right-hand (unitarity) cut is the only singularity in "E". Then Cauchy's theorem permits us to write $t_\ell(q',q;\omega)$ for arbitrary complex ω (not on $(0,\infty)$), in terms of the discontinuity across the cut. It will be more convenient to use the fact (most easily derived from (1.35d) and (1.35e) that the half-shell t-matrix carries the on-shell phase. That is

$$t_\ell(q,q_E;E^+) = -\ (\pi\mu_{12}q_E)^{-1}\ \exp\ (i\delta_\ell)\ \sin\ \delta_\ell f_\ell(q,q_E) \quad , \tag{1.43}$$

where $f_\ell(q,q_E)$ is a real function - the Kowalski-Noyes half-shell function [1.9]. Now Cauchy's theorem implies

$$t_\ell(q',q;\omega) = V_\ell(q',q) + (2\pi i)^{-1} \int_0^\infty dE \; \mathrm{disc}\,\{[t_\ell(q',q;E)]\}/(\omega - E) \quad , \tag{1.44}$$

where V_ℓ is the limit as $\omega \to \infty$ of $t_\ell(\omega)$ - it is the potential when that exists. Using (1.43) in (1.39) and substituting in (1.44) one finds

$$t_\ell(q',q,\omega) = V_\ell(q',q) + (\pi^2\mu_{12}^2)^{-1} \int_0^\infty \frac{dq_E \sin^2 \delta_\ell(q_E) f_\ell(q',q_E) f_\ell(q,q_E)}{\omega - q_E^2/2\mu_{12}} \quad . \tag{1.45}$$

The unwanted potential can be removed by one subtraction [1.10, 11]. Finally, we observe that this procedure is easily extended to include a bound state - in which case one needs the bound state wave function.

1.1.5 The Separable Potential

The technical definition of a one-term separable potential in terms of the partial wave expansion is simply

$$V_\ell(q',q) = g_\ell(q')\lambda_\ell g_\ell(q) \quad . \tag{1.46}$$

Here $\lambda_\ell = -1(+1)$ defines an attractive (repulsive) interaction. For the sake of simplicity of generalization we prefer Dirac notation, so that (1.46) becomes

$$V_\ell = |g_\ell > \lambda_\ell < g_\ell| \quad . \tag{1.47}$$

Such a potential is rather pleasant to use, reducing the two-body problem to algebra, and (as we shall see in Sect.1.8) reducing the three-body problem to a numerically solvable form.

Consider the Lippmann-Schwinger equation (1.27a) (after partial wave expansion) with this potential

$$t_\ell(E) = |g_\ell>\lambda_\ell<g_\ell| + |g_\ell>\lambda_\ell<g_\ell|G_0 t_\ell(E) \quad . \tag{1.48}$$

This has a solution of the form

$$t_\ell(E) = |g_\ell>\tau_\ell(E)<g_\ell| \quad , \tag{1.49}$$

$$\tau_\ell^{-1}(E) = \lambda_\ell^{-1} - <g_\ell|G_0(E)|g_\ell> \quad , \tag{1.50}$$

$$<g_\ell|G_0(E)|g_\ell> = \int_0^\infty dq\, q^2 g_\ell(q)(E^+ - q^2/2\mu_{12})^{-1} g_\ell^+(q) \quad . \tag{1.51}$$

Thus in order to calculate the phase shifts for this potential one need only evaluate a principal value integral.

The most clear cut physical justification for a separable t-matrix is when one is close to a bound state or resonance. To see this, using (1.25) in (1.24a) one can show that $t_3G_0 = V_3G_3$, and hence (using (1.24b))

$$t_3 = V_3 + t_3G_0V_3 = V_3 + V_3G_3V_3 \quad . \tag{1.52}$$

All that remains is to realize that G_3 is diagonal in the eigenstates of H. Then replacing G_3 by its spectral representation one gets

$$t_3(E^+) = V_3 + \sum_n \frac{V_3|\psi_n\rangle\langle\psi_n|V_3}{E^+ - \varepsilon_n} + \int \frac{d\varepsilon V_3|\psi_\varepsilon^{(+)}\rangle\langle\psi_\varepsilon^{(+)}|V_3}{E^+ - \varepsilon} \quad . \tag{1.53}$$

Clearly as E approaches a bound state energy (ε_n), this pole term will dominate - and it is separable.

This close relationship between separability and the presence of bound states or resonances was first emphasized by LOVELACE [1.6]. Physically this idea is apparent in (1.49), where there are independent amplitudes for the formation "$\langle g|$", propagation "τ", and decay "$|g\rangle$" of some state, with no memory of the formation carrying over to the decay. That such interactions can arise in field theory has been clearly described by AMADO [1.12].

It is also possible that a separable t-matrix may be realistic even if there is no nearby pole. To see this, we observe that it has been shown (e.g., by MONGAN [1.11]) that (two-term) separable potentials fitted to N-N data over several hundred MeV produce half-shell functions close to those of the "realistic" potentials. In view of (1.45), one expects that for use in many-body problems, the separable interactions should be quite acceptable.

So far there has been only fleeting mention of the bound state problem. When one partial wave has a bound state, there is a very simple relation between the bound state wave function and the form-factor of the corresponding separable potential. A bound state wave function satisfies the homogeneous Lippmann-Schwinger equation

$$|\psi_\ell^B\rangle = -(B + H_0)^{-1}V_\ell|\psi_\ell^B\rangle \quad , \tag{1.54}$$

where $-B<0$ is the bound state energy. Taking an attractive potential ($\lambda_\ell = -1$ in (1.47)) in this case, one can solve (1.54).

$$|\psi_\ell^B\rangle = -(B + H_0)^{-1}|g_\ell\rangle F_\ell \quad . \tag{1.55}$$

For a one-term potential "F_ℓ" is a normalization constant. Formally it is defined as the solution of

$$\left[\lambda_\ell^{-1} - \langle g_\ell | G_0(-B) | g_\ell \rangle\right] F_\ell = 0 \quad . \tag{1.56}$$

Thus, if one knows the bound state wave function (e.g., from e-scattering), one can immediately define a one-term separable potential with form-factor

$$g_\ell(q) \equiv \langle q | g_\ell \rangle = -(B + q^2/2\mu_{12})\psi_\ell^B(q) F_\ell^{-1} \quad , \tag{1.57}$$

which produces this function exactly. The resultant t-matrix is called the unitary pole approximation (UPA) to whatever true t-matrix describes the system. For a thorough discussion of this aspect of separable interactions (and their successful application to the 3N bound state problem) we refer to the review by LEVINGER [1.13].

To complete this discussion it is interesting to note that if one assumes an underlying one-term separable potential, the inverse scattering problem has a unique solution. Briefly $\tau(E)$ is constructed from the knowledge that the discontinuity across the rigth-hand cut is entirely given by the phase shifts. Indeed $\tau_\ell(E^+)$ can be shown [1.14] to be simply exp $[i\Delta(E^+)]$, with $\Delta(E)$ given by (3.19). Substituting this into the on-shell relation (1.32) with a separable form on the left uniquely determines $g^2(q_E)$. Potentials constructed in this way for pion-nucleon scattering have been quite successfully applied in a description of pion-nucleus scattering [1.15].

So far we have almost exclusively considered rank one potentials. For higher rank [1.16], $|g_\ell\rangle$ can be considered a row matrix of length equal to the rank of the potential. Then λ_ℓ becomes a square symmetric matrix of strengths - which can be diagonal. Eq.(1.47-51) for scattering, and (1.54-57) for the bound state are then all valid.

1.2 The Simplest Three-Body Problem

The purpose of this rather brief section is two fold. First, it is a convenient place to summarize the notation used in most of the book to describe three interacting particles. Second, the discussion of Subsection 1.1.4 is extended to two-body scattering in the presence of a third non interacting particle - since this forms the input to the three-body problem. Even this elementary discussion introduces some unexpected features of three-body systems.

1.2.1 Notation

Let us consider three distinguishable, spinless particles of masses (m_1, m_2, m_3). Each of these has kinetic energy K_i, and the total kinetic energy is H_0. As a general rule, all expressions (in this book) will be written in the three-body CM. That is, the momenta $\underline{p}_i$ satisfy the the condition that $\sum_i \underline{p}_i$ be zero. Thus, there are only two independent momenta which are usually taken to be: one of the $\underline{p}_i$ and the relative momentum of the remaining pair, labelled $\underline{q}_i$. Adopting the convention that (i,j,k) be always cyclic, $\underline{q}_i$ is defined as

$$\underline{q}_i = (m_k \underline{p}_j - m_j \underline{p}_k)/(m_j + m_k) \quad . \tag{1.58}$$

The inverse transformation to (1.58) is

$$\underline{p}_j = \underline{q}_i - (m_j/m_{jk}) \underline{p}_i \, ; \underline{q}_j = -(m_i/m_{ik}) \underline{q}_i - (m_k M/m_{ik} m_{jk}) \underline{p}_i \quad , \tag{1.59a}$$

$$\underline{p}_k = -\underline{q}_i - (m_k/m_{jk}) \underline{p}_i \, ; \underline{q}_k = -(m_i/m_{ij}) \underline{q}_i + (m_j M/m_{ij} m_{jk}) \underline{p}_i \quad , \tag{1.59b}$$

where we use M for $(\sum_i m_i)$, and m_{ij} for $(m_i + m_j)$, etc.

In terms of these momenta, the total kinetic energy is

$$H_0 = p_i^2/2\mu_i + q_i^2/2\mu_{jk} \quad . \tag{1.60}$$

Here μ_{jk} is the reduced mass of j and k, and μ_i the reduced mass of particle i and the pair (jk)

$$\mu_{jk} = m_j m_k/m_{jk} \, ; \; \mu_i = m_i m_{jk}/M \quad . \tag{1.61}$$

Clearly we now have three sets of variables which describe the same state. That is, $|\underline{p}_1 \underline{p}_2 \underline{p}_3\rangle$ is equally well described by any $|\underline{p}_i \underline{q}_i\rangle$ with i = 1,2,3. These states are normalized (cf. (1.17)) as

$$\langle \underline{p}_i' \underline{q}_i' | \underline{p}_i \underline{q}_i \rangle = \delta(\underline{p}_i - \underline{p}_i') \delta(\underline{q}_i - \underline{q}_i') \quad , \tag{1.62}$$

which means that the coordinate space representation of a free three-body (plane wave) state is

$$\langle \underline{\rho}_i , \underline{r}_{jk} | \underline{p}_i \underline{q}_i \rangle = (2\pi)^{-3} \exp(i\underline{p}_i \cdot \underline{\rho}_i) \exp(i\underline{q}_i \cdot \underline{r}_{jk}) \quad . \tag{1.63}$$

Here $\underline{r}_{jk}$ is the separation of j and k $(\underline{r}_j - \underline{r}_k)$, and $\underline{\rho}_i$ is the position of particle i, with respect to the CM of j and k

$$\underline{\rho}_i = \underline{r}_i - (m_j \underline{r}_j + m_k \underline{r}_k)/m_{jk} \quad . \tag{1.64}$$

In the three-body CM (defined by $\sum_i m_i \underline{r}_i = 0$) $\underline{\rho}_i$ becomes simply $(M/m_{jk})\underline{r}_i$.

While the inclusion of three-body forces introduces no formal difficulties, we assume that there are only pairwise interactions. As in Section 1.1, these are labelled by the spectator particle, e.g., V_1 is the interaction between particles 2 and 3. Hence the total Hamiltonian is

$$H = H_0 + V \equiv H_0 + \sum_{i=1}^{3} V_i = H_0 + V_j + V^j \quad , \tag{1.65}$$

where for convenience we introduce V^i which is the sum of all interactions other than V_i.

The channel Hamiltonian H_i, in which only one pair interacts, is simply

$$H_i = H_0 + V_i \quad , \tag{1.66}$$

which is not the same as "H" in Section 1.1 since (1.66) also includes the kinetic energy of particle i, and the CM motion of the pair (jk) (i.e., $p_i^2/2\mu_i$). Further, one can define the full and channel Green's functions by

$$G(z) = (z - H)^{-1}; \; G_i(z) = (z - H_i)^{-1} \quad . \tag{1.67}$$

(Defining $V_0 \equiv 0$, the latter can include the free three-body Green's function too (i = 0)).

Equations (1.62) and (1.63) define one possible asymptotic state - with all particles free. However, the most common experimental situation is to have initially a projectile (say) i, incident on a two-body bound state (j and k, in a state "c").[2] In this case the asymptotic state $|\chi_{i;c}\rangle$ is

$$\langle \underline{\rho}_i, \underline{r}_{jk} | \chi_{i;c}\rangle = (2\pi)^{-3/2} \exp\,(i\underline{p}_i \cdot \underline{\rho}_i)\tilde{\psi}_{i;c}(\underline{r}_{jk}) \quad , \tag{1.68a}$$

$$\langle \underline{p}_i', \underline{q}_i' | \chi_{i;c}\rangle = \delta(\underline{p}_i - \underline{p}_i')\psi_{i;c}(\underline{q}_i') \quad , \tag{1.68b}$$

where $\psi(\tilde{\psi})$ is the normalized bound state wave function.

1.2.2 The Two-Body t-Matrix in the Three-Body System

In any real "two-body" scattering experiment there are many spectator particles. Nevertheless, for short-range interactions, a careful choice of experimental conditions usually enables one to isolate the two-body interaction. (The exceptional case of Coulomb scattering is well described in Ref.1.3, p.263). Let us consider

---

[2] Indeed, we shall see in Chapter 6 that a many-body problem can often be reduced to an effective three-body problem, as in the construction of an optical model for particle-nucleus scattering.

the scattering of particles j and k in the presence of a noninteracting spectator i (with momentum $\underline{p}_i$ which cannot change). Since we work in the three-body CM the pair (jk) has total momentum $-\underline{p}_i$, and the two-body sub-energy (E_{jk}) is related to the total three-body energy (E) by

$$E = E_{jk} + p_i^2/2\mu_i \quad . \tag{1.69}$$

Suppose the initial and final states are $|\underline{p}_i\underline{q}_i>$ and $|\underline{p}_i'\underline{q}_i'>$, respectively, with $q_i = q_i' = (2\mu_{jk}E_{jk})^{1/2}$ - corresponding to real elastic scattering. Then the entire discussion of Section 1.1 carries through. In particular, all the scattering information is contained in a t-matrix satisfying the Lippmann-Schwinger equation

$$t_i(E^+) = V_i + V_i G_0(E^+) t_i(E^+) \quad . \tag{1.70}$$

Note that we consider it unnecessary to discriminate between operators in two- and three-body Hilbert space (cf. the "hat" notation of LOVELACE [1.6]). One need only remember that the three-body matrix elements of V_i and $G_0(E)$ are

$$<\underline{p}_i'\underline{q}_i'|V_i|\underline{p}_i\underline{q}_i> = \delta(\underline{p}_i - \underline{p}_i')V_i(\underline{q}_i',\underline{q}_i) \quad , \tag{1.71a}$$

$$<\underline{p}_i'\underline{q}_i'|G_0(E^+)|\underline{p}_i\underline{q}_i> = \delta(\underline{p}_i - \underline{p}_i')\delta(\underline{q}_i - \underline{q}_i')[E^+ - p_i^2/2\mu_i - q_i^2/2\mu_{jk}]^{-1} \quad . \tag{1.71b}$$

Putting (1.71) into (1.70), one finds that it has a solution of the form $\delta(\underline{p}_i-\underline{p}_i')\tilde{t}_i(\underline{q}_i',\underline{q}_i;E;p_i)$, where $\tilde{t}_i$ satisfies a purely two-body Lippmann-Schwinger equation, but with energy $(E^+-p_i^2/2\mu_i)$. Thus, (1.70) has the solution

$$\delta(\underline{p}_i - \underline{p}_i')\ t_i\ (\underline{q}_i',\underline{q}_i;E^+ - p_i^2/2\mu_i) \quad , \tag{1.72}$$

which is exactly what one expects physically. Note that the mathematical reason for this expression is the separation of p_i and q_i in G_0. In Chapter 5 it will be seen that this separation is not so trivial in the relativistic case.

While (1.72) was derived for on-shell two-body scattering (E_{jk} related to q_i,q_i'), (1.70) defines the two-body t-matrix in the three-body system for any energy E, and for arbitrarily large spectator momentum p_i. If $p_i^2>2\mu_i E$, the interacting pair has negative energy, which does not correspond to a real scattering situation. Nevertheless, virtual processes in which a spectator has arbitrarily large momentum occur in all three-body calculations - one needs the two-body t-matrix at negative energy to describe even positive energy three-body phenomena. That this particular energy dependence is related to three-body unitarity we establish in Section 1.6. For the moment we remark that in (1.45) the t-matrix at arbitrary energy was related to the phase shifts and the half-shell function (the latter being essentially the scattering wave functions) at positive energy. Thus the appearance of negative energy t-matrices is not so mysterious, being related to the analytic properties of the scattering amplitude. (A similar point, in somewhat different language, has been made in [1.17]).

To conclude this section, we describe an important problem in three-body scattering theory, which is already visible in (1.72). In Subsection 1.1.3 it was observed that a sufficient condition for a unique solution of an integral equation is that the kernel be L^2. However, because of the third particle, Trace $[(G_0V)^{+}G_0V]$ includes $\int d\underline{p}_i \int d\underline{p}_i'[\delta(\underline{p}_i - \underline{p}_i')]^2$, which is infinite - whatever the two-body subenergy. Clearly the problem arises because the third particle does not interact - this is known as the "disconnectedness" problem.

This problem may seem artificial, since one can always factor out the dangerous δ-function. But in the true three-body problem the kernel is G_0V (recall $V = \sum_i V_i$), and there is no common term to remove. Finally we recall that the L^2 condition is sufficient, not necessary, for a compact kernel. In the next section it is shown that even this lesser condition does not hold. Then the reason for Faddeev's work will become clear.

1.3 Three Distinguishable Particles - Wave Function Approach

In this section all pairs are allowed to interact. While for the present we consider distinguishable particles, this restriction is easily lifted (as shown in Sec.1.7). The omission of spin is also for pedagogical reasons, since the spinless case illustrates most of the interesting physics. This omission too will be remedied later (Sec.1.8), when we discuss fully the case of separable interactions with arbitrary spin. As with the two-body problem, we introduce the topic through the wave function (usually in coordinate space). The concept of the coordinate space wave function is basic to the development of the boundary condition method in Chapter 4.

Let us briefly outline the contents of this section. First we repeat essentially Faddeev's explanation of what is wrong with a Lippmann-Schwinger type of equation for three particles. Next we simply assume the Faddeev decomposition of G(z), and show that this leads to a subdivision of the wave function which solves the non-uniqueness problem. The question of actually proving the Faddeev form for G(z) is left for Section 1.4.

1.3.1 The Non-Uniqueness Problem

To be definite, consider the (physically unlikely) situation where both the initial and final states contain three free particles (this is the $3 \to 3$ amplitude). The initial state (ϕ_0) corresponds to momenta $(\underline{p}_i, \underline{q}_i)$, the final state ($\phi_f$) to $(\underline{p}_i', \underline{q}_i')$

- both have the form (1.63). Now the basic formula which we take without further justification from ECKSTEIN and GERJUOY [1.18] is that the scattering amplitude from an initial state Φ_i to a final state Φ_f is

$$T_{fi} = \langle\Phi_f|V^f|\Psi_i^{(+)}\rangle \quad , \tag{1.73a}$$

$$= \langle\Psi_f^{(-)}|V^i|\Phi_i\rangle \quad . \tag{1.73b}$$

Here V^f is the part of V not included in Φ_f, and $\Psi_i^{(+)}$ is the full scattering solution corresponding to an initial state Φ_i. In the 3 → 3 case under consideration, V^f is V and Φ_f is ϕ_f. Thus the whole problem reduces to finding a meaningful integral equation for $\Psi_0^{(+)}$.

Of course, $\Psi_0^{(+)}$ satisfies the Schrödinger equation, with the boundary condition that it tends asymptotically to ϕ_0, plus outgoing spherical waves. (One gets exactly the same results by working with $\Psi_f^{(-)}$; however we place less emphasis there because it corresponds to physically unrealizable ingoing boundary conditions). As we saw in Subsection 1.1.3 for the two-body case (cf. (1.21)), these two conditions can be formally combined [1.18] into one equation

$$\Psi_i^{(+)} = \lim_{\varepsilon\to 0+} i\varepsilon G(E + i\varepsilon)\Phi_i \quad , \tag{1.74a}$$

$$\Psi_f^{(-)} = \lim_{\varepsilon\to 0+} i\varepsilon G(E - i\varepsilon)\Phi_f \quad . \tag{1.74b}$$

Indeed, using the operator identity (cf. (1.24))

$$G(z) = G_0(z) + G_0(z)VG(z) \tag{1.75}$$

in (1.74a) - with $[\Psi_0^{(+)},\phi_0]$ replacing $[\Psi_i^{(+)},\Phi_i]$ - one finds

$$\begin{aligned}\Psi_0^{(+)} &= \lim_{\varepsilon\to 0+} i\varepsilon(E + i\varepsilon - H_0)^{-1}\phi_0 + G_0V \lim_{\varepsilon\to 0+} i\varepsilon G(E + i\varepsilon)\phi_0 \\ &= \phi_0 + G_0(E^+)V\Psi_0^{(+)} \quad ,\end{aligned} \tag{1.76}$$

since ϕ_0 is an eigenstate of H_0 with eigenvalue E.

This inhomogeneous integral equation looks like the end of the story. However, the considerations of Subsection 1.2.2, showing that $G_0(z)V$ is not L^2 even for "z" complex, might raise suspicions that the kernel of (1.76) is not compact. To confirm this, consider the problem where the initial state has one particle (say 1) incident on a bound state "c" of the other two (with binding energy ε_c). The initial wave function is then $\chi_{1;c}$ (cf. (1.68)), which satisfies

$$(H_0 + V_1)\chi_{1;c} = E\chi_{1;c} \quad , \tag{1.77}$$

and the projectile has momentum $p_1^2 = 2\mu_1(E+\varepsilon_c)$.(Since the bound state wave function satisfies $[-\nabla_{23}^2/2\mu_{23}+V_1]\tilde{\psi}_{1;c} = -\varepsilon_c\tilde{\psi}_{1;c}$).

Using (1.74a) and (1.75), one can write the corresponding full solution $\Psi_{1;c}^{(+)}$ as

$$\Psi_{1;c}^{(+)} = \lim_{\varepsilon\to 0+} i\varepsilon G_0(E + i\varepsilon)\chi_{1;c} + G_0(E^+)V \lim_{\varepsilon\to 0+} i\varepsilon G(E + i\varepsilon)\chi_{1;c} \quad . \tag{1.78}$$

Now the first term on the right tends to zero, since

$$\begin{aligned} G_0(E + i\varepsilon)\chi_{1;c} &= \int d\underline{p}_1' d\underline{q}_1' \frac{|\underline{p}_1'\underline{q}_1'><\underline{p}_1'\underline{q}_1'|\chi_{1;c}>}{E + i\varepsilon - p_1'^2/2\mu_1 - q_1'^2/2\mu_{23}} \\ &= \int \frac{d\underline{q}_1' |\underline{p}_1\underline{q}_1'>\psi_{1;c}(\underline{q}_1')}{-\varepsilon_c + i\varepsilon - q_1'^2/2\mu_{23}} \end{aligned} \tag{1.79}$$

is finite. The difference from the previous case is here the energy denominator is always finite (for finite ε_c), while $G_0(E+i\varepsilon)\phi_0$ is $(i\varepsilon)^{-1}\phi_0$. Thus (1.78) becomes

$$\Psi_{1;c}^{(+)} = G_0(E^+)V\Psi_{1;c}^{(+)} \quad . \tag{1.80}$$

Comparing (1.80) and (1.76), one sees that $\Psi_{1;c}^{(+)}$ is a solution of the homogeneous version of (1.76). Thus, if one has any solution $\Psi_{0\alpha}^{(+)}$ of (1.76), then there is an infinite number of different solutions $[\Psi_{0\alpha}^{(+)} + \lambda\Psi_{1;c}^{(+)}]$ - with λ arbitrary. That is, this equation does not satisfy the first criterion of a useful integral equation, namely that it should have a unique solution. (Incidentally this proves "G_0V" is not compact).

To conclude this section, we observe that it is possible to write an inhomogeneous equation for $\Psi_{1;c}$, by starting with a different relation for $G(z)$

$$G(z) = G_1(z) + G_1(z)V^1G(z) \tag{1.81a}$$

$$= G_1(z) + G(z)V^1G_1(z) \quad . \tag{1.81b}$$

This is derived using (1.23) with $B = G$ and $A = G_1$ (or $B = G_1$, $A = G$ for (1.81b)). Eq. (1.81a) leads to an inhomogeneous equation because $\chi_{1;c}$ is an eigenstate of H_1 with eigenvalue E. However this equation also suffers from non-uniqueness problems as one can show by constructing (for example) $\Psi_{2;d}^{(+)}$ corresponding to an initial state $\chi_{2;d}$. In fact, $\Psi_{2;d}^{(+)}$ satisfies the homogeneous version of the equation for $\Psi_{1;c}^{(+)}$, and so on. Clearly there are very fundamental problems in this approach to finding integral equations.

1.3.2 The Solution (3 → 3)

Let us now assume the Faddeev decomposition of G(z) and show how it cures the non-uniqueness problem. This form for G(z), which will be proven in Section 1.4, is

$$G(z) = G_0(z) + G^1(z) + G^2(z) + G^3(z) \quad , \tag{1.82}$$

where the three-body operators G^i satisfy the set of three coupled equations

$$G^i = (G_i - G_0) + G_0 t_i (G^j + G^k) \quad . \tag{1.83}$$

(All operators in (1.83) have the same argument, and $i \neq j \neq k \neq i$).

Accepting (1.82) and (1.83) for the present, we again construct $\Psi_0^{(+)}$. Using (1.82) in (1.74) yields

$$\Psi_0^{(+)} = \phi_0 + \sum_{i=1}^{3} \lim_{\varepsilon \to 0+} i\varepsilon G^i(E + i\varepsilon)\phi_0 \equiv \phi_0 + \sum_{i=1}^{3} \psi_0^i \quad . \tag{1.84}$$

(Note that one really means $\psi_0^{(+)i}$, with the "plus" indicating the outgoing boundary conditions - we omit it only for ease of printing). Now using (1.83) for G^i leads to a set of coupled integral equations for ψ_0^i,

$$\begin{aligned} \psi_0^i &= \lim_{\varepsilon \to 0+} i\varepsilon G^i(E + i\varepsilon)\phi_0 \\ &= (\phi_i - \phi_0) + G_0 t_i (\psi_0^j + \psi_0^k) \quad , \end{aligned} \tag{1.85}$$

where $\phi_i = \lim_{\varepsilon \to 0+} i\varepsilon G_i(E+i\varepsilon)\phi_0$. One often writes these equations in the equivalent matrix form

$$\begin{pmatrix} \psi_0^1 \\ \psi_0^2 \\ \psi_0^3 \end{pmatrix} = \begin{pmatrix} \phi_1 - \phi_0 \\ \phi_2 - \phi_0 \\ \phi_3 - \phi_0 \end{pmatrix} + G_0 \begin{pmatrix} 0 & t_1 & t_1 \\ t_2 & 0 & t_2 \\ t_3 & t_3 & 0 \end{pmatrix} \begin{pmatrix} \psi_0^1 \\ \psi_0^2 \\ \psi_0^3 \end{pmatrix} \quad . \tag{1.86}$$

The kernel of these equations is now a 3 x 3 matrix, with zeros on the diagonal, which is the crucial point. Let us call the product of G_0 and this 3 x 3 matrix of t-matrices "$\underset{\approx}{F}$". Then forming Trace ($\underset{\approx}{F}^\dagger \underset{\approx}{F}$), one sees immediately that there are terms like $(G_0 t_1)^\dagger G_0 t_1$, which leads to an infinite Schmidt norm (cf. Subsec.1.2.2). Essentially this is because $\underset{\approx}{F}$ is still disconnected.

However, iterating (1.86) once implies

$$\underset{\sim}{\psi}_0 = \underset{\sim}{\phi}_0 + \underset{\approx}{F}\underset{\sim}{\phi}_0 + \underset{\approx}{F}\underset{\approx}{F}\underset{\sim}{\psi}_0 \quad , \tag{1.87}$$

where $\underset{\sim}{\phi}_0$ and $\underset{\sim}{\psi}_0$ are column vectors, in an obvious notation. The kernel of (1.87) is

$$\underset{\approx}{F}^2 = G_0 \begin{pmatrix} t_1G_0t_2 + t_1G_0t_3 & t_1G_0t_3 & t_1G_0t_2 \\ t_2G_0t_3 & t_2G_0t_1 + t_2G_0t_3 & t_2G_0t_1 \\ t_3G_0t_2 & t_3G_0t_1 & t_3G_0t_1 + t_3G_0t_2 \end{pmatrix} \quad , \tag{1.88}$$

which is completely connected - there are no δ-functions in the elements. It is straightforward to show that Trace $[(\underset{\approx}{F}^2)^\dagger \underset{\approx}{F}^2]$ is finite for E off the positive real axis (for reasonable two-body interactions) [1.19]. Thus, as for the two-body problem, the kernel is compact for any complex E, and (1.87) has a unique solution which may be found by Fredholm methods. This *suggests* that there is also a unique solution in the limit E+i0.

Indeed, Faddeev has established that under rather weak conditions on the t-matrices the fifth iterate of (1.86) has a kernel ($\underset{\approx}{F}^5$) which is compact in a particular Banach space of continuous functions [1.1]. Somewhat more powerful results can be proved by working in momentum space (as in Chap.2) [1.20]. For present purpose we need only the result that the coupled integral equations (1.86) have a unique solution (a direct consequence of $\underset{\approx}{F}^5$ being compact).

Having solved the mathematical problem, we ask what the resultant equations mean physically. Clearly (1.84) means that $\Psi_0^{(+)}$ is the sum of the incident wave ϕ_0, and some scattered waves ψ_0^i. Comparing the inhomogeneous term $(\phi_i - \phi_0)$ with (1.21) and (1.26), one sees that it is the scattered wave resulting from the interaction of j and k in the presence of i (noninteracting). In coordinate space

$$\begin{aligned}(\phi_i - \phi_0)(\underline{\rho}_i, \underline{r}_{jk}) = &- \frac{\mu_{jk}}{(2\pi)^4} \exp(i\underline{p}_i \cdot \underline{\rho}_i) \int d\underline{r}'_{jk} d\underline{r}''_{jk} \\ &\times \frac{\exp iq_i |\underline{r}_{jk} - \underline{r}'_{jk}|}{|\underline{r}_{jk} - \underline{r}'_{jk}|} t_i(\underline{r}'_{jk}, \underline{r}''_{jk}; E - p_i^2/2\mu_i) \\ &\times \exp(i\underline{q}_i \cdot \underline{r}''_{jk}) \quad ,\end{aligned} \tag{1.89}$$

while in momentum space it has the simple form

$$(\phi_i - \phi_0)\,(\underset{\sim}{p}_i',\underset{\sim}{q}_i') = \delta(\underset{\sim}{p}_i - \underset{\sim}{p}_i')(E^+ - p_i'^2/2\mu_i - q_i'^2/2\mu_{jk})^{-1}$$
$$\times \quad t_i(\underset{\sim}{q}_i',\underset{\sim}{q}_i;E^+ - p_i^2/2\mu_i) \quad , \tag{1.90}$$

involving the half-off-shell two-body t-matrix.

To next order one has terms like $G_0 t_i(\phi_j-\phi_0)(i\neq j)$, corresponding to (ik) scattering first, and then k interacting with j through t_i, and so on. Such an expansion of (1.85) clearly shows that it makes physical as well as mathematical sense.

1.3.3 Scattering from a Two-Body Bound State

To be specific, suppose the initial state is $\chi_{1;c}$ (cf. (1.68) and (1.77)). Once again the full scattering wave function is given by (1.74a), but with the Faddeev form for G. That is

$$\psi_{1;c}^{(+)} = \sum_{i=1}^{3} \psi_{1;c}^{i}; \ \psi_{1;c}^{i} = \lim_{\varepsilon\to 0+} G^i(E + i\varepsilon)\chi_{1;c} \tag{1.91}$$

where G_0 does not contribute (as explained below (1.78)). Using these same arguments, and (1.83) for G^i, one finds

$$\psi_{1;c}^{i} = \lim_{\varepsilon\to 0+} i\varepsilon[G_i(E + i\varepsilon) - G_0(E + i\varepsilon)]\chi_{1;c} + G_0(E^+)t_i(E^+)[\psi_{1;c}^{j} + \psi_{1;c}^{k}]$$
$$= \delta_{1;i}\chi_{1;c} + G_0 t_i(\psi_{1;c}^{j} + \psi_{1;c}^{k}) \tag{1.92}$$

with $i\neq j\neq k\neq i$. (The only nonzero first term results from $i\varepsilon G_1(E+i\varepsilon)\chi_{1;c}$ because $\chi_{1;c}$ is an eigenstate of H_1). In matrix form this is

$$\underset{\sim}{\psi}_{1;c} = \underset{\sim}{\chi}_{1;c} + \underset{\approx}{F}\underset{\sim}{\psi}_{1;c} \quad , \tag{1.93}$$

where the inhomogeneous term is $(\chi_{1;c},0,0)^T$. Similar equations may be derived for particle 2 (or 3) incident on a bound state of the other two. The sole difference is that then only the second (or third) position in $\underset{\sim}{\chi}_{i;c}$ is nonzero. Because the same kernel $\underset{\approx}{F}$ occurs here, identical arguments to Subsection 1.3.2 show that $\underset{\sim}{\psi}_{1;c}$, and hence $\psi_{1;c}^{(+)}$, is unique.

It seems worthwhile to comment on why the inhomogeneous term of (1.92) is so much simpler than that of (1.85). Briefly, in the latter case any two particles could scatter and produce an outgoing wave without ever approaching the third. However in the former case, in order for any scattering to occur the projectile had to hit the bound pair. Thus even the simplest process producing a scattered wave must involve all three particles together at some time.

While one can now evaluate the interesting scattering amplitudes by using (1.73a), and even eliminate all reference to potentials in favor of t-matrices [see Ref.1.1, JETP article], it is preferable to wait for Section 1.5 where we write integral equations directly for the scattering amplitude. Finally we note that in going from the Lippmann-Schwinger to the Faddeev equations we have tripled the number of equations for a given initial state. This increase was necessary to permit one to incorporate a complete set of boundary conditions. In fact, the failure of the Lippmann-Schwinger theory can be attributed to its failure to satisfy the complete set of boundary conditions in a single integral equation.

1.4 The Faddeev Equations ($3 \to 3$)

Here our first task is to give a rigorous derivation of (1.83), which was simply assumed in the last section. Faddeev's original proof of this fundamental equation, as well as that of most texts, relies on a rearrangement of the perturbation expansion for $G(z)$. Since we prefer not to rely on the convergence of an unknown series, we present instead the rigorous derivation based on operator algebra.

Having firmly established the Faddeev equations for the Green's function, one can define the three-body t-matrix T (which is closer to experiment), and write Faddeev-type equations for its components ($T = \sum_{i=1}^{3} T^i$). Although T purports to describe only the situation of three particles in and three out ($3 \to 3$), by looking the residues of T at the two-body bound state poles, one can see how bound state elastic and rearrangement scattering is included.

1.4.1 A Rigorous Derivation of Faddeev Equations for G(z)

Let us begin by dividing the integral equation (1.75) for the resolvent $G(z)$ into four components

$$G(z) = G_0(z) + \sum_{i=1}^{3} G^i(z); \quad G^i(z) = G_0(z)V_iG(z) \quad . \tag{1.94}$$

We notice also the obvious extension of (1.81)

$$G(z) = G_i(z) + G_i(z)V^iG(z) \tag{1.95a}$$

$$= G_i(z) + G(z)V^iG_i(z) \tag{1.95b}$$

where G_i is the channel Green's function of (1.67). Substituting (1.95a) into the definition of $G^i(z)$ - (1.94) - implies

$$G^i(z) = G_0(z)V_iG_i(z) + G_0(z)V_iG_i(z)V^iG(z) \quad . \tag{1.96}$$

Now using the extension of (1.24a) to the three-body system $[G_0(z)V_iG_i(z) = G_i(z) - G_0(z)]$ implies that

$$G^i(z) = [G_i(z) - G_0(z)] + [G_i(z) - G_0(z)]G_0^{-1}(z)[G^j(z) + G^k(z)] \quad , \tag{1.97}$$

where we have inserted $G_0G_0^{-1}$ between G_i and V^i in the last term of (1.96), and identified G_0V_jG with G^j. This is essentially the Faddeev form of (1.83). All that remains is to substitute $G_0t_iG_0$ for (G_i-G_0) (cf. (1.25)) in the last term.

In conclusion, it is worth noting that this derivation, and particularly the definition (1.94) makes clear what G^i means physically. That is, G is a propagator which describes the development through (1.74) (or better its time-dependent equivalent) of an symptotic state, say ϕ_0, into a fully interacting wave function, $\psi_0^{(+)}$. After splitting off the part of G with no interaction (i.e., G_0) there remain only three possibilities. One of the three pairs must have interacted last. The expression $G^i = G_0V_iG$ makes clear that G^i is that part of the full propagator in which the pair (jk) interacts last.

1.4.2 The Three-Body t-Matrix

In analogy with the two-body case, it is usual to define the three-body t-matrix by

$$G(z) = G_0(z) + G_0(z)T(z)G_0(z) \quad . \tag{1.98}$$

Then following the steps which led to (1.52), we have

$$T(z) = V + VG(z)V \tag{1.99a}$$

$$= \sum_{i,j} [V_i\delta_{ij} + V_iGV_j] \quad , \tag{1.99b}$$

and hence

$$T = \sum_{i,j} M_{ij}; \; M_{ij} = V_i\delta_{ij} + V_iG(z)V_j \quad . \tag{1.100}$$

Substituting for G from (1.75) $(G = G_0+G_0[\sum_k V_k]G)$, and rearranging terms, (1.100) becomes

$$\begin{aligned} M_{ij} &= V_i\delta_{ij} + V_iG_0 \sum_k [\delta_{kj}V_k + V_kGV_j] \\ &= V_i\delta_{ij} + V_iG_0 \sum_k M_{kj} \quad . \end{aligned} \tag{1.101}$$

This is now readily brought to Faddeev form by bringing the term $V_iG_0M_{ij}$ to the left, and multiplying throughout by $[1-V_iG_0]^{-1}$ (to the left). This yields

$$M_{ij} = t_i\delta_{ij} + \sum_{k\neq i} t_i G_0 M_{kj} \quad , \tag{1.102}$$

where the last step involves identifying $[1-V_iG_0(z)]^{-1}V_i$ as t_i - from (1.23a). Eq. (1.102) has the obvious matrix representation

$$\underset{\approx}{M} = \underset{\approx}{t} + \underset{\approx}{F}'\underset{\approx}{M} \quad , \tag{1.103}$$

with $\underset{\approx}{F}'$ having the same essential structure as the kernel $\underset{\approx}{F}$ of Section 1.3 (i.e, $\underset{\approx}{F}' = G_0^{-1}\underset{\approx}{F}G_0$). Indeed all the conclusions about the uniqueness of the solution of (1.86) also apply to (1.103).

These equations, symmetric in the initial and final states, are what is usually meant by "the Faddeev equations". Physically what they describe is the 3→3 scattering process, (i.e., three free particles initially and finally). The most natural representation of these equations is the momentum representation

$$\begin{aligned}\langle \underline{p}_i'\underline{q}_i'|M_{ij}(E)|\underline{p}_j\underline{q}_j\rangle &= \delta_{ij}\delta(\underline{p}_i' - \underline{p}_i)t_i(\underline{q}_i',\underline{q}_i;E - p_i^2/2\mu_i) \\ &+ \sum_{k\neq i}\int d\underline{p}_k'' d\underline{q}_k''\delta(\underline{p}_i' - \underline{p}_i'')t_i(\underline{q}_i',\underline{q}_i'';E - p_i'^2/2\mu_i) \\ &\times [E^+ - p_k''^2/2\mu_k - q_k''^2/2\mu_{ij}]^{-1}\langle \underline{p}_k''\underline{q}_k''|M_{kj}(E)|\underline{p}_j\underline{q}_j\rangle \quad .\end{aligned} \tag{1.104}$$

Note that the same physical state is described by $|\underline{p}_1\underline{q}_1\rangle$, $|\underline{p}_2\underline{q}_2\rangle$ or $|\underline{p}_3\underline{q}_3\rangle$, while $|\underline{p}_i'\underline{q}_i'\rangle$ is different (as described in Subsec.1.2.1). Eq. (1.104) results in a set of three-dimensional integral equations (after the δ-function is taken into account). To remove the δ-function in the inhomogeneous term one simply defines $W_{ij} = M_{ij} - \delta_{ij}t_i$, which satisfies

$$W_{ij} = \bar{\delta}_{ij}t_iG_0t_j + t_iG_0\sum_{k\neq i} W_{kj} \tag{1.105}$$

with $\bar{\delta}_{ij} = (1-\delta_{ij})$.

Although (1.105) gives the 3→3 amplitude, which is not usually measured experimentally, it can be used to derive equations for the usual situation in which a projectile hits a two-body bound state. To extract such an amplitude, we suppose that t_j (in the inhomogeneous term) has a bound state pole at $(E-p_j^2/2\mu_j) = -\varepsilon_{j;c}$. Then by taking the residue of the equation at this pole, one can extract an equation for break-up reactions. One can also extract the elastic and rearrangement amplitudes by taking the residue at the appropriate pole in the final state. Because the next section is devoted to an alternative derivation of equations for these amplitudes, this discussion will not be taken further. For a more detailed discussion of this procedure, and its relation to the next section, the reader is referred to the paper of OSBORN and KOWALSKI [1.21].

Equation (1.102) has been used as the starting point for deriving r-matrix (or K-matrix) equations for the three-body system. However, the more complex analytic structure of the three-body amplitude leads to more than one set of integral equations. For details the reader is referred to the work of CAHILL [1.22], KOWALSKI [1.23] and SASAKAWA [1.24].

To conclude this section we write an alternative set of three-body equations, which was given in Faddeev's first paper, and is used in Chapter 3. If we define T^i as $\sum_j M_{ij}$, then clearly (cf. (1.100)) T is $\sum_i T^i$, and from (1.102) these T^i satisfy the Faddeev-type equations

$$T^i = t_i + t_i G_0 (T^j + T^k) \quad . \tag{1.106}$$

Finally, substituting this subdivision of T into (1.98) and taking $\lim_{\varepsilon \to 0+} i\varepsilon G(E+i\varepsilon)\phi_0$, one finds an expression for the full scattering wave function which is needed in Chapter 4

$$\Psi_0^{(+)} = \phi_0 + \sum_i G_0(E^+) T^i \phi_0 \quad . \tag{1.107}$$

1.5 Bound State Scattering

As we have already stated, the processes most easily studied experimentally are elastic and rearrangement scattering of a projectile from a bound state. For practical purposes, it is convenient to have Faddeev-like equations for these amplitudes directly.

1.5.1 The Amplitudes U_{ji}

Consider the process

$$i + (jk)_c \longrightarrow j + (ki)_d \tag{1.108}$$

where $(\ )_c$ and $(\ )_d$ are bound states. From (1.74) and (1.81) the total three-body wave function is

$$\Psi_{i;c}^{(\pm)} = \chi_{i;c} + G(E^\pm) V^i \chi_{i;c} \quad . \tag{1.109}$$

Now applying (1.33) to G(E) we find

$$G(E^\pm) = G(E^\mp) \mp 2\pi i \delta(E - H) \quad , \tag{1.110}$$

and hence $\Psi_{i;c}^{(+)}$ is related to $\Psi_{i;c}^{(-)}$ by

$$\Psi^{(\pm)}_{i;c} = \Psi^{(\mp)}_{i;c} \mp 2\pi i\delta(E - H)\chi_{i;c} \quad , \tag{1.111}$$

and the S-matrix is then given by

$$S_{j(d);i(c)} \equiv \langle\Psi^{(-)}_{j;d}|\Psi^{(+)}_{i;c}\rangle$$

$$= \delta_{ji}\delta_{dc} - 2\pi i\delta(E_j - E_i)\langle\chi_{j;d}|V^j|\Psi^{(+)}_{i;c}\rangle \tag{1.112a}$$

$$= \delta_{ji}\delta_{dc} - 2\pi i\delta(E_j - E_i)\langle\Psi^{(-)}_{j;d}|V^i|\chi_{i;c}\rangle \quad . \tag{1.112b}$$

One therefore has two expressions for the on-shell three-body t-matrix

$$T_{j(d);i(c)} = \langle\chi_{j;d}|V^j|\Psi^{(+)}_{i;c}\rangle \equiv \langle\chi_{j;d}|U^{(+)}_{ji}|\chi_{i;c}\rangle \tag{1.113a}$$

$$= \langle\Psi^{(-)}_{j;d}|V^i|\chi_{i;c}\rangle \equiv \langle\chi_{j;d}|U^{(-)}_{ji}|\chi_{i;c}\rangle \quad . \tag{1.113b}$$

The operators $U^{(\pm)}_{ji}$ were first introduced by LOVELACE [1.6]. Using (1.109) and (1.73) one can write

$$U^{(+)(-)}_{ji}(z) = V^{j(i)} + V^j G(z) V^i \quad , \tag{1.114}$$

from which one sees that although $U^{(\pm)}_{ji}$ are equal on-shell, they differ off-energy-shell. Now, substituting $V^i = \sum_{k\neq i} V_k$, and (1.95b) for G in terms of G_k and V^k, into (1.114) leads to

$$\begin{aligned} U^{(+)}_{ji} &= V^j + \sum_{k\neq i} \{V^j G_k V_k + V^j G V^k G_k V_k\} \\ &= V^j + \sum_{k\neq i} U^{(+)}_{jk} G_0 t_k \quad , \end{aligned} \tag{1.115a}$$

where the last line is a consequence of (1.114) for $U^{(+)}_{ji}$ and the equality $G_k V_k = G_0 t_k$ (cf. Subsec.1.1.4). Analogous steps for $U^{(-)}_{ji}$, lead to the expression

$$U^{(-)}_{ji} = V^i + \sum_{k\neq j} t_k G_0 U^{(-)}_{ki} \quad . \tag{1.115b}$$

While (1.115) are of the Faddeev type, they have the double disadvantage of introducing two possible off-shell extensions of the three-body scattering amplitude, as well as satisfying equations which involve *both* the two-body potentials and t-matrices (unlike (1.102)).

To resolve both these problems, we introduce a new operator "U_{ji}", defined by

$$G(z) = G_j(z)\delta_{ij} + G_j(z) U_{ji}(z) G_i(z) \quad . \tag{1.116}$$

Then, substituting (1.95b) into (1.95a) - with j replacing i in the latter - one finds

$$G = G_j + G_jV^jG_i + G_jV^jGV^iG_i = G_j + G_jU_{ji}^{(+)}G_i \quad , \tag{1.117}$$

and hence (by comparison with (1.116))

$$U_{ji}(z) = \bar{\delta}_{ji}(z - H_i) + U_{ji}^{(+)}(z) \quad . \tag{1.118a}$$

Similarly, substituting (1.95a) into (1.95b) leads to

$$U_{ji}(z) = \bar{\delta}_{ji}(z - H_j) + U_{ji}^{(-)}(z) \quad . \tag{1.118b}$$

To obtain an integral equation for U_{ji}, let us substitute (1.118a) into (1.115a):

$$U_{ji} = \bar{\delta}_{ji}G_i^{-1} + V^j - \sum_{\ell \neq i} \bar{\delta}_{j\ell}G_\ell^{-1}G_0t_\ell + \sum_{\ell \neq i} U_{j\ell}G_0t_\ell \quad . \tag{1.119}$$

Next we show that the sum of the first three terms on the rhs of (1.119) is just $\bar{\delta}_{ji}(z-H_0)$. To do so, we note (from (1.25)) that $G_\ell^{-1}G_0t_\ell$ equals $(G_0^{-1}-G_\ell^{-1})$. Hence this sum becomes

$$\sum_\ell \bar{\delta}_{j\ell}(z - H_\ell) - \Big[\sum_{\ell \neq i} \bar{\delta}_{j\ell}(z - H_0) - \sum_\ell \bar{\delta}_{j\ell}V_\ell\Big] \quad , \tag{1.120}$$

and adding and subtracting $\bar{\delta}_{ji}(z-H_0)$ to the term in brackets proves the result. Therefore we have established that U_{ji} satisfies the Faddeev-type equation

$$U_{ji}(z) = \bar{\delta}_{ji}(z - H_0) + \sum_{\ell \neq i} U_{j\ell}(z)G_0(z)t_\ell(z) \quad . \tag{1.121a}$$

By substituting (1.118b) into (1.115b) and following analogous steps, one can easily show now that

$$U_{ji}(z) = \bar{\delta}_{ji}(z - H_0) + \sum_{\ell \neq j} t_\ell(z)G_0(z)U_{\ell i}(z) \quad . \tag{1.121b}$$

Clearly U_{ji} is a much more symmetric operator than $U_{ji}^{(\pm)}$, with none of their drawbacks. The integral equations (1.121) were first derived by ALT et al. [1.25], and are often referred to as the AGS equations. Clearly, since the matrix elements of $[z-H_{i(j)}]$ vanish on-shell, U_{ji} predicts the correct physical scattering amplitudes

$$T_{j(d);i(c)}(E^+) = \langle \chi_{j;d}|U_{ji}(E^+)|\chi_{i;c}\rangle \quad . \tag{1.122}$$

1.5.2 Break-Up in Terms of U_{ji}

Although the main purpose of this section was to derive (1.121) for elastic and rearrangement scattering, in most experiments break-up is another important possibility. Since it is very easy to calculate break-up once one has solved for U_{ji}, we give the expression here.

None of the steps leading to (1.114) for $U_{ji}^{(\pm)}$ relied on the final state having a bound pair. In fact, if the asymptotic final state is a plane wave ϕ_0, the expression for the break-up (2→3) matrix element is given by

$$T_{0;i(c)}(E) = \langle\phi_0|U_{0i}^{(\pm)}(E)|\chi_{i;c}\rangle \quad , \tag{1.123}$$

where $U_{0i}^{(\pm)}$ is given by (1.114) with $V^0 \equiv V$. Now, using (1.118b) to express $U_{0i}^{(-)}$ in terms of U_{0i} implies

$$\begin{aligned} U_{0i}^{(-)}(E) &= (H_0 - E) + U_{0i} \\ &= \sum_{\ell=1}^{3} t_\ell G_0 U_{\ell i} \quad , \end{aligned} \tag{1.124}$$

where the last step involves substituting (1.121b), with j = 0, for U_{0i}. Hence the break-up amplitude is

$$T_{0,i(c)} = \langle\phi_0|\sum_\ell t_\ell G_0 U_{\ell i}|\chi_{i;c}\rangle \quad , \tag{1.125}$$

and all that is required in order to calculate a physical break-up cross section is a three-dimensional integration ($\int d\underline{q}_\ell''$).

It is rather important to note that (1.125) is only correct when both $\chi_{i;c}$ and ϕ_0 are on-shell. If the three-particle state is an intermediate step in some more complicated calculation and can therefore have arbitrary energy, one is no longer free to use $U_{0j}^{(\pm)}$ interchangeably. In such a case one should use the form $U_{0i}^{(+)}$, which from (1.114) differs from $U_{0i}^{(-)}$ by V_i. Thus, with ϕ_0 off-energy shell the correct break-up amplitude is given by (1.125) plus $\langle\phi_0|V_i|\chi_{i;c}\rangle$, which can be written without specific mention of V_i as $\langle\phi_0|(E-H_0)|\chi_{i;c}\rangle$.

1.5.3 Calculation of Cross-Sections

With the normalization of momentum states that is used here (cf. (1.17)), and assuming bound state wave functions normalized to unity, the t-matrices for elastic and rearrangement scattering (1.122), and break-up (1.125) are the same as those defined by GOLDBERGER and WATSON [1.3]. Therefore one can use their formulas for the scattering cross section. That is, we assume a two-body initial state $|i\rangle$ (e.g., $\chi_{i;c}$) (since in practice one cannot study scattering with three particles incident), in which the particles have relative velocity $v_{in}(\equiv p_i/\mu_i)$. The final state $|f\rangle$ contains N_f particles (for us N_f is two for $\chi_{j;d}$, or three for ϕ_0), with momenta $\underline{p}_j$ (j = 1, ..., N_f). Then the differential cross section in the three-body CM is

$$d\sigma_{fi} = (2\pi)^4 v_{in}^{-1} |T_{fi}|^2 \delta(E_f - E_i)\delta(\sum_{j=1}^{N_f} \underline{p}_j) \prod_{j=1}^{N_f} d\underline{p}_j \quad , \tag{1.126}$$

[cf. Ref.1.3, Eq. (142) p.92]. (In our notation T_{fi} is $T_{j(d);i(c)}$ or $T_{0;i(c)}$). The reader should verify that this expression leads to (1.6) and (1.30) for two-body elastic scattering.

For comparison with experiment one must calculate the cross section in the laboratory system. This transformation is also completely described in Ref.1.3, pp.221 to 225. One could not do better than refer the reader to that discussion. Finally, when the particles have spin, the labels f and i include spin projections. If the spin projections are not known initially or finally, the differential cross section is obtained by summing the rhs of (1.126) over the final spins, and averaging over the initial spins.

1.6 Unitarity

1.6.1 Formal Expression for disc $\{U_{ji}\}$

In Subsection 1.1.4 we showed that the discontinuity of the on-shell two-body t-matrix across its right-hand cut was proportional to the total cross section at that energy. It is our purpose here to show that there is a similar condition on the three-body scattering amplitudes. In order to show this we follow essentially the method of [1.26]. Consider (1.121b) as a matrix equation

$$\underset{\approx}{U} = \underset{\approx}{U}^{(0)} + \underset{\approx}{U}^{(0)}\underset{\approx}{\Gamma}\underset{\approx}{U} \quad , \tag{1.127}$$

$$[\underset{\approx}{U}]_{ij} = U_{ij};[\underset{\approx}{U}^{(0)}]_{ij} = \bar{\delta}_{ij}G_0^{-1} \quad , \tag{1.128a}$$

$$[\underset{\approx}{\Gamma}]_{ij} = \delta_{ij}G_0 t_i G_0 \quad . \tag{1.128b}$$

Let us rewrite (1.127) in the equivalent form

$$\underset{\approx}{U}^{-1}(E^+) = \underset{\approx}{U}^{(0)-1}(E^+) - \underset{\approx}{\Gamma}(E^+) \quad . \tag{1.129}$$

Then, subtracting the same expression at E^- gives

$$\underset{\approx}{U}^{-1}(E^-) - \underset{\approx}{U}^{-1}(E^+) = [\underset{\approx}{U}^{(0)-1}(E^-) - \underset{\approx}{U}^{(0)-1}(E^+)] + [\underset{\approx}{\Gamma}(E^+) - \underset{\approx}{\Gamma}(E^-)] \quad . \tag{1.130}$$

The only quantity here which has not been written explicitly is $\underset{\approx}{U}^{(0)-1}$. One can readily verify, by writing both (1.128a) and $\underset{\approx}{U}_0^{-1}$ as 3x3 matrices, that $[\underset{\approx}{U}^{(0)-1}]_{ij}$ is simply $(^1/2 - \delta_{ij})G_0$. Using this, plus the "disc" notation of (1.38), one can express (1.130) as

$$\text{disc}\ \{[\underset{\approx}{U}^{-1}(E)]_{ij}\} = \frac{1}{2}\ \text{disc}\ \{G_0(E)\} - \delta_{ij}\ \text{disc}\ \{G_i(E)\} \quad . \tag{1.131}$$

From (1.67) and (1.33) one finds that the discontinuity of $G_i(E)$ is $-2\pi i\delta(E-H_i)$ (including i=0). In order to see more clearly what this means, we multiply it by a unit operator - expressed in terms of a complete set of eigenfunctions of H_i (cf. (1.53)). This includes two-body bound states ($\chi_{i;c}$) and scattering states (ϕ_i - see Subsec.1.3.2) in the presence of a noninteracting spectator.

$$\begin{aligned}\text{disc}\{G_i(E)\} &= \sum_c \int d\underline{p}_i(-2\pi i)|\chi_{i;c}(\underline{p}_i)\rangle\delta(E - p_i^2/2\mu_i - \varepsilon_{i;c})\langle\chi_{i;c}(\underline{p}_i)| \\ &+ \int d\underline{p}_i d\underline{q}_i(-2\pi i)|\phi_i(\underline{p}_i,\underline{q}_i)\rangle\delta(E - p_i^2/2\mu_i - q_i^2/2\mu_{jk})\langle\phi_i(\underline{p}_i,\underline{q}_i)| \quad .\end{aligned} \tag{1.132}$$

Of course, $|\phi_0(\underline{p}_i,\underline{q}_i)\rangle$ is just $|\underline{p}_i\underline{q}_i\rangle$.

Let us now make the definitions

$$\Delta_{i;c} = \int d\underline{p}_i|\chi_{i;c}(\underline{p}_i)\rangle\delta(E - p_i^2/2\mu_i - \varepsilon_{i;c})\langle\chi_{i;c}(\underline{p}_i)| \quad , \tag{1.133}$$

$$\Delta_0 = \int d\underline{p}_i d\underline{q}_i|\underline{p}_i\underline{q}_i\rangle\delta(E - p_i^2/2\mu_i - q_i^2/2\mu_{jk})\langle\underline{p}_i\underline{q}_i| \quad . \tag{1.134}$$

Then, identifying $\phi_i = (1+G_0t_i)\phi_0$ (from (1.26)) gives

$$\begin{aligned}\text{disc}\ \{[\underset{\approx}{U}^{-1}(E)]_{ij}\} &= -\pi i\Delta_0(E) + 2\pi i\delta_{ij}\ \{\sum_c \Delta_{i;c}(E) \\ &+ [1 + G_0(E^+)t_i(E^+)]\Delta_0(E)[1 + t_i(E^-)G_0(E^-)]\} \quad .\end{aligned} \tag{1.135}$$

Multiplying to the left by $\underset{\approx}{U}(E^+)$ and the right by $\underset{\approx}{U}(E^-)$ gives

$$\begin{aligned}\text{disc}\{U_{ji}(E)\} &= \pi i \sum_{k,\ell} U_{jk}(E^+)\Delta_0 U_{\ell i}(E^-) \\ &- 2\pi i\{\sum_{k,c} U_{jk}(E^+)\Delta_{k;c}(E)U_{ki}(E^-) \\ &+ \sum_k U_{jk}(E^+)[1 + G_0(E^+)t_k(E^+)]\Delta_0(E)[1 + t_k(E^-)G_0(E^-)]U_{ki}(E^-)\} \quad .\end{aligned} \tag{1.136}$$

Next, we sum (1.121b) over j, to find that $\sum_k U_{ki}$ is just $2U_{0i}$. Also, adding $t_jG_0U_{ji}$ to both sides of (1.121b) one finds that

$$(1 + t_kG_0)U_{ki} = [\bar{\delta}_{ki} - 1](E - H_0) + U_{0i} \quad . \tag{1.137}$$

Finally we may use these results to eliminate U_{ki} in the first and last terms of (1.136) in favor of U_{0i}. (The $(E-H_0)$ term in (1.137) gives zero acting on Δ_0, and so may be dropped). Thus, the complete expression of three-body unitarity is

$$\mathrm{disc}\{U_{ji}(E)\} = -2\pi i\left\{\sum_{k,c} U_{jk}(E^+)\Delta_{k;c}(E)\,U_{ki}(E^-) + U_{j0}(E^+)\Delta_0(E)U_{0i}(E^-)\right\} \quad . \tag{1.138}$$

1.6.2 Meaning for Elastic Bound State Scattering

Consider the discontinuity of the forward, on-shell, elastic scattering amplitude for particle i hitting a bound state "c". Using (1.122) and (1.125) this is

$$\begin{aligned}\mathrm{disc}\{T_{i(c);i(c)}(E,\theta = 0)\} = -2\pi i\Big\{\sum_{d,j} \int d\underline{p}_i\,|T_{j(d);i(c)}(E^+)|^2 \\ \times\,\delta(E - p_j^2/2\mu_j - \varepsilon_{j;d}) + \int d\underline{p}_i d\underline{q}_i\,|T_{0;i(c)}(E^+)|^2 \\ \times\,\delta(E - p_i^2/2\mu_i - q_i^2/2\mu_{jk})\Big\} \quad .\end{aligned} \tag{1.139}$$

Now using (1.126) one sees that the imaginary part of the forward scattering amplitude is proportional to the total cross section for all possible reactions. In particular, one finds that

$$\mathrm{Im}\{T_{i(c);i(c)}(E,\theta=0)\} = -\pi(2\pi)^{-4}v_{in}\left\{\sum_{j,d} \sigma_{j(d)\leftarrow i(c)} + \sigma_{b\text{-}up}\right\} \quad , \tag{1.140}$$

where the "σ's" are the total cross sections. This is the three-body extension of the simple optical theorem found in Section 1.1.

1.6.3 Remarks

Equation (1.138) shows that the cut structure of the three-body amplitude is rather more complicated than the two-body case. There is a cut starting at each $\varepsilon_{j;c}$, which corresponds to the right-hand cut for elastic (or rearrangement) scattering. In addition there is the cut corresponding to break-up. Notice how important in this structure is the form of the effective two-body interaction energy in the three-body system - as discussed in Subsection 1.2.2.

To conclude, we recall that in the two-body system one can remove the unitarity cut and work entirely in terms of a real r-matrix. The extension of this procedure to the three-body system is discussed in [1.22-24]. Essentially one must remove the cuts described above one at a time, until only a real quantity remains. The resulting equations have been applied to n-d scattering [1.27].

1.7 Identical Particles

The inclusion of identical particles is a topic that is often glossed over, with little more than a reference to Lovelace's discussion of n-d scattering [1.6]. In this section we describe a systematic procedure for treating identical particles. Indeed with a few basic rules, one can easily write the correct combination of amplitudes needed in any case of interest. Our discussion relies heavily on the more detailed discussion in Ref.1.3, Chapter 7.

The essence of the problem is to guarantee that if in any intermediate state there are two (or three) identical particles, then that state must have the appropriate symmetry. It is a pleasant fact that if one has a transition operator which is symmetric under interchange of these particles, using initial and final states of the appropriate symmetry guarantees that only intermediate states of the correct symmetry can contribute. Thus, one can do the whole calculation with (unsymmetrized) plane wave intermediate states, and only symmetrize the initial and final states!

As a simple example, consider the Lippmann-Schwinger equation (1.23a) for the scattering of identical bosons. Then the procedure just outlined tells one to place t between symmetric initial and final states - written $|12+21\rangle$. (If "12" represents $\exp(i\underline{q}\cdot\underline{r}_{12})$, then "21" is $\exp(-i\underline{q}\cdot\underline{r}_{12})$). It is left as an exercise for the reader to show that if (1.27a) is evaluated between such states, then the contribution from antisymmetric intermediate states (in VG_0t) vanishes[3]. Consequently the physical scattering amplitude for bosons (fermions) is then given by

$$2^{-1/2}\langle 12 \overset{+}{(-)} 21|t|12 \overset{+}{(-)} 21\rangle 2^{-1/2} = \langle 12|t|12\rangle \overset{+}{(-)} \langle 12|t|21\rangle \quad , \tag{1.141}$$

which shows the origin of the "direct plus (minus) exchange" recipe (as used in [1.6]).

These same fundamental ideas apply in the three-body problem. One can use plane waves throughout, with no worries about symmetry, provided that the physical scattering amplitudes are evaluated using correctly normalized initial and final states of the appropriate symmetry. Consider the scattering of three identical fermions (bosons), in the case where there is only one two-body bound state. One then has scattering from, and break-up of, this bound state. Assuming a normalized bound state ($\langle\psi|\psi\rangle=1$) of the correct symmetry [$\psi(12)=(\mp)\psi(21)$], an appropriately normalized asymptotic state is[4]

---

[3] One needs to realize that replacing 1 by 2, and vice-versa, throughout a matrix element of V (or t) does not change its value, i.e., $\langle 12|V|12\rangle = \langle 21|V|21\rangle$ and $\langle 12|V|21\rangle = \langle 21|V|12\rangle$.

[4] On this occasion the time-independent theory is not strictly correct. The factor $3^{-1/2}$ arises because one assumes $\langle 1(23)|2(31)\rangle = 0$, which is only justifiable for wave packet states.

$$3^{-1/2}|1(23) + 2(31) + 3(12)\rangle \quad . \tag{1.142}$$

Here the bound pair is shown in brackets. (The cyclic ordering means that 1(23) is related to 2(31) by two interchanges - which is why (1.142) is correct for both bosons and fermions).

The physical scattering amplitude "$T_{2\leftarrow 2}$" is now found by placing the U operator between states (1.142)

$$\begin{aligned} T_{2\leftarrow 2} &= 3^{-1/2}\langle 1(23) + 2(31) + 3(12)|U|1(23) + 2(31) + 3(12)\rangle 3^{-1/2} \\ &= \langle\chi|U_{11} + U_{21} + U_{31}|\chi\rangle \\ &= \langle\chi|U^D + 2U^E|\chi\rangle \equiv \langle\chi|\tilde{U}|\chi\rangle \quad . \end{aligned} \tag{1.143}$$

Here, $|\chi\rangle$ represents a three-body state containing the bound state $|\psi\rangle$ (cf. (1.86)). As in the two-body case the identity of the particles has led us to equate $(U_{11}+U_{21}+U_{31})$ with $(U_{22}+U_{12}+U_{32})$, and U_{21} with U_{31}, etc. In order to calculate cross sections, one needs only U, for which one can derive a single integral equation. In fact, adding the equations (1.121b) for $U_{i1}(i = 1,2,3)$, one finds

$$\tilde{U} = 2(z - H_0) + 2tG_0\tilde{U} \quad , \tag{1.144}$$

which is a threefold reduction in the number of equations.

One would also like to evaluate the physical break-up amplitude $T_{3\leftarrow 2}$. Suppose the three final momenta are $(\underline{p}_a, \underline{p}_b, \underline{p}_c)$. Then with the notation that $(123) \equiv [\underline{p}_a(1), \underline{p}_b(2), \underline{p}_c(3)]$, $(312) \equiv [\underline{p}_a(3), \underline{p}_b(1), \underline{p}_c(2)]$ and so on, the correctly normalized and symmetrized final state for fermions (bosons) is

$$6^{-1/2}|123\ (\bar{+})\ 132 + 231\ (\mp)\ 213 + 312\ (\bar{+})\ 321\rangle \quad . \tag{1.145}$$

To find $T_{3\leftarrow 2}$, one places the break-up transition operator $\sum_\ell t_\ell G_0 U_{\ell -}$[5] (cf. (1.124)) between the states (1.142) and (1.145). After some algebra this becomes

$$\begin{aligned} T_{3\leftarrow 2} = 2^{-1/2}\{&\langle\underline{p}_a\underline{q}_a|(t^D\ (\bar{+})\ t^E)G_0\tilde{U} \\ &+ \langle\underline{p}_b\underline{q}_b|(t^D\ (\mp)\ t^E)G_0\tilde{U} + \langle\underline{p}_c\underline{q}_c|(t^D\ (\bar{+})\ t^E)G_0\tilde{U}\}|\chi\rangle \quad , \end{aligned} \tag{1.146}$$

where t^D, t^E are the direct and exchange matrix elements - corresponding to $\langle 12|t|12\rangle$ and $\langle 21|t|12\rangle$, respectively.

---

[5] The "-" is determined by the spectator label in the initial state, i.e., 1(23) gives $U_{\ell 1}$, etc.

It is a long but instructive exercise, which we leave to the motivated reader, to verify that the unitarity relation for the forward elastic scattering amplitude is once again (1.140). Now, however, $T_{2\leftarrow 2}$ is given by (1.143), and the total cross section for break-up is

$$\sigma_{b\text{-}up} = \frac{(2\pi)^4}{3!v_{in}} \int d\underline{p}_i' d\underline{q}_i' |T_{3\leftarrow 2}(\underline{p}_i',\underline{q}_i';\underline{p}_{in})|^2 \delta\left(E - \frac{p_i'^2}{2\mu_i} - \frac{q_i'^2}{2\mu_{jk}}\right) , \tag{1.147}$$

where the 2→3 amplitude is given by (1.146). We emphasize that the factor of 1/3! is physically crucial, since there are six ways of arranging the identical particles among the three final state momenta. Without this factor, one would overcount the total cross section by a factor of six. Finally, we note that in calculating the differential cross section for break-up, one simply squares the expression (1.146) for $T_{3\leftarrow 2}$ and uses it in (1.126) with no statistical factors.

To conclude this section we emphasize that the procedure outlined here is always correct, even though this need not be obvious. To illustrate this, consider the expansion to second order of U_{33}, describing the elastic scattering of a distinguishable projectile "3" (e.g., a pion) from a bound pair of identical fermions "12" (e.g., a deuteron). This is $(t_1+t_2+t_1G_0t_2+t_2G_0t_1)$, in which there is no obvious symmetry ensuring that only antisymmetric intermediate states (for 12) can contribute. However, if one expands the t_i to second order (as $v_i+v_ig_0v_i$), this can be rewritten (to second order in the v's now) as $(v_1+v_2)+(v_1+v_2)G_0(v_1+v_2)$ - which does have the required symmetry. A similar proof can be carried on to all orders. Clearly then, one of the prices we have paid in replacing the potentials by t-matrices in the Faddeev approach is to obscure some underlying symmetry properties.

1.8 Three-Body Scattering With Separable Interactions

In this section we present the Faddeev equations in their most practical form. The majority of applications of three-body formalism have used separable two-body interactions. This has the great advantage of leading to quasi-two-body equations for the physical amplitudes (i.e., T rather than U). Moreover, after angular momentum reduction these equations become one-dimensional integral equations (see Sec.1.9).

1.8.1 General Formulation

The t-matrix for a separable two-body potential was discussed in Subsection 1.1.5. For conciseness, we choose to write such a t-matrix in three-body Hilbert space as (cf. (1.49))

$$t_i(E) = \sum_a \int d\underline{p}_i |i\underline{p}_i;a\rangle \tau_{ia}(E - p_i^2/2\mu_i)\langle i\underline{p}_i;a| . \tag{1.148}$$

For spinless particles the form-factor has the momentum representation

$$\langle \underline{p}_i' \underline{q}_i' | i\underline{p}_i;a\rangle = \delta(\underline{p}_i - \underline{p}_i') g_\ell(q_i') Y_{\ell m}(\hat{\underline{q}}_i) \quad , \tag{1.149}$$

where clearly the label a is just "ℓm" here. The energy dependence of "τ" was discussed in detail in Subsection 1.2.2.

Let us first substitute (1.148) for t_ℓ in (1.121b) for U_{ji}, and then take matrix elements of the resultant equation between the products $\langle j\underline{p}_j';b|G_0$ and $G_0|i\underline{p}_i;a\rangle$. This gives

$$\begin{aligned} \langle j\underline{p}_j';b|G_0 U_{ji} G_0|i\underline{p}_i;a\rangle &= \bar{\delta}_{ji} \langle j\underline{p}_j';b|G_0|i\underline{p}_i;a\rangle \\ &+ \sum_{\ell,c} \int d\underline{p}_\ell'' \bar{\delta}_{j\ell} \langle j\underline{p}_j';b|G_0|\ell\underline{p}_\ell'';c\rangle \tau_{\ell c}(E - p_\ell''^2/2\mu_\ell) \\ &\times \langle \ell\underline{p}_\ell'';c|G_0 U_{\ell i} G_0|i\underline{p}_i;a\rangle \quad , \end{aligned} \tag{1.150}$$

where all three-body operators correspond to energy E. If one makes the definitions

$$X_{jb;ia}(\underline{p}_j',\underline{p}_i;E) \equiv \langle j\underline{p}_j';b|G_0(E)U_{ji}(E)G_0(E)|i\underline{p}_i;a\rangle \quad , \tag{1.151}$$

$$Z_{jb;ia}(\underline{p}_j',\underline{p}_i;E) \equiv \langle j\underline{p}_j';b|G_0(E)|i\underline{p}_i;a\rangle \quad , \tag{1.152}$$

(1.150) takes the form

$$\begin{aligned} X_{jb;ia}(\underline{p}_j',\underline{p}_i;E) &= Z_{jb;ia}(\underline{p}_j',\underline{p}_i;E) \\ &+ \sum_{\ell,c} \int d\underline{p}_\ell'' \, Z_{jb;\ell c}(\underline{p}_j',\underline{p}_\ell'';E)\tau_{\ell c}(E - p_\ell''^2/2\mu_\ell) X_{\ell c;ia}(\underline{p}_\ell'',\underline{p}_i;E) \quad . \end{aligned} \tag{1.153}$$

This is a set of coupled three-dimensional integral equations, which look very much like the Lippmann-Schwinger equation for a two-body coupled channels problem. Here "τ" is the propagator, and "Z" an energy-dependent potential. What exactly are the amplitudes "X"?

To understand the X amplitudes, we recall (Subsec.1.1.5) that the bound state wave function is very closely related to the form-factor for a separable potential. Indeed, if the pair (jk) has a bound state in channel a at energy $-\varepsilon_{i;a}$, then for $\underline{p}_i$ corresponding to an on-shell projectile bound pair system (i.e., $p_i^2 = 2\mu_i[E+\varepsilon_{i;a}]$), (1.57) leads to

$$\langle \underline{p}_i'\underline{q}_i'|G_0(E)|i\underline{p}_i;a\rangle = \delta(\underline{p}_i - \underline{p}_i')(\varepsilon_{i;a} + p_i^2/2\mu_i)^{-1}g_a(q_i')Y_a(\hat{\underline{q}}_i')$$
$$\equiv \langle \underline{p}_i'\underline{q}_i'|\chi_{i;a}(\underline{p}_i)\rangle \quad , \tag{1.154}$$

where $\chi_{i;a}$ should be familiar by now as the wave function for a two-body bound state in three-body Hilbert space.

In writing (1.154) we have assumed that the normalization constant (F_ℓ in (1.57)) is included into the definition of the form-factor - this can only be done for a one term potential. Now, from (1.154) it is readily seen that in the case where both $\underline{p}_j'$ and $\underline{p}_i$ correspond to on-shell scattering (as defined above), the X amplitudes are

$$X_{jb;ia}(\underline{p}_j',\underline{p}_i;E) = \langle \chi_{j;b}(\underline{p}_j')|U_{ji}(E)|\chi_{i;a}(\underline{p}_i)\rangle \quad . \tag{1.155}$$

Therefore, from (1.122), they are the physical scattering amplitudes $T_{j(b);i(a)}$, for elastic or rearrangement scattering.

The matrix element for break-up of a bound state is now very easy to write down. Using a separable t-matrix in (1.125) one sees at once that (on-shell)

$$T_{0\leftarrow i(a)}(\underline{p}',\underline{q}';\underline{p}_i) = \sum_{\ell,c} \langle \underline{q}_\ell'|\ell;c\rangle\tau_{\ell c}(E - p_\ell'^2/2\mu_\ell)X_{\ell c;ia}(\underline{p}_\ell',\underline{p}_i;E) \quad , \tag{1.156}$$

which is an algebraic relation involving the half-shell scattering amplitude X, and the two-body form-factor $\langle \underline{q}_\ell'|\ell;c\rangle[\equiv g_c(q_\ell')Y_c(\hat{\underline{q}}_\ell')]$.

For completeness we note that there is a well-defined procedure, due to AGS [1.25], for including residual nonseparable parts of the t-matrices. Unfortunately we have no space to elaborate that here. Finally we note that one can write unitarity relations satisfied by these amplitudes. In fact, taking the matrix element of (1.138) between $\langle j\underline{p}_j';b|G_0$ and $G_0|i\underline{p}_i;a\rangle$, and using the definitions of $\Delta_{k;c}$, Δ_0 and U_{0i}, this unitarity relation is just

$$\begin{aligned} \mathrm{disc}\{X_{jb;ia}(\underline{p}_j',\underline{p}_i;E)\} = -2\pi i \Big\{ &\sum_{k,c} \int d\underline{p}_k'' X_{jb;kc}(\underline{p}_j',\underline{p}_k'';E^+) \\ &\times \delta(E - p_k''^2/2\mu_k + \varepsilon_{k;c})X_{kc;ia}(\underline{p}_k'',\underline{p}_i;E^-) \\ &+ \int d\underline{p}''d\underline{q}''[\sum_{k,c} X_{jb;kc}(\underline{p}_j',\underline{p}_k'';E^+)\tau_{kc}(E^+ - p_k''^2/2\mu_k)\langle kc|\underline{q}_k''\rangle] \\ &\times \delta(E - p_1''^2/2\mu_1 - q_1''^2/2\mu_{23})[\sum_{\ell,d}\langle \underline{q}_\ell''|\ell;d\rangle\tau_{\ell d}(E^- - p_\ell''^2/2\mu_\ell)X_{\ell d;ia}(\underline{p}_\ell'',\underline{p}_i;E^-)]\Big\} \quad . \end{aligned} \tag{1.157}$$

1.8.2 Three Identical Particles

All that is involved here is an application of the principles described in Section 1.7. That is, one takes the matrix elements of X between states of the type (1.142) - for elastic or rearrangement processes. Perhaps the easiest way to get the final equations is just to take matrix elements of (1.144) with $\langle b|G_0$ and $G_0|a\rangle$, to find

$$X_{b,a}(\underline{p}',\underline{p};E) = 2Z_{b,a}(\underline{p}',\underline{p};E) + \sum_c \int d\underline{p}'' 2Z_{b,c}(\underline{p}',\underline{p}'';E)\tau_c(E - p''^2/2\mu_1)X_{c,a}(\underline{p}'',\underline{p};E) \quad , \tag{1.158}$$

where $\mu_1 = 2m/3$ (m the mass of each particle), and[6]

$$Z_{b,c}(\underline{p}',\underline{p}'';E) = \langle 2\underline{p}';b|G_0(E)|1\underline{p}'';c\rangle \quad , \tag{1.159}$$

$$X_{b,a}(\underline{p}',\underline{p};E) = X_{1b,1a}(\underline{p}',\underline{p};E) + X_{2b,1a}(\underline{p}',\underline{p};E) + X_{3b,1a}(\underline{p}',\underline{p};E) \quad . \tag{1.160}$$

The number of coupled equations is now just equal to the number of channels in the two-body t-matrix.

From the discussion of Section 1.7, it should be clear that in order to calculate a differential cross section for elastic scattering, one simply identifies $X_{a,a}$ as $T_{f,i}$ in (1.126). The physical matrix element for break-up can also be written down from (1.146). To simplify things we assume that only symmetric (antisymmetric) two-body channels are included in (1.148) for the identical boson (fermion) case - so that $\langle \underline{p}_i'\underline{q}_i'|i\underline{p}_i;c\rangle$ is $(\pm)$ $\langle \underline{p}_i'(-\underline{q}_i')|i\underline{p}_i;c\rangle$. Then the break-up amplitude is (on-shell)

$$T_{0\leftarrow a}(\underline{p}',\underline{q}';\underline{p}) = \sqrt{2} \sum_{i,c} \langle \underline{q}_i'|c\rangle\tau_c(E^+ - p_i'^2/2\mu_1)X_{c,a}(\underline{p}_i',\underline{p};E^+) \quad , \tag{1.161}$$

for both bosons and fermions (although the contributing states "c" are clearly different in the two cases).

1.9 Rotationally Invariant Equations for Separable Interactions

The reduction of the quasi-two-body equations (1.153) to rotationally invariant form is an essential step in most practical three-body calculations. While this is well described in many places for the three-nucleon case with s-wave interactions, one has to search the literature to find the result in the case of arbitrary spin (and angular momentum). In view of the variety of problems now being tackled by

---

[6] As in Section 1.7 we realize that interchanging any indices throughout a matrix element does not alter its value, so that $\langle 1(23)|X|2(31)\rangle \equiv \langle 3(12)|X|1(23)\rangle$, etc. Also τ_{ic} is independent of i, and is written "τ_c".

three-body methods (as described in Chap.6), it seems worthwhile to present the general result here. The motivated reader should check that the standard (spinless and three-nucleon) results follow from the general formulas presented here.

To begin, recall that nowhere in the derivation of (1.153) did one need to assume that the particles were spinless. In fact, with the appropriate interpretation these equations are suitable for particles of arbitrary spin. In such a case, the form factor $|i\underline{p}_i;a\rangle$ is completely specified (in the three-body CM) by the momentum $\underline{p}_i$ of the spectator i, and its internal quantum numbers (e.g., parity, spin and spin projection (π_i,j_i,m_i)), as well as the properties of the interacting pair $(\bar{\pi}_i,\bar{J}_i,\bar{m}_i)$. Thus, in the general case, the label "a" in this form factor can be interpreted as $(\bar{m}_i,m_i,n_a)$, where n_a includes all the other quantum numbers needed to uniquely specify the channel (e.g., $n_a = (\bar{\pi}_i,\bar{J}_i,\pi_i,j_i)$).

The two-body t-matrix in this case can still be written in the form (1.148), but it should be remembered that "τ" is independent of the spin projections $(\bar{m}_i,m_i)$, and also of the internal quantum numbers of the spectator. In fact, let us write the form factor as

$$\langle\underline{p}_i'\underline{q}_i'|i\underline{p}_i;a\rangle = \delta(\underline{p}_i - \underline{p}_i')\chi_{j_i m_i} g_{\bar{J}_i\bar{\pi}_i}(q_i')\mathcal{Y}_{\bar{J}_i\bar{m}_i}(\hat{q}_i') \quad , \tag{1.162}$$

where $\chi_{j_i m_i}$ is the intrinsic wave function of i, and $\mathcal{Y}_{\bar{J}_i\bar{m}_i}$ is an eigenstate of the pair's total angular momentum and projection. Then the function "$\tau_{\bar{J}_i\bar{\pi}_i}$" is given by (1.50) and (1.51) with $g_{\bar{J}_i\bar{\pi}_i}$ replacing g_ℓ.

Now, showing the spin projections explicitly, the matrix element for which one solves (1.153) is

$$\langle\bar{m}_j m_j n_b;\underline{p}_j'|X|\bar{m}_i m_i n_a;\underline{p}_i\rangle \quad . \tag{1.163}$$

We choose to express this in terms of rotationally invariant amplitudes written in the channel spin coupling scheme. That is, we couple the pair spin $\bar{J}$ to the spectator spin j to give the channel spin S (with z-projection Σ), which is then coupled to the angular momentum of the spectator with respect to the pair CM L, to give the total angular momentum of the system J. To summarize

$$\bar{J} + j \to S;\ L + S \to J \quad . \tag{1.164}$$

In this coupling scheme (1.163) may be written[7]

$$\langle\bar{m}_j m_j n_b;\underline{p}_j|X|\bar{m}_i m_i n_a;\underline{p}_i\rangle \tag{1.165}$$

---

[7] We apologize for the awkward j-sub-j notation. This is the reason for using (α, β,γ) as particle labels in the Appendix.

$$= \sum_{S_i\Sigma_i S_j\Sigma_j} \sum_{L_iM_iL_jM_j} \sum_{JM} \langle \bar{J}_j\bar{m}_j j_j m_j | S_j\Sigma_j \rangle$$

$$\times \langle L_jM_jS_j\Sigma_j | JM \rangle Y_{L_jM_j}(\hat{\underline{p}}'_j) \langle L_jS_jn_b;p'_j | X^J | L_iS_in_a;p_i \rangle \tag{1.165}$$

$$\times Y^*_{L_iM_i}(\hat{\underline{p}}_i) \langle L_iM_iS_i\Sigma_i | JM \rangle \langle \bar{J}_i\bar{m}_i j_i m_i | S_i\Sigma_i \rangle \quad ,$$

where the Clebsch-Gordan coefficients and spherical harmonics are as defined in [1.4]. The amplitude X is diagonal in (JM), because the Hamiltonian that we use commutes with (J,J_z). In addition, the fact $\langle L_jS_jn_b;p'_j|X^J|L_iS_in_a;p_i\rangle$ is independent of M relies on the Wigner-Eckart theorem (Ref.1.4, p.73). This independence of M is why the amplitude is referred to as rotationally invariant. (The effect of a rotation of a frame in which a given state is $|JM\rangle$ is to make it a linear combination of states $|JM'\rangle$ in the new coordinates). One can write an identical expression for the driving term Z in (1.153) in terms of the rotationally invariant amplitude $\langle L_jS_jn_b;p'_j|Z^J|L_iS_in_a;p_i\rangle$.

The next step is to substitute (1.165) and the corresponding expression for Z into the three-body equations (1.153). Using the well-known orthogonality relations of the vector coupling coefficients and spherical harmonics [1.4], one can show (after considerable algebra), that these equations reduce to

$$X^J_{N_j,N_i}(p'_j,p_i;E) = Z^J_{N_j,N_i}(p'_j,p_i;E)$$
$$+ \sum_{N_\ell,\ell} \int_0^\infty dp''_\ell p''^2_\ell Z^J_{N_j,N_\ell}(p'_j,p''_\ell;E)\tau_{N_\ell}(E - p''^2_\ell/2\mu_\ell) X^J_{N_\ell,N_i}(p''_\ell,p_i;E) \quad , \tag{1.166}$$

where $N_i \equiv (L_i,S_i,n_a)$, etc., and

$$X^J_{N_j,N_i}(p'_j,p_i;E) \equiv \langle L_jS_jn_b;p'_j|X^J(E)|L_iS_in_a;p_i\rangle \quad , \tag{1.167}$$

with a similar relation for $Z^J_{N_j,N_i}$.

Equation (1.166) is the final form of the quasi-two-body equations, which one uses in practical calculations. As promised, it is a one-dimensional integral equation for the rotationally invariant scattering amplitude. Note that there is one set of coupled equations to be solved for each value of J. The input to these equations consists of the functions τ and Z^J. Both of these are completely determined by the two-body form factors ($g_{\bar{J}\pi}$ in (1.162)). Explicit expressions for Z^J in the case of arbitrary spin are to be found in the Appendix to this chapter.

1.10 Conclusion

In the preceding sections, we have described the basic ideas of two- and three-body scattering theory. Naturally, in such a brief space, there are many developments which we can merely touch on. In this section we mention some of the most important aspects that have been omitted, and give references to more detailed treatments.

In principle, with the rather complete treatment of the separable interaction case which we have given - culminating in (1.166) and (1.182) - the reader could tackle a real problem. However, no mention has been made of the technical problems which arise in the numerical solution of (1.166). In fact, as it stands this equation cannot be solved without enormous numerical effort [1.28]. The simplest technique for solving it is to rotate the contour of integration ($\int_0^\infty dq_\ell''$) into the complex plane. For elastic scattering this was first described by HETHERINGTON and SCHICK [1.29], while the correct procedure for break-up is due to CAHILL and SLOAN [1.30]. The justification for the whole procedure was provided by BRAYSHAW [1.31]. A rather elegant proof of this procedure is given in Subsection 2.2.6. For an up-to-date description of this and other numerical methods (such as Padé approximants, variational methods, etc.) the reader is referred to [1.26].

Of course, the separable interaction is not the only possible input to a three-body calculation. In fact, there are a number of cases (particularly for the three-nucleon system) in which the Faddeev equations for a nonseparable interaction have been solved. Such a calculation involves the solution of coupled *two*-dimensional equations (in p and q, not just p), after angular momentum decomposition. It is in this sort of calculation that the Padé techniques can be particularly valuable [1.32].

So far, we have made no mention of alternative formulations of three-body theory - such as that of WEINBERG or SUGAR and BLANKENBECLER [1.33]. All these approaches aim to produce equations with a connected kernel. Some of these formulations introduce spurious poles, which do not correspond to bound states or resonances in the physical system. (This is known as "FEDERBUSH disease" [1.34]). However, NOBLE has shown that there are an indefinite number of valid alternatives to Faddeev theory which do not have this problem [1.35]. For a slightly longer discussion of this topic the reader is referred to [1.19] and [1.36], but the only really satisfactory explanation is in the original papers. With regard to the possible ways of rewriting Faddeev theory, we have already mentioned the work of [1.17] and [1.21]. It is possible that such approaches will deepen our understanding of three-body processes in the future.

Naturally, there are also a number of topics which have been left for later chapters. In particular, the analytic properties of the amplitudes X are discussed in Chapter 2. Applications of the theory are left for Chapter 6, where there is a careful explanation of how this relates to many-body problems (e.g., in the construction of an optical potential). Of course, our considerations have concerned solely the nonrelativistic problem. The state of our understanding in relativistic three-body systems is explained in Chapter 5.

Finally, in Chapters 3 and 4 there are two attempts, in unrelated areas, to do away with the theory we have outlined. The former deals with the problem of incorporating important constraints such as unitarity in the analysis of three-hadron final states, without the enormous effort of solving the full problem. The latter is a suggestion on how three-body scattering theory can be reformulated in terms of a boundary condition.

Appendix

Here we sketch the derivation of an explicit expression for $Z^J_{N_\alpha,N_\beta}$ - the driving term in (1.166) - in the case of arbitrary spin and angular momentum. This is all one needs in order to carry out a practical three-body calculation with separable interactions.

In order to avoid nasty notational accidents (such as j-sub-j), we use (α,β,γ) as (cyclic) particle labels. The relative angular momentum of an interacting pair (e.g., $(\beta\gamma)$) is ℓ_α (z-projection λ_α), and we choose to couple in the ℓ-s scheme in such a state. (That is $j_\beta+j_\gamma\to s_\alpha$ (with z-projection σ_α), and $\ell_\alpha+s_\alpha\to\bar{J}_\alpha$). The definition of the driving term is

$$Z^J_{N_\beta,N_\alpha}(p_\beta,p_\alpha;E) = \langle N_\beta;p_\beta;J|G_0(E)|N_\alpha;p_\alpha;J\rangle \quad , \tag{1.168}$$

where according to the coupling scheme we have outlined

$$|N_\alpha;p_\alpha;J\rangle = |\{L_\alpha[(\ell_\alpha(j_\beta j_\gamma)s_\alpha)\bar{J}_\alpha j_\alpha]S_\alpha\}JM;p_\alpha\rangle \quad . \tag{1.169}$$

Using this scheme we can write $[\hat{J} = (2J+1)^{1/2}]$

$$\langle N_\beta;p_\beta;J|G_0(E)|N_\alpha;p_\alpha;J\rangle = \hat{J}^{-2}\sum_{\text{all projections}}$$

$$\langle j_\gamma m_\gamma j_\alpha m_\alpha|s_\beta\sigma_\beta\rangle\langle j_\beta m_\beta j_\gamma m_\gamma|s_\alpha\sigma_\alpha\rangle \tag{1.170}$$

$$\times \langle \ell_\beta\lambda_\beta s_\beta\sigma_\beta|\bar{J}_\beta\bar{m}_\beta\rangle\langle \ell_\alpha\lambda_\alpha s_\alpha\sigma_\alpha|\bar{J}_\alpha\bar{m}_\alpha\rangle$$

$$\times <\bar{j}_\beta\bar{m}_\beta j_\beta m_\beta|S_\beta\Sigma_\beta><\bar{j}_\alpha\bar{m}_\alpha j_\alpha m_\alpha|S_\alpha\Sigma_\alpha>$$

$$\times <L_\beta M_\beta S_\beta\Sigma_\beta|JM><L_\alpha M_\alpha S_\alpha\Sigma_\alpha|JM> \tag{1.170}$$

$$\times <\lambda_\beta M_\beta L_\beta n_\beta p_\beta|G_0(E)|\lambda_\alpha M_\alpha L_\alpha n_\alpha p_\alpha> \quad ,$$

where the sum is over all angular momentum projections, and we have used the fact that $Z^J_{N_\beta,N_\alpha}$ is independent of M to sum over this too - dividing by (2J+1) to compensate. The last term in (1.170) is

$$<\lambda_\beta M_\beta L_\beta n_\beta p_\beta|G_0(E)|\lambda_\alpha M_\alpha L_\alpha n_\alpha p_\alpha>$$

$$= \int d\hat{\underline{p}}_\beta d\hat{\underline{p}}_\alpha d\underline{p}'_\beta d\underline{q}'_\beta Y_{L_\beta M_\beta}(\hat{\underline{p}}_\beta)<n_\beta\lambda_\beta\underline{p}_\beta|\underline{p}'_\beta\underline{q}'_\beta> \tag{1.171}$$

$$\times (E - p'^2_\beta/2\mu_\beta - q'^2_\beta/2\mu_{\gamma\alpha})^{-1}<\underline{p}'_\alpha\underline{q}'_\alpha|n_\alpha\lambda_\alpha\underline{p}_\alpha>Y_{L_\alpha M_\alpha}(\hat{\underline{q}}_\alpha) \quad ,$$

where as we have explained before, the form factor is

$$<\underline{p}'_\alpha\underline{q}'_\alpha|n_\alpha\lambda_\alpha\underline{p}_\alpha> = \delta(\underline{p}_\alpha - \underline{p}'_\alpha)g_{n_\alpha}(q'_\alpha)Y_{\ell_\alpha\lambda_\alpha}(\hat{\underline{q}}'_\alpha) \quad , \tag{1.172}$$

and n_α is a complete set of quantum numbers describing the interacting pair ($\beta\gamma$). (Eq. (1.170) and (1.171) are the inverse of (1.165)).

The integration over the δ-functions in (1.171) is carried out by changing variables from $(\underline{p}'_\beta,\underline{q}'_\beta)$ to $(\underline{p}'_\beta,\underline{p}'_\alpha)$ using

$$\underline{q}'_\alpha = \underline{p}'_\beta + \rho_\alpha\underline{p}'_\alpha ; \ \underline{q}'_\beta = - \underline{p}'_\alpha - \rho_\beta\underline{p}'_\beta \quad , \tag{1.173}$$

with $\rho_\alpha = m_\beta/m_{\beta\gamma}$ and $\rho_\beta = m_\alpha/m_{\alpha\gamma}$, which follows from the definition (1.58) and $\sum_i\underline{p}_i = 0$. (Note also that (1.173) implies that $d^3p_\beta d^3q_\beta = d^3p_\beta d^3p_\alpha$). In order to do the remaining angular integrals ($\hat{\underline{p}}_\alpha$ and $\hat{\underline{p}}_\beta$), one must expand the spherical harmonics in $\hat{\underline{q}}_\alpha$ and $\hat{\underline{q}}_\beta$ using the relation [1.37]

$$q_\alpha^{\ell_\alpha}Y_{\ell_\alpha\lambda_\alpha}(\hat{\underline{q}}_\alpha) = \sum_{abm_a m_b}\delta_{\ell_\alpha,a+b}\{4\pi(2\ell_\alpha + 1)!/(2a + 1)!(2b + 1)!\}^{1/2}$$

$$\times (\rho_\alpha p_\alpha)^a p_\beta^b Y_{am_a}(\hat{\underline{p}}_\alpha)Y_{bm_b}(\hat{\underline{p}}_\beta)<am_a bm_b|\ell_\alpha\lambda_\alpha> \quad . \tag{1.174}$$

In addition one needs to expand the product of form factors and Green's function in terms of spherical harmonics in $\hat{\underline{p}}_\alpha$ and $\hat{p}_\beta$. That is

$$\frac{q_\beta^{-\ell_\beta} g^\dagger_{n_\beta}(q_\beta) g_{n_\alpha}(q_\alpha) q_\alpha^{-\ell_\alpha}}{E-p_\alpha^2/2m_\alpha - p_\beta^2/2m_\beta - (\underline{p}_\alpha+\underline{p}_\beta)^2/2m_\gamma} = 4\pi \sum_{LM} F^L_{n_\beta,n_\alpha}(p_\beta,p_\alpha;E) Y^*_{LM}(\hat{\underline{p}}_\beta) Y_{LM}(\hat{\underline{p}}_\alpha) \quad , \tag{1.175}$$

with

$$F^L_{n_\beta,n_\alpha}(p_\beta,p_\alpha;E) = \frac{1}{2}\int_{-1}^{+1} dx \, \frac{q_\beta^{-\ell_\beta} g^\dagger_{n_\beta}(q_\beta) g_{n_\alpha}(q_\alpha) q_\alpha^{-\ell_\alpha} P_L(x)}{E-p_\alpha^2/2m_\alpha - p_\beta^2/2m_\beta - (\underline{p}_\alpha+\underline{p}_\beta)^2/2m_\gamma} \quad , \tag{1.176}$$

and $x = \hat{\underline{p}}_\alpha \cdot \hat{\underline{p}}_\beta$. Using (1.175) and (1.174) in (1.171) leads to an expression with four spherical harmonics in each of $\hat{\underline{p}}_\alpha$ and $\hat{\underline{p}}_\beta$. This can be reduced, using Eq.(4.6.5) of Ref.1.4 to a product of two spherical harmonics in each variable, for which one can use the orthogonality of spherical harmonics. The resultant expression involves a product of vector coupling coefficients and 3-j symbols. Summing over all the m-quantum numbers with the help of Eqs. (2.20) and (3.21) of Ref.1.38, one finds

$$\begin{aligned}
&\langle \lambda_\beta M_\beta L_\beta n_\beta p_\beta | G_0(E) | \lambda_\alpha M_\alpha L_\alpha n_\alpha p_\alpha \rangle \\
&= (-)^{\ell_\alpha+\lambda_\alpha+M_\beta} \hat{\ell}_\alpha \hat{\ell}_\beta \hat{L}_\alpha \hat{L}_\beta \sum_L (-)^L \hat{L}^2 F^L_{n_\beta,n_\alpha}(p_\beta,p_\alpha;E) \\
&\times \sum_{aa'bb'} B^{\ell_\beta \ell_\alpha}_{a'b';ab}(p_\beta,p_\alpha) \sum_{\Lambda\Lambda' f} [\hat{f}\hat{\Lambda}\hat{\Lambda}']^2 \begin{pmatrix} a & a' & \Lambda \\ 0 & 0 & 0 \end{pmatrix} \begin{pmatrix} b & b' & \Lambda' \\ 0 & 0 & 0 \end{pmatrix} \\
&\times \begin{pmatrix} \Lambda & L & L_\alpha \\ 0 & 0 & 0 \end{pmatrix} \begin{pmatrix} \Lambda' & L & L_\beta \\ 0 & 0 & 0 \end{pmatrix} \begin{Bmatrix} L_\alpha & L_\beta & f \\ \Lambda' & \Lambda & L \end{Bmatrix} \begin{Bmatrix} \ell_\alpha & \ell_\beta & f \\ a & a' & \Lambda \\ b & b' & \Lambda' \end{Bmatrix} \sum_{M_f} \begin{pmatrix} \ell_\alpha & \ell_\beta & f \\ -\lambda_\alpha & \lambda_\beta & M_f \end{pmatrix} \begin{pmatrix} L_\alpha & L_\beta & f \\ M_\alpha & M_\beta & M_f \end{pmatrix} \quad ,
\end{aligned} \tag{1.177}$$

where the 3-j, 6-j symbols are as defined in [1.4], and

$$\begin{aligned}
B^{\ell_\beta \ell_\alpha}_{a'b';ab} &= \delta_{\ell_\alpha, a+b} \delta_{\ell_\beta, a'+b'} \rho_\alpha^a \rho_\beta^{b'} p_\alpha^{a+a'} p_\beta^{b+b'} \\
&\times \left\{ \frac{(2\ell_\alpha+1)!(2\ell_\beta+1)!}{(2a)!(2b)!(2a')!(2b')!} \right\}^{1/2} \quad .
\end{aligned} \tag{1.178}$$

Now, to obtain the driving term we substitute (1.177) into (1.170), and replace all Clebsch-Gordan coefficients by 3-j symbols. The resultant expression has an m-sum over ten 3-j symbols. This can be simplifed using the relations

$$\sum_{M_\alpha M_\beta M} (-)^{M_\beta - 2M} \begin{pmatrix} L_\alpha & L_\beta & f \\ M_\alpha & -M_\beta & M_f \end{pmatrix} \begin{pmatrix} L_\alpha & S_\alpha & J \\ M_\alpha & \Sigma_\alpha & -M \end{pmatrix} \begin{pmatrix} L_\beta & S_\beta & J \\ M_\beta & \Sigma_\beta & -M \end{pmatrix}$$

$$= (-)^{J-f+\Sigma_\alpha} \begin{pmatrix} S_\alpha & S_\beta & f \\ -\Sigma_\alpha & \Sigma_\beta & M_f \end{pmatrix} \begin{Bmatrix} S_\alpha & S_\beta & f \\ L_\beta & L_\alpha & J \end{Bmatrix} \quad , \tag{1.179}$$

and

$$\sum_{\text{all projections}} (-)^{-\Sigma_\alpha - \bar{m}_\alpha - \bar{m}_\beta - \sigma_\alpha - \sigma_\beta + \lambda_\alpha} \begin{pmatrix} S_\alpha & S_\beta & f \\ -\Sigma_\alpha & \Sigma_\beta & M_f \end{pmatrix} \begin{pmatrix} \ell_\alpha & \ell_\beta & f \\ -\lambda_\alpha & \lambda_\beta & M_f \end{pmatrix}$$

$$\times \begin{pmatrix} \bar{j}_\alpha & j_\alpha & S_\alpha \\ \bar{m}_\alpha & m_\alpha & -\Sigma_\alpha \end{pmatrix} \begin{pmatrix} \bar{j}_\beta & j_\beta & S_\beta \\ \bar{m}_\beta & m_\beta & -\Sigma_\beta \end{pmatrix} \begin{pmatrix} \ell_\alpha & s_\alpha & \bar{j}_\alpha \\ \lambda_\alpha & \sigma_\alpha & -\bar{m}_\alpha \end{pmatrix} \begin{pmatrix} \ell_\beta & s_\beta & \bar{j}_\beta \\ \lambda_\beta & \sigma_\beta & -\bar{m}_\beta \end{pmatrix}$$

$$\times \begin{pmatrix} j_\beta & j_\gamma & s_\alpha \\ m_\beta & m_\gamma & -\sigma_\alpha \end{pmatrix} \begin{pmatrix} j_\gamma & j_\alpha & s_\beta \\ m_\gamma & m_\alpha & -\sigma_\beta \end{pmatrix} = (-)^{2f+2S_\beta+2\ell_\alpha} \tag{1.180}$$

$$\times (-)^{2\bar{j}_\beta - s_\alpha + 2s_\beta - j_\alpha + 2j_\beta} \begin{Bmatrix} j_\alpha & S_\alpha & S_\beta & j_\beta \\ \bar{j}_\alpha & f & \bar{j}_\beta & j_\gamma \\ s_\alpha & \ell_\alpha & \ell_\beta & s_\beta \end{Bmatrix} \quad ,$$

which are most easily proved using the diagrammatic procedures of [1.39]. (The 12-j symbol is that defined by ORD-SMITH [1.40]).

Collecting all these pieces, one now has a general expression for the rotationally invariant Born term

$$Z^J_{N_\beta,N_\alpha}(p_\beta,p_\alpha;E) = \bar{\delta}_{\alpha\beta} p_\alpha^{\ell_\beta} p_\beta^{\ell_\alpha} \sum_L F^L_{N_\beta,N_\alpha}(p_\beta,p_\alpha;E)$$

$$\times \sum_{a=0}^{\ell_\alpha} \sum_{b=0}^{\ell_\beta} A^{L,a,b}_{N_\beta,N_\alpha} (p_\alpha/p_\beta)^{a-b} \quad , \tag{1.181}$$

with all the dynamical information contained in $F^L_{N_\beta,N_\alpha}$ (which is given by (1.176)), and

$$A^{L,a,b}_{N_\beta,N_\alpha} = (-)^R \hat{\ell}_\alpha \hat{\ell}_\beta \hat{L}_\alpha \hat{L}_\beta \hat{S}_\alpha \hat{S}_\beta \hat{\bar{j}}_\alpha \hat{\bar{j}}_\beta \hat{s}_\alpha \hat{s}_\beta \hat{L}^2$$

$$\times \rho_\alpha^a \rho_\beta^b \{(2\ell_\alpha + 1)!(2\ell_\beta + 1)!/(2a)!(2b)!(2\ell_\alpha - 2a)!(2\ell_\beta - 2b)!\}^{1/2} \tag{1.182}$$

$$\times \sum_{f\Lambda\Lambda'} [\hat{f}\hat{\Lambda}\hat{\Lambda}']^2 \begin{Bmatrix} S_\alpha & S_\beta & f \\ L_\beta & L_\alpha & J \end{Bmatrix} \begin{Bmatrix} j_\alpha & S_\alpha & S_\beta & j_\beta \\ \bar{j}_\alpha & f & \bar{j}_\beta & j_\gamma \\ s_\alpha & \ell_\alpha & \ell_\beta & s_\beta \end{Bmatrix}$$

$$\times \begin{Bmatrix} L_\alpha L_\beta f \\ \Lambda' \Lambda \; L \end{Bmatrix} \begin{Bmatrix} \ell_\alpha & \ell_\beta & f \\ a & \ell_\beta - b & \Lambda \\ \ell_\alpha - a & b & \Lambda' \end{Bmatrix} \begin{pmatrix} \Lambda' & L & L_\beta \\ 0 & 0 & 0 \end{pmatrix} \begin{pmatrix} \Lambda & L & L_\alpha \\ 0 & 0 & 0 \end{pmatrix} \begin{pmatrix} \ell_\alpha - a & b & \Lambda' \\ 0 & 0 & 0 \end{pmatrix} \begin{pmatrix} a & \ell_\beta - b & \Lambda \\ 0 & 0 & 0 \end{pmatrix} \quad , \tag{1.182}$$

$$R = -J + L_\alpha + L_\beta + S_\alpha + S_\beta + \bar{J}_\alpha + \bar{J}_\beta + s_\alpha + \ell_\beta - j_\alpha \quad . \tag{1.183}$$

Essentially the same result, using noncyclic ordering, was first found by STINGL and RINAT [1.41], to wich we refer for a little more detail. In the form given here, this result may be found in Appendix B of Ref.1.42.

Note that while we have derived the result only for $\alpha \rightarrow \beta$, the case $\alpha \rightarrow \gamma$ is easily obtained from it. Finally, we observe that in most cases of practical interest the particles have isospin (i_α for particle α, $\bar{i}_\alpha$ for the pair $(\beta\gamma)$). In that case one follows the procedure outlined in Section 1.9, but also couples to a total isospin I. (This can be done independently of the spin, angular momentum reduction). The result is that Z is now labelled by both I and J, and the coefficient $A^{L,a,b}_{N_\beta,N_\alpha}$ is multiplied by the recoupling coefficient

$$\begin{aligned} &\langle [(i_\gamma i_\alpha)\bar{i}_\beta i_\beta]I | [(i_\beta i_\gamma)\bar{i}_\alpha i_\alpha]I \rangle \\ &= (-)^{i_\alpha + i_\gamma - \bar{i}_\beta + 2I} \hat{\bar{i}}_\alpha \hat{\bar{i}}_\beta \begin{Bmatrix} i_\beta & i_\gamma & \bar{i}_\alpha \\ i_\alpha & I & \bar{i}_\beta \end{Bmatrix} \quad . \end{aligned} \tag{1.184}$$

References

1.1 L.D. Faddeev: *Mathematical Aspects of the Three-Body Problem* (Daniel Davey and Co., Inc., New York 1965);
L.D. Faddeev: Soviet Physics JETP 12, 1014 (1961); Doklady 6, 384 (1961); 7, 600 (1963)

1.2 I.E. McCarthy: *Introduction to Nuclear Theory*, 1st. ed. (John Wiley & Sons, Inc., New York, London, Sydney, Toronto 1968) p.18

1.3 M.L. Goldberger, K.M. Watson: *Collision Theory*, 1st. ed. (John Wiley & Sons, Inc., New York, London, Sydney 1964)

1.4 A.R. Edmonds: *Angular Momentum in Quantum Mechanics*, Revised Printing (Princeton University Press, Princeton 1968)

1.5 A. Messiah: *Quantum Mechanics*, 1st. ed. (North Holland Publishing Company, Amsterdam 1961) p.489

1.6 C. Lovelace: Phys. Rev. B135, 1225 (1964);
C. Lovelace: *Strong Interactions and High Energy Physics*, ed. by R.G. Moorhouse (Oliver and Boyd, London 1964)

1.7 M.K. Srivastava, D.W.L. Sprung: *Advances in Nuclear Physics*, Vol. 8 (Plenum Press, New York, London 1975)

1.8 P.A.M. Dirac: *The Principles of Quantum Mechanics*, 4th ed. (Oxford University Press 1958)

1.9 K.L. Kowalski: Phys. Rev. Letters 15, 798 (1965)
H.P. Noyes: Phys. Rev. Letters 15, 538 (1965)

1.10 M. Baranger, B. Giraud, S.K. Mukhopadhyay, P.U. Sauer: Nuclear Phys. A138, 1 (1969)

1.11 R.T. Mongan: Phys. Rev. 189, 1888 (1969)

1.12 R.D. Amado: Phys. Rev. 132, 485 (1963)

1.13 J.S. Levinger: *Springer Tracts in Modern Physics*, Vol. 71 (Springer-Verlag, Berlin, Heidelberg, New York 1974) pp.88-240

1.14 F. Tabakin: Phys. Rev. 177, 1443 (1969)

1.15 R.H. Landau, S.C. Phatak, F. Tabakin: Ann. Phys. (N.Y.) 78, 299 (1973)

1.16 R. Stagat: Nuclear Phys. A125, 654 (1969); I.R. Afnan, D.M. Clement, F.J.D. Serduke: Nuclear Phys. A170, 625 (1970)

1.17 B.R. Karlsson, E.M. Zeiger: Phys. Rev. D11, 939 (1975)

1.18 H. Eckstein: Phys. Rev. 101, 880 (1956); E. Gerjuoy: Ann. Phys. (N.Y.) 5, 58 (1958)

1.19 K.M. Watson, J. Nuttall: *Topics in Several Particle Dynamics*, 1st. ed. (Holden-Day, Inc., San Francisco, Cambridge, London, Amsterdam 1967)

1.20 M. Rubin, R. Sugar, G. Tiktopoulos: Phys. Rev. 146, 1130 (1966); 159, 1348 (1967); 162, 1555 (1967)

1.21 T.A. Osborn, K.L. Kowalski: Ann. Phys. (N.Y.) 68, 361 (1971)

1.22 R.T. Cahill: Nuclear Phys. A194, 599 (1972)

1.23 K.L. Kowalski: Phys. Rev. D10, 1271 (1974)

1.24 T. Sasakawa: Nuclear Phys. A203, 496 (1973)

1.25 E.O. Alt, P. Grassberger, W. Sandhas: Nuclear Phys. B2, 167 (1967)

1.26 E.W. Schmid, H. Ziegelmann: *The Quantum Mechanical Three-Body Problem* (Pergamon Press, Oxford, Edinburgh, New York, Toronto, Sydney 1974)

1.27 P.C. Tandy, R.T. Cahill, I.E. McCarthy: Phys. Letters 41B, 241 (1972)

1.28 N.M. Larson, J.H. Hetherington: Phys. Rev. C9, 699 (1974)

1.29 J.H. Hetherington, L.H. Schick: Phys. Rev. 137B, 935 (1965)

1.30 R.T. Cahill, I.H. Sloan: Nuclear Phys. A165, 161 (1971)

1.31 D.D. Brayshaw: Phys. Rev. 176, 1855 (1968)

1.32 R.A. Malfliet, J.A. Tjon: Ann. Phys. (N.Y.) 61, 425 (1970)

1.33 S. Weinberg: Phys. Rev. 133, B232 (1964); R. Sugar, R. Blankenbecler: Phys. Rev. 136B, 472 (1964)

1.34 P.G. Federbush: Phys. Rev. 148, 1556 (1966)

1.35 J. Noble: Phys. Rev. 161, 1495 (1967)

1.36 R.D. Amado: *Brandeis University Summer Institute*, Vol.2 (Gordon and Breach, New York 1970) p.1

1.37 M. Moshinsky: Nuclear Phys. 13, 104 (1959)

1.38 M. Rotenberg, R. Bivins, N. Metropolis, J.K. Wooten, Jr.: *The 3-j and 6-j Symbols* (The Technology Press, MIT 1959)

1.39 A.P. Yutsis, I.B. Levinson, V.V. Vanagas: *Mathematical Apparatus of the Theory of Angular Momentum* (Israel Program for Scientific Translations, Jerusalem 1962)

1.40 R.J. Ord-Smith: Phys. Rev. 94, 1227 (1954)

1.41 M. Stingl, A.S. Rinat: Nuclear Phys. A154, 613 (1970)

1.42 I.R. Afnan, A.W. Thomas: Phys. Rev. C10, 109 (1974)

2. Analytic Structure of On-Shell Three-Body Amplitudes

L. R. Dodd

With 8 Figures

2.1 Background

There are several possible answers to the question of why one should study the analytic structure of three-body amplitudes.

A general aim is to prove the existence of three-body on-shell amplitudes and to apply dispersion methods to calculate them. In this chapter work is reviewed which shows that the on-shell N/D method [2.1] is feasible in nonrelativistic problems and in some circumstances should be considered as a worthwhile alternative to the solution of the off-shell Faddeev equations [2.2]. A knowledge of analytic properties is also essential in studying and parameterizing threshold behavior [2.3] and final state interactions [2.4,5], and in the S-matrix approach to statistical mechanics [2.6-8]. Although numerical solution of the Faddeev equations is by now a standard procedure, a study of the analytic properties of the amplitudes involved has several practical consequences. Often interesting properties of three-particle scattering are not easily revealed in numerical solutions. For example the Efimov effect [2.9-11] and coherence in final-state interactions [2.4,5] were only understood when a proper analysis was carried out, some time after numerical solutions were available. Another example is provided by the rotation of contours transformation [2.12] used in numerical work with the Faddeev equations. In order to prove the validity of this transformation, the singularities of the amplitudes must be located. BRAYSHAW [2.13] was the first to do this in his study of analytic properties of half-off-shell amplitudes.

As in the case of potential scattering, the application of on-shell methods to nonrelativistic three-body problems, where exact amplitudes are calculable, may extend our understanding of the implementation of unitarity and analyticity in relativistic problems.

It is clear that in three-body problems, where there are a number of degrees of freedom, analytic continuation in many different variables is possible. This review will be confined to a discussion of the analytic properties in the total energy variable of on-shell three-body amplitudes, the scattering angles being held fixed. This choice is advantageous for the formulation of partial wave dispersion relations [2.14]. However, analytic properties in pair sub-energy variables are needed in the analysis of final-state interactions [2.4,5]. Also there has been some consideration of continuation in angular momentum variables [2.15,16].

Our discussion is based on a series of papers by RUBIN et al.[2.17], who studied the Fredholm representation of scattering amplitudes for three particles, each pair interacting through a Yukawa potential. STELBOVICS [2.18] has made a similar analysis for separable potentials. The essential properties of the amplitudes are still present when separable interactions are considered. Separable interactions have the advantage that the break-up and free-scattering amplitudes are expressible in terms of the elastic amplitude. Also the Faddeev equations are more easily solved than with local potentials. Therefore for definiteness and simplicity of presentation, the system of three spinless bosons, interacting through separable Yamaguchi potentials [2.19,20] will be used as an illustrative example in this review. The methods and techniques discussed are, however, easily generalized to more complex systems and interactions.

The mathematical tool used is the Fredholm theory of integral equations with L^2 kernels as presented by SMITHIES [2.21]. In Section 2.2 the Fredholm representations of the solutions of the Faddeev equations for the above model are considered. The Fredholm series solution may be regarded as a rearrangement of the terms in the iterative solution, i.e., the multiple scattering series generated when the three-body equations are iterated. Each term in the series is a multiple integral over momentum variables. The convergence of the series is guaranteed if the kernel of the integral equations is square integrable. As in the case of two-body scattering, the kernel is singular in the physical on-shell limit, which is needed to compute cross sections. The technique used to construct the on-shell amplitudes is to transform, without changing its value, each term in the Fredholm solution by making contour deformations [2.22,23]. The new series is identified with the solution of a modified integral equation, which has a square integrable kernel, so that the convergence of the original Fredholm series solution for the on-shell amplitude is proved.

In Section 2.3 the Fredholm representations of the on-shell amplitudes are continued to complex energies, holding the scattering angles fixed. It is proved that, apart from the Born term in the amplitude associated with the scattering of three free particles in both the initial and final states, all the singularities in the complex energy plane are confined to the real axis. Consequently the amplitudes are suitable for the application of dispersion relations. Also in Section 2.3 positions of important classes of singularities are found, using the Landau rules familiar from the analysis of Feynman diagrams in field theory [2.24,25]. An important result is that in "minimal" three-body scattering [2.26], where the vertex function is structureless, the whole contribution of the left-hand cut may be calculated from single and double particle exchange terms of the multiple scattering series.

The analytic properties of the on-shell amplitudes are summarized in Subsection 2.3.5.

On-shell calculations of nonrelativistic three-body amplitudes are reviewed in Section 2.4. Dispersion relations are formulated for the three-boson problem and the N/D method of solution outlined. In Subsection 2.4.2, N/D calculations of neutron-deuteron scattering are discussed, with particular emphasis on the comparison of N/D solutions with solutions of the off-shell Faddeev equations. Applications to other systems are mentioned in Subsection 2.4.3 and our conclusions are presented in Section 2.5.

2.2 The Fredholm Representation of On-Shell Amplitudes

2.2.1 Definition of Amplitudes in the Separable Potential Model

For simplicity in this and the next section we consider a system of three, identical spinless bosons, the attractive interaction of each pair being described by a separable two-particle t-matrix

$$t(\underline{q},\underline{q}';W) = g(q)g(q')\tau(W) \tag{2.1}$$

with simple s-wave Yamaguchi vertex functions

$$g(q) = (\beta^2 + q^2)^{-1} \quad . \tag{2.2}$$

The two-particle propagator [2.19]

$$\tau(w) = -\lambda[1 + \lambda\pi^2/\beta(w^{1/2} + i\beta)^2]^{-1} \tag{2.3}$$

has a single bound state pole at energy $W = -B$ with B related to the potential strength λ and the range parameter β by

$$\lambda = \beta(B^{1/2} + \beta)^2/\pi \quad . \tag{2.4}$$

The following considerations may be extended in a straightforward way to include spin, isospin and more complicated separable potentials.

The basic three-body equation for the off-shell elastic amplitude $X(\underline{p}',\underline{p};w)$ describing the scattering of a particle of momentum $\underline{p}$ from the bound state of the other pair is [2.27] (cf.Sec.1.8)[1]

$$X(\underline{p}',\underline{p};W) = Z(\underline{p}',\underline{p};W) + \int Z(\underline{p}',\underline{p}'';W)\tau(W - \tfrac{3}{4}p''^2)X(\underline{p}'',\underline{p};W)d^3p \quad . \tag{2.5}$$

Here the "potential" is defined in terms of the vertex functions of (2.2) and an energy denominator representing the free propagation of the particles,

$$Z(\underline{p}',\underline{p};W) = g(\underline{p} + 1/2\underline{p}')g(-\underline{p}' - 1/2\underline{p})(W - p'^2 - p^2 - \underline{p}\cdot\underline{p}')^{-1} \quad . \tag{2.6}$$

Eq. (2.5) in operator form will be written as

$$X(W) = Z(W) + K(W)X(W) \tag{2.7}$$

with the kernel

$$K(\underline{p}',\underline{p};W) = Z(\underline{p}',\underline{p};W)\tau(W - \tfrac{3}{4}p^2) \quad . \tag{2.8}$$

The on-shell elastic amplitude is defined by

$$X_{BB}(\underline{p}_B',\underline{p}_B;E + i0) = \lim_{\varepsilon\to 0+} X(\underline{p}_B',\underline{p}_B;E + i\varepsilon) \tag{2.9}$$

with

$$p_B'^2 = p_B^2 = 4(E + B)/3 \quad .$$

An advantage of the separable potential model is that the physical break-up and the three free particle amplitudes can be found from the solution of (2.5). The break-up amplitude is proportional to (cf. (1.161))

$$X_{0B}(\underline{p}_1',\underline{q}_1',\underline{p}_B;W) = \sum_{i=1}^{3} g(q_i')\tau(W - \tfrac{3}{4}p_i'^2)X(\underline{p}_i',\underline{p}_B;W) \quad , \tag{2.10}$$

where

$$4E = 3p_i'^2 + 4q_i'^2 = -4B + 3p_B^2 \quad , \tag{2.11}$$

the limit $\varepsilon\to 0+$ is taken in $W = E+i\varepsilon$, and the amplitude for the scattering of three free particles is proportional to

$$X_{00}(\underline{p}_1',\underline{q}_1',\underline{p}_1,\underline{q}_1;W) = \sum_{i=1}^{3} \delta(\underline{p}_i - \underline{p}_i')g(q_i')\tau(W - \tfrac{3}{4}p_i'^2)g(q_i)$$

$$+ \sum_{i=1}^{3}\sum_{j=1}^{3} g(q_i')\tau(W - \tfrac{3}{4}p_i'^2)X(\underline{p}_i',\underline{p}_j;W)\tau(W - \tfrac{3}{4}p_j^2)g(q_j) \quad , \tag{2.12}$$

---

[1] Since we only deal with three equal mass particles here, we set $\hbar = c = m = 1$, and the reduced masses (cf.Chap.1) become numerical constants.

where

$$4E = 3p_i'^2 + 4q_i'^2 = 3p_j^2 + 4q_j^2 \tag{2.13}$$

and the limit $\varepsilon \to 0+$ is taken in $W = E+i\varepsilon$.

For complex energies (Im $W \neq 0$) it can be verified that the kernel of (2.5) is L^2, i.e.,

$$\| K(W) \| = \int d^3p \int d^3p' |K(\underline{p},\underline{p}';W)|^2 < \infty \quad , \tag{2.14}$$

and also that $Z(\underline{p}',\underline{p};W)$ is an L^2 function of $\underline{p}'$,

$$\int |Z(\underline{p}',\underline{p};W)|^2 \, d^3p < \infty \quad . \tag{2.15}$$

Thus the conditions for the SMITHIES [2.21] form of the Fredholm solution are satisfied,

$$X(W) = Z(W) + R(W)Z(W) \quad . \tag{2.16}$$

The resolvent $R(W)$ is an L^2 kernel satisfying the equation

$$R(W) = K(W) + K(W)R(W) \tag{2.17}$$

and is expressed as the ratio of two power series

$$R(\underline{p}',\underline{p};W) = N(\underline{p}',\underline{p};W)/D(W) \tag{2.18}$$

with $D(W) = \Sigma_n D_n(W)$ and $N(\underline{p}',\underline{p};W) = \Sigma_n N_n(\underline{p}',\underline{p};W)$. The series for D and N are absolutely convergent. Also the kernel $N(\underline{p}',\underline{p};W)$ is L^2 and its series representation is uniformly convergent in $\underline{p}'$ and $\underline{p}$. Explicit forms for N_n and D_n in terms of powers of the kernel K and $\sigma_n = Tr(K^n)$ are given in the Appendix. In this way the three-body amplitude X for complex W may be constructed.

In the on-shell limit, (2.9), the kernel K is no longer L^2 and the L^2 theory does not establish the convergence of the Fredholm series for the physical amplitude. This is a problem which is very familiar from two-body potential scattering, where a number of different ways of establishing the existence of solutions have been considered [2.28]. In this section a method [2.22,23] of constructing the physical amplitudes based on the analytic properties of the terms in the iterative solution of (2.5) will be outlined. The details are given in the unpublished thesis of STELBOVICS [2.18], which makes extensive use of the techniques of RUBIN et al.[2.17].

2.2.2 Singularities of Multiple Scattering Terms

The n-th order term $T_n(W) = K^n(W)Z(W)$, obtained by iteration of (2.5), written explicitly is

$$T_n(\underline{p},\underline{p}';W) = \int d^3p_1 \dots \int d^3p_n Z(\underline{p},\underline{p}_1;W)\tau(W - \tfrac{3}{4}p_1^{\,2}) \times Z(\underline{p}_1,\underline{p}_2;W)\tau(W - \tfrac{3}{4}p_2^{\,2}) \dots Z(\underline{p}_n,\underline{p}';W) \quad . \tag{2.19}$$

The immediate aim of this section is to show that the integration paths over $\underline{p}_1,\underline{p}_2,\dots,\underline{p}_n$ can be rotated away from the energy-dependent singularities of the integrand without changing the value of the integral. We will then be able to show that the terms in the Fredholm solution are also invariant under a rotation of contours, and that the sum of the transformed terms converges in the on-shell limit.

From (2.2) and (2.6) it can be seen that the integrand of T_n depends on the magnitudes of the momenta and a set of n+1 parameters u_i, defined by

$$\begin{aligned} u_1 &= \hat{\underline{p}} \cdot \hat{\underline{p}}_1 \\ u_{i+1} &= \hat{\underline{p}}_i \cdot \hat{\underline{p}}_{i+1} \qquad i = 1,\dots,n-1 \\ u_{n+1} &= \hat{\underline{p}}' \cdot \hat{\underline{p}}_n \quad . \end{aligned} \tag{2.20}$$

Rather than using the cartesian coordinates of $\underline{p}_i = (p_{i1},p_{i2},p_{i3})$, it is advantageous to transform to a system of spherical polar coordinates,

$$p_{ij} = P \cos\theta_{3i+j-4} \prod_{k=3i+j-3}^{3n-1} \sin\theta_k, \qquad \begin{array}{l} i = 1,2,\dots,n \\ j = 1,2,3 \end{array} \tag{2.21}$$

with

$$P > 0,\ \theta_0 = 0,\ 0 < \theta_1 \le 2\pi,\ 0 \le \theta_k \le \pi \qquad k = 2,\dots,3n-1 \quad .$$

Since $p_i^{\,2} = \Sigma_j p_{ij}^{\,2} \le \Sigma_{ij} p_{ij}^{\,2} = P^2$, we set $p_i = z_i P$ with $0 \le z_i \le 1$, $i=1,\dots,n$. The parameters z_i and u_i of (2.20) are independent but both are functions of $\theta_1 \dots \theta_{3n-1}$. In this coordinate system the integral for T_n takes the form

$$T_n(\underline{p},\underline{p}';W) = \int_0^\infty dP P^{3n-1} \int_{\Omega_{3n}} d\hat{P} I_n(p,p',P;z_1\dots z_n,u_1\dots u_{n+1}) \quad . \tag{2.22}$$

As the angular integrations are carried out, each z_i ranges over the finite interval [0,1] and u_i over [-1,1]. The singularities of the integrand in the complex P plane are poles due to the vanishing of denominators in Z and τ, and square root branch points from the argument of τ. Consider first the denominators of W. From (2.2) and (2.6) the form factor denominators in $Z(\underline{p},\underline{p}_1;W)$ vanish when

$$\beta^2 + p^2 + P^2 z_1^{\,2}/4 + pPu_1z_1 = 0 \tag{2.23a}$$

and

$$\beta^2 + p^2/4 + P^2 z_1^{\,2} + pPu_1z_1 = 0 \quad . \tag{2.23b}$$

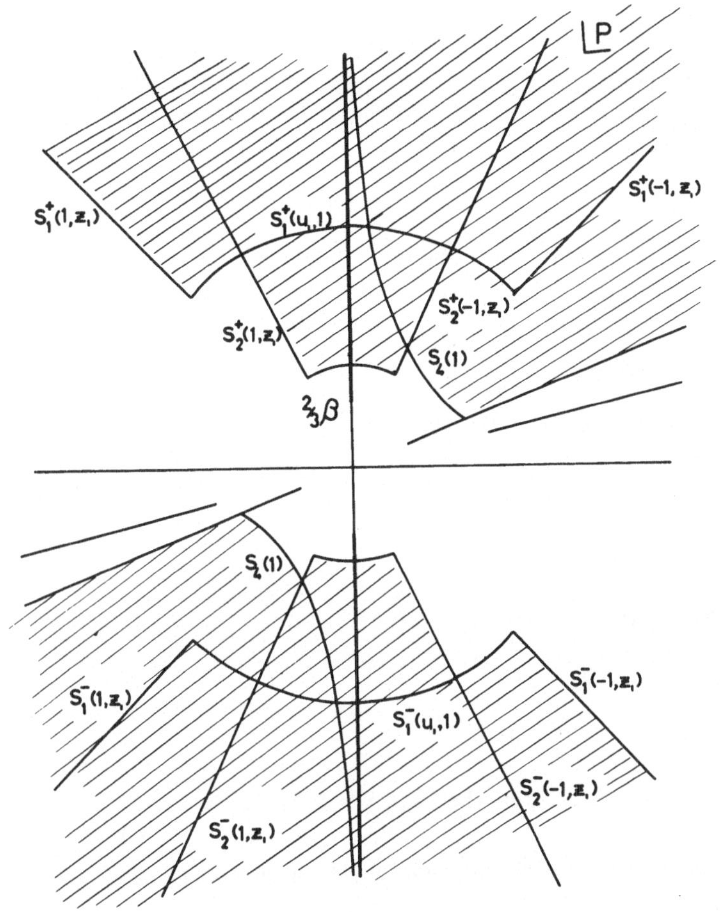

Fig. 2.1. Singularity structure of integrand of T_n, excluding the free-particle propagators containing external momenta, in the complex P plane. The integrand is analytic in the unshaded region for all values of the parameters u and z

We denote the solutions of (2.23a) and (2.23b) by $s_1^{\pm}(u_1,z_1)$ and $s_2^{\pm}(u_1,z_1)$, respectively, the superscript referring to the two possible solutions for each combination of u_1 and z_1. As u_1 and z_1 take all values in their range, $s_1^{\pm}$ and $s_2^{\pm}$ trace out families of curves, which cover the shaded regions indicated in Fig.2.1. Similar considerations apply to the other term involving an external momentum $Z(\underline{p}_n,\underline{p}';W)$. The vertex function denominators of the other terms $Z(\underline{p}_i,\underline{p}_{i+1};W)$ vanish when either

$$\beta^2 + p^2 z_i^{\,2} + p^2 z_{i+1}^2/4 + p^2 z_i z_{i-1} u_{i+1} = 0 \tag{2.24a}$$

or

$$\beta^2 + P^2z_i^2/4 + P^2z_{i+1}^2 + P^2z_iz_{i+1}u_{i+1} = 0 \quad . \tag{2.24b}$$

The associated solutions $s_3^{\pm}(z_i,z_{i+1},u_{i+1})$ and $s_3^{\pm}(z_{i+1},z_i,u_{i+1})$ lie on the imaginary P axis with $|\mathrm{Im}\{P\}| \geq 2\beta/3$.

The energy-dependent denominators in the potential functions $Z(p_i,p_{i+1};W)$ vanish when

$$W - P^2z_i^2 - P^2z_{i+1}^2 - P^2z_iz_{i+1}u_{i+1} = 0 \quad , \tag{2.25}$$

and the denominator of the propagator τ vanishes when

$$W + B - 3P^2z_i^2/4 = 0 \quad . \tag{2.26}$$

The loci of solutions of (2.25) and (2.26) are the lines

$$\arg P = \pm 1/2 \arg W, \ |P| \geq |W/3|^{1/2}$$

and

$$\arg P = \pm 1/2 \arg (W + B), \ |P| \geq |4(W + B)/3|^{1/2} \quad ,$$

respectively. They are shown in Fig.2.1 for the case $W = E+i\varepsilon$, $E>0$.

The branch cut of the square root in the propagator is taken along arg $(W-3P^2z_i^2/4) = 0$ whose locus,

$$3/2 \ \mathrm{Im}\{P\}\mathrm{Re}\{P\} = \mathrm{Im}\{W/z_i^2\}, \ (\mathrm{Im}\{P\})^2 \geq 2(|W| - \mathrm{Re}\{W\})/3z_i^2$$

is that part of the hyperbola shown in Fig.2.1 as $S_4(z_i)$.

The remaining singularities arise from the Green's function denominators in the Z functions containing the external momenta. For example $Z(\underline{p},\underline{p}_i;W)$ has a pole whenever

$$W - p^2 - P^2z_1^2 - pPu_1z_1 = 0 \tag{2.27}$$

leading in the limit $W = E+i0$ to the solution

$$P = \{- u_1p \pm [p^2(u_1^2 - 4) + 4E]\}/2z_1$$

which may be real or complex depending on p. The different possibilities are shown in Fig.2.2.

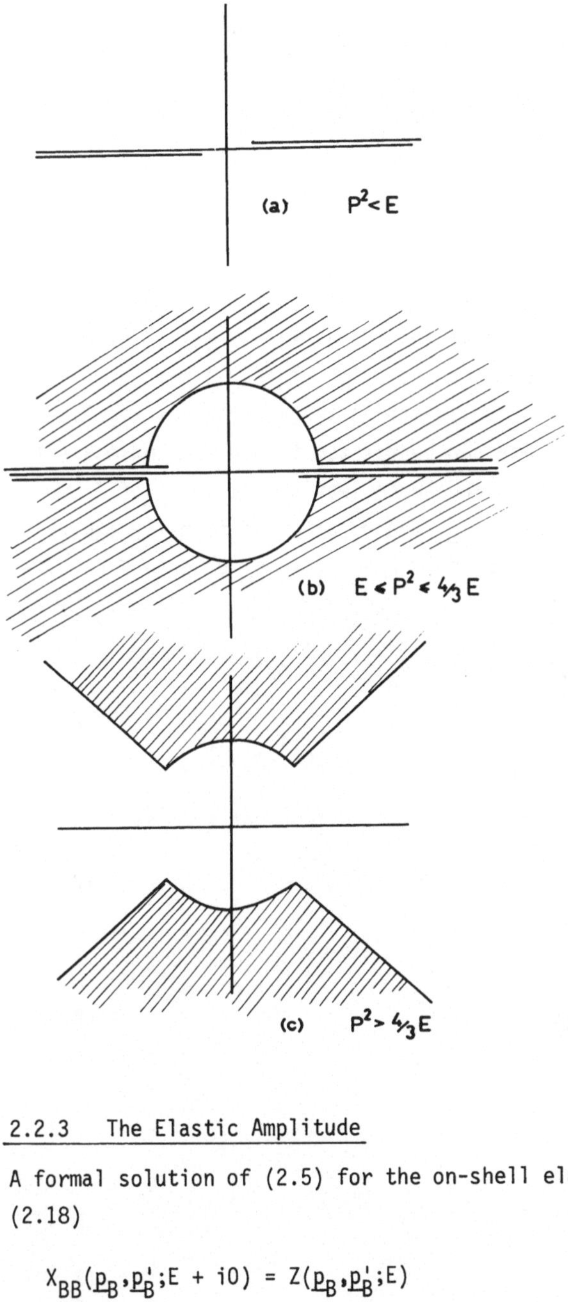

Fig.2.2a-c. Singularity structure in the variable P of the Green's function in $Z(\underline{p},\underline{p}_1;E+i0)$ containing the external momentum $\underline{p}$

2.2.3 The Elastic Amplitude

A formal solution of (2.5) for the on-shell elastic amplitude is from (2.16) and (2.18)

$$X_{BB}(\underline{p}_B,\underline{p}_B';E + i0) = Z(\underline{p}_B,\underline{p}_B';E)$$

$$+ \lim_{\varepsilon\to 0+} \sum_n D(E + i\varepsilon)^{-1} \int d\underline{p}_1 N_n(\underline{p}_B,\underline{p}_1;E + i\varepsilon)Z(\underline{p}_1,\underline{p}_B', E + i\varepsilon) \quad . \tag{2.28}$$

To obtain an alternative series for X_{BB}, which unlike the above representation can be proved to converge, we make use of the analytic properties of N_nZ and D_n. First we note from (2.75) and (2.76) that N_nZ and D_n are expressible as linear combinations of $T_1 \dots T_n$ and $\sigma_1 \dots \sigma_n$, where $\sigma_i = \mathrm{Tr}(K^i)$. For $\underline{p} = \underline{p}_B$ and $\underline{p}' = \underline{p}_B'$, it follows from Subsection 2.2.2 that the integrand of T_i is free of singularities in a sector of the P plane defined by

$$- \theta_m < \arg P < 0 \quad ,$$

where

$$\theta_m = \min \left[\mathrm{atan}\,(\beta/p_B),\ \mathrm{atan}\,(2B^{1/2}/p_B)\right] \quad .$$

The angle $\theta = \mathrm{atan}\,(\beta/p_B)$ is the polar angle of the radial line $S^-(-1,z)$ of Fig.2.1, while $\theta = \mathrm{atan}\,(2B^{1/2}/p_B)$ is associated with the solution of (2.27) with $u_1 = -1$ and $p^2 > 4E/3$, corresponding to the radial line in the fourth quadrant of Fig.2.2c. Thus the integration contour in the P plane can be rotated through an angle θ with $-\theta_m < \theta < 0$ without crossing any singularities and hence leaving the value of T_n unchanged. In terms of the intermediate momenta $\underline{p}_i$ the rotation is formally equivalent to $\underline{p}_i \to \underline{p}_i e^{i\theta}$ [2.22], i.e., the term T_n is unchanged when $\underline{p}_1$ to $\underline{p}_n$ undergo a simultaneous rotation through an angle θ. Thus

$$\begin{aligned} T_n(\underline{p}_B,\underline{p}_B';E + i0) &= \lim_{\varepsilon\to 0} \int d^3p_n K^n(\underline{p}_B,\underline{p}_n;E + i\varepsilon)Z(\underline{p}_n,\underline{p}_B';E + i0) \\ &= \int d^3p_1 \int d^3p_n e^{3i\theta} K(\underline{p}_B,\underline{p}_1 e^{i\theta};E)K^{n-1}(\underline{p}_1 e^{i\theta},\underline{p}_n e^{i\theta};E)Z(\underline{p}_n e^{i\theta},\underline{p}_B';E) \quad . \end{aligned} \tag{2.29}$$

To save space the last equation will be written in an obvious notation as

$$(T_n)_{BB} = K_{B\theta} K_{\theta\theta}^{n-1} Z_{\theta B} \quad . \tag{2.30}$$

Similarly, one can show that

$$\lim_{\varepsilon\to 0+} \sigma_n(E + i\varepsilon) = \int d^3p e^{3i\theta} K_n(\underline{p}e^{i\theta},\underline{p}e^{i\theta}) \equiv \mathrm{Tr}(K_{\theta\theta}^n) \tag{2.31}$$

for

$$0 < -\theta < \pi/2 \quad .$$

It follows that

$$\lim_{\varepsilon\to 0+} D(E + i\varepsilon) = D_\theta(E) \tag{2.32}$$

where D_θ is the Fredholm determinant of the kernel $K_{\theta\theta}$. The physical elastic amlitude now has the representation

$$X_{BB} = Z_{BB} + K_{B\theta}Z_{\theta B} + K_{B\theta}R_{\theta\theta}Z_{\theta B} \quad , \tag{2.33}$$

where $R_{\theta\theta} = \Sigma_n(N_n)_{\theta\theta}/\Sigma_n(D_n)_\theta$ is the resolvent kernel of the operator $K_{\theta\theta}(E)$. Moreover since $\|K_{\theta\theta}(W)\|^2<\infty$ unless arg W = 2θ or arg (W+B) = 2θ, $R_{\theta\theta}$ is an L^2 kernel. Also $K_{B\theta} = K(\underline{p}_B,\underline{p}_1e^{i\theta},E)$ and $Z_{\theta B} = Z(\underline{p}_2e^{i\theta},\underline{p}_B')$ are square integrable functions of $\underline{p}_1$ and $\underline{p}_2$, so that the Fredholm representation (2.33) is uniformly convergent in $\underline{p}_B$ and $\underline{p}_B'$ and since $p_B = p_B' = [4/3(E + B)]^{1/2}$ in E also.

2.2.4 The Break-up Amplitude

To construct the break-up amplitude for physical energies it is sufficient to prove that the sum for $X(\underline{p}_0,\underline{p}_B,E+i0)$ converges. Because $p_0<(4E/3)^{1/2}$, the energy denominator (2.27) may have singularities along the positive real axis, as shown in Fig.2.2b, preventing the rotation used for the elastic amplitude. However, a simple change in variable in the integration over the momenta $\underline{p}$ in T_n, before the introduction of generalized spherical polar coordinates avoids this difficulty:

$$\begin{aligned} \underline{p}_1 \to \underline{q}_1 &= \underline{p}_1 + {}^{1/2}\underline{p}_0 \\ \underline{p}_i \to \underline{q}_i &= \underline{p}_i \qquad i = 2,\dots,n \end{aligned} \tag{2.34}$$

The discussion then proceeds as before, yielding the following representation for the half-off-shell elastic amplitude

$$X(\underline{p}_0,\underline{p}_B,E + i0) = Z_{0B} + K_{0\bar\theta}Z_{\bar\theta B} + K_{0\bar\theta}K_{\bar\theta\theta}Z_{\theta B} + K_{0\bar\theta}K_{\bar\theta\theta}R_{\theta\theta}Z_{\theta B} \quad , \tag{2.35}$$

where for example $K_{0\bar\theta}K_{\bar\theta\theta}$ stands for

$$\int d^3q_1e^{3i\theta}K(p_0,q_1e^{i\theta} - {}^{1/2}\underline{p}_0;E)K(q_1e^{i\theta} - {}^{1/2}\underline{p}_0,q_2e^{i\theta};E) \quad ,$$

the barred subscript indicating translation before rotation. The maximum angle of rotation is now atan $(2B^{1/2}/|\underline{p}_B-\underline{p}_0|)$. It is not difficult to show that $K_{0\bar\theta}K_{\bar\theta\theta}$ is an L^2 function of $\underline{q}_2$ and we have already mentioned that $Z_{\theta B}$ is L^2, so that the resolvent term is uniformly convergent in $\underline{p}_0$ and $\underline{p}_B$ and as $p^2<4E/3$ also with respect to energy. Of the remaining terms neither $(T_2)_{0B} = K_{0\bar\theta}K_{\bar\theta\theta}Z_{\theta B}$ nor $(T_1)_{0B} = K_{0\bar\theta}Z_{\bar\theta B}$ is singular since the same rotation may be performed. Finally Z_{0B} does not have any singularities for real $\underline{p}_0$ and $\underline{p}_B$.

2.2.5 The Free-Particle Amplitude

Since both $\underline{p}_i$ and $\underline{p}_j<(4E/3)^{1/2}$ in the free-particle amplitude (2.12), it is necessary to use the translation (2.34) at both ends of each term of the iterative expansion. This leads to the following expression for the fully off-shell elastic amplitude

$$X = Z_{00} + (T_1)_{00} + (T_2)_{00} + K_{0\bar{\theta}}K_{\bar{\theta}\theta}K_{\theta\bar{\theta}}Z_{\bar{\theta}\theta} + K_{0\bar{\theta}}K_{\bar{\theta}\theta}R_{\theta\theta}K_{\theta\bar{\theta}}Z_{\bar{\theta}\theta} \quad . \tag{2.36}$$

Since $R_{\theta\theta}$, $K_{0\bar{\theta}}K_{\bar{\theta}\theta}$ and $K_{0\bar{\theta}}Z_{\bar{\theta}\theta}$ are all L^2 in the momentum variables denoted by θ, the Fredholm series is once again uniformly convergent. The rotation of variables can be carried out for T_2 and this term is not singular. But in the case of T_1, the coordinate transformations fail to remove the singularities from the integration path; this term contains the well-known triangle scattering singularity. The Born term also gives rise to a pole singularity for appropriate real values of $\underline{p}_0$ and $\underline{p}_0'$.

Hence the free scattering on-shell amplitude, (2.12), can be expressed as the sum of three singular terms, the disconnected part containing momentum conserving δ-functions, the Born term, and T containing the rescattering singularity, together with a uniformly convergent Fredholm series.

2.2.6 Remark on the Rotation of Contours Method

Fredholm theory may be used to establish the validity of the rotation of contours technique, used extensively in numerical solution of the off-shell Faddeev equations above the break-up threshold. The contour of integration is deformed away from the nearby momentum-dependent singularities of the kernel. The validity of the contour deformation, originally suggested by HETHERINGTON and SCHICK [2.12], was assumend in early calculations, but later confirmed by BRAYSHAW [2.13]. By studying the analyticity of the half-off-shell amplitudes in the complex momentum plane, he showed that the continued amplitude $X(\underline{p}e^{i\theta},\underline{p}';E)$ satisfied the equation

$$X_{\theta B} = Z_{\theta B} + K_{\theta\theta}X_{\theta B} \quad . \tag{2.37}$$

Since $K_{\theta\theta}$ is L^2 and $Z(\underline{p}e^{i\theta},\underline{p}_B';E)$ is an L^2 function of $\underline{p}$, this equation is readily solved numerically. The required on-shell amplitude is then determined by

$$X_{BB}(\underline{p}_B,\underline{p}_B',E + i0) = Z_{BB} + K_{B\theta}X_{\theta B} \quad . \tag{2.38}$$

Alternatively, by means of the Fredholm solution it is possible to justify the rotation without studying the properties of the half-off-shell amplitude.

For let Y be the solution of the equation

$$Y = Z_{\theta B} + K_{\theta\theta}Y, \quad -\theta_m < \theta < 0 \quad . \tag{2.39}$$

Since this equation is Fredholm

$$Y = Z_{\theta B} + R_{\theta\theta}Z_{\theta B} \tag{2.40}$$

which combined with (2.33) yields

$$X_{BB} = Z_{BB} + K_{B\theta}Y \quad . \tag{2.41}$$

2.3 Analytic Continuation in Energy of the On-Shell Amplitudes

2.3.1 The Elastic Amplitude

The Fredholm representation may be used to define a continuation of the physical on-shell amplitude $X_{BB}(\underline{p}_B,\underline{p}_B';E+i0)$. Let

$$p_B^{\,2} = p_B'^{\,2} = 4(W + B)/3 \tag{2.42}$$

for complex W, then

$$X(\underline{p}_B,\underline{p}_B';W) = Z(\underline{p}_B,\underline{p}_B';W) + \int d^3p'K(\underline{p}_B,\underline{p}';W)Z(\underline{p}',\underline{p}_B';W)$$
$$+ \sum_{n=0}^{\infty} D(W)^{-1} \int d^3p' \int d^3p''K(\underline{p}_B,\underline{p}';W)N_n(\underline{p}',\underline{p}'';W)Z(\underline{p}'',\underline{p}_B;W) \quad , \tag{2.43}$$

defines a continuation $X(\underline{p}_B,\underline{p}_B',W)$ of the on-shell amplitude to complex energies, provided

i) the energy-dependent singularities of the integrands of the terms in the series do not cross any integration contours and

ii) the series for all such energies is uniformly convergent, i.e.,

$$\int d^3p'|K(\underline{p}_B,\underline{p}';W)|^2<\infty \quad , \tag{2.44a}$$

$$\int d^3p'|Z(\underline{p}_B,\underline{p}';W)|^2<\infty \quad , \tag{2.44b}$$

and

$$\|K(W)\|<\infty \quad . \tag{2.44c}$$

It is important to remember that because of the constraint (2.42), $X(\underline{p}_B,\underline{p}_B';W)$ can be regarded as a function of complex energy W and four real parameters $\hat{n}$ and $\hat{n}'$, specifying the angles of the momentum vectors $\underline{p}_B$ and $\underline{p}_B'$, respectively. It is easier to work with the variable $p = |\underline{p}_B| = |\underline{p}_B'|$ and to transform the results to the physical sheet of the complex energy plane by the mapping (2.42).

Examination of the explicit forms of the integrands of (2.43) shows that the integrals are finite for

$$0 < \mathrm{Im}\{p\} < 2B^{1/2},\ \mathrm{Im}\{W^{1/2}\} > 0 \quad . \tag{2.45}$$

As discussed in Subsection 2.2.1 the kernel K(W) is L^2 provided Im p does not vanish. Hence the series for X(W), excluding the Born term, is uniformly convergent in the region of the complex p plane defined by (2.45). The analyticity of the amplitude follows immediately since the integrands of the terms T_n (and hence N_n) are analytic in the region (2.45) and therefore so are the integrals. The Born term $Z(\underline{p}_B,\underline{p}_B';W)$ has poles at

$$p = \pm i\beta(5/4 + u)^{-1/2} \quad \text{and} \quad p = \pm iB^{1/2}(5/4 + u)^{-1/2} \quad , \tag{2.46}$$

with

$$u = \hat{n} \cdot \hat{n}', \quad -1 \le u \le 1 \quad .$$

Assuming that the bound state wave number $B^{1/2}$ is smaller than the inverse range parameter β, we conclude that X(W) of (2.43) defines an analytic continuation of the elastic scattering amplitude in the strip d_0 defined by

$$0 < \mathrm{Im}\{p\} < 2B^{1/2}/3 \quad . \tag{2.47}$$

Our next task is to extend the domain of analyticity beyond d_0 by considering a rotated amplitude. Let

$$Y = Z_{BB} + K_{B\theta}Z_{\theta B} + \Sigma_n D(W)^{-1}K_{B\theta}(N_n)_{\theta\theta}Z_{\theta B} \quad ,$$

where it is understood that the terms are evaluated for complex energy W and that condition (2.42) holds. A region is sought where $\int d^3p'|K(\underline{p}_B,\underline{p}'e^{i\theta};W)|^2$, $\int d^3p'|Z(\underline{p}_B,\underline{p}'e^{i\theta};W)|^2$ and $\|K(W)\|$ exist, and the Born term is nonsingular.

After a straightforward but tedious examination of the relevant integrals, it is found that Y(W) is analytic in the region d_θ defined by

$$\begin{aligned} &0 < \mathrm{Im}\{(pe^{-i\theta})\} < 2B^{1/2}\cos\theta/3 \quad , \\ &\mathrm{Im}\{(pe^{-i\theta})\notin\ell_\theta\},\ \mathrm{Im}\{(W^{1/2}e^{-i\theta})\} > 0 \quad , \end{aligned} \tag{2.48}$$

where ℓ_θ is the set of points on the rectangular hyperbola

$$(4B\sin 2\theta/3) + 2\mathrm{Re}\{(pe^{-i\theta})\}\mathrm{Im}\{(pe^{-i\theta})\} = 0$$

such that

$$|\mathrm{Im}\{(pe^{-i\theta})\}| \le (4B/3)^{1/2}|\sin\theta| \quad .$$

Note that the endpoints of ℓ_θ, $p = \pm(4B/3)^{1/2}$, p being real, are independent of the angle θ.

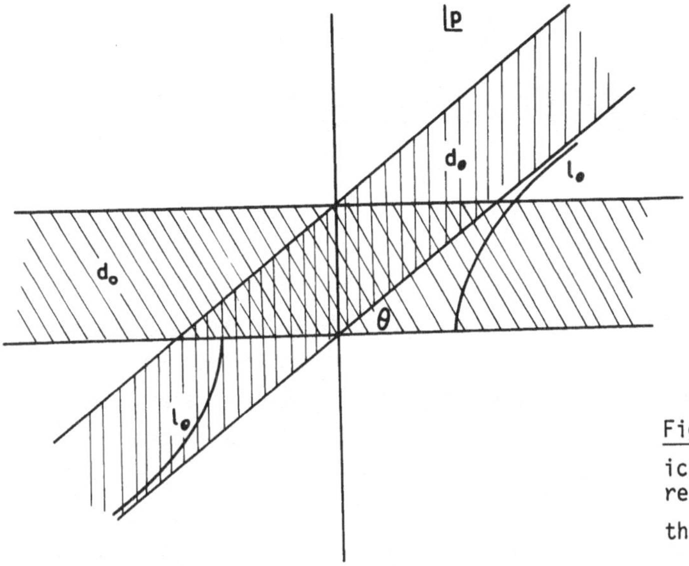

Fig.2.3. Region d_0 of analyticity of the amplitude X, and region d_θ of analyticity of the rotated ampliude Y

The domains of analyticity of X(W) and Y(W) are shown in Fig.2.3. Part of the imaginary axis $0<\mathrm{Im}\{p\}<2B^{1/2}/3$ is always in the intersection of d_0 and d_θ. Furthermore it can be verified that Y(W) and X(W) are equal on this line so that Y(W) is the analytic continuation of X(W) to the region d_θ.

As θ is varied from $-\pi/2$ to $\pi/2$ analytic continuation of X(W) is established in the entire p plane with the exception of

i) the imaginary axis with $\mathrm{Im}\{p\}>2B^{1/2}/3$, $\mathrm{Im}p\leq 0$. (2.49)

ii) $p = \pm (4B/3)^{1/2}$.

Consequently on the first sheet ($\mathrm{Im}W^{1/2}>0$) of the W plane, the singularities of the on-shell amplitude are confined to the real axis.

2.3.2 Location of Singularities of the Elastic Amplitude

In this subsection the positions of the singularities in the complex W plane are discussed. In general a singularity arises when the integration contour in a term of the Fredholm expansion is trapped between two or more singularities of the integrand. Rules for locating such singularities are well known from the analysis of Feynman graphs in field theory [2.24,25]. Pinch singularities may be generated from the simultaneous divergence of the following functions in the integrand of a general term T_n,

i) Green's function $(W-p_i^2-p_j^2-\underline{p}_i\cdot\underline{p}_j)^{-1}$ in $Z(\underline{p}_i,\underline{p}_j;W)$ describing propagation of three free particles in intermediate states.

ii) Green's function $(W-\underline{p}_B^2-\underline{p}_i^2-\underline{p}_i\cdot\underline{p}_B)^{-1}$ in $Z(\underline{p}_B,\underline{p}_i;W)$ describing propagation in initial or final states.

iii) Functions $\tau(W-3/4p_i^2)$ describing the propagation of a bound state of a pair together with a free particle.

iv) Vertex functions $[\beta^2+(\underline{p}_i+1/2\underline{p}_j)^2]^{-1}$, $[\beta^2+(\underline{p}_j+1/2\underline{p}_i)^2]^{-1}$ in $Z(\underline{p}_i,\underline{p}_j,W)$ describing the structure of a bound pair in intermediate states.

v) Vertex functions $[\beta^2+(\underline{p}_B+1/2\underline{p}_i)^2]^{-1}$, $[\beta^2+(1/2\underline{p}_B+\underline{p}_i)^2]$ in $Z(\underline{p}_B,p_i;W)$ describing the structure of a bound pair in initial or final states.

Apart from the vertex functions which give rise to singularities depending on the form of the separable potential, the remaining denominators are always present irrespective of the type of potential, and also appear in the work of RUBIN et al. for local Yukawa interactions [2.17]. STELBOVICS [2.18] has made a careful analysis of the pinches which arise between functions i), ii) and iii). The important result is that *there is only one singularity in the elastic amplitude due to pinches between Green's functions and bound state propagators*. It occurs in $T_1(\underline{p}_B,\underline{p}_B';W)$ and is the result of a pinch between the Green's functions in $Z(\underline{p}_B,\underline{p}_1;W)$ and $Z(\underline{p}_1,\underline{p}_B';W)$. This is proved by applying the Landau rules to each perturbation term T_n. Here we shall only give the analysis for T_1 as an example. To determine the singularities of

$$T_1(\underline{p}_B,\underline{p}_B';W) = \int d^3p_1 Z(\underline{p}_B,\underline{p}_1;W)\tau(W - \tfrac{3}{4}p_1^2)Z(\underline{p}_1,\underline{p}_B';W)$$

due to pinches between energy denominators, the denominator $d_1 = W-(3/4)p_B^2-(\underline{p}+1/2\underline{p}_B)^2$ of $Z(\underline{p}_B,\underline{p}_1;W)$, $d_3 = W-(3/4)p_B'^2-(\underline{p}_1+1/2\underline{p}_B')$ of $Z(\underline{p}_1,\underline{p}_B';W)$ and $d_2 = p_1^2-p_B^2$ of $\tau(W-(3/4)p_1^2)$ are combined to form a single denominator $D = \alpha_1 d_1+\alpha_2 d_2+\alpha_3 d_3$, using Feynman's identity

$$(d_1d_2d_3)^{-1} = \int_0^1 d\alpha_1 \int_0^1 d\alpha_2 \int_0^1 d\alpha_3 \frac{\delta(\alpha_1+\alpha_2+\alpha_3-1)}{(\alpha_1d_1+\alpha_2d_2+\alpha_3d_3)^3} \quad .$$

The integrand of T_1 now contains a hypersurface $D = 0$, where the integrand is singular. The necessary conditions that this hypersurface is trapped in the domain of integration, and hence that T_1 is singular are obtained from the Landau rules,

$$D = 0 \quad ,$$
$$\frac{\partial D}{\partial \alpha_i} = 0 \quad , \qquad (2.50)$$
$$\frac{\partial D}{\partial \underline{p}_1} = 0 \quad .$$

The condition $\partial D/\partial \underline{p}_1 = 0$ yields the equation

$$2(\alpha_1+\alpha_2+\alpha_3)\underline{p}_1 + \alpha_1\underline{p}_B + \alpha_3\underline{p}_B' = 0 \qquad (2.51a)$$

where $\alpha_i \geq 0$. The three subsidiary equations $\partial D/\partial\alpha_i = 0$ (which also imply $D = 0$) are

$$\alpha_1 = 0 \quad \text{or} \quad (\underline{p}_1 + 1/2\underline{p}_B)^2 = -B \tag{2.51b}$$

$$\alpha_2 = 0 \quad \text{or} \quad \underline{p}_1{}^2 = \underline{p}_B{}^2 \tag{2.51c}$$

$$\alpha_3 = 0 \quad \text{or} \quad (\underline{p}_1 + 1/2\underline{p}_B')^2 = -B \quad . \tag{2.51d}$$

Since the singularities are known to lie on the imaginary axis, let $p_B = iP_B$ and $p_B' = iP_B'$; P_B and P_B' are real. With the transformation $\underline{x}_1 = i\underline{p}_1$ (2.50) become

$$2\underline{x}_1(\alpha_1 + \alpha_2 + \alpha_3) + \alpha_1\underline{P}_B + \alpha_3\underline{P}_B' = 0 \tag{2.52a}$$

$$\alpha_1 = 0 \quad \text{or} \quad (\underline{x}_1 + 1/2\underline{P}_B)^2 = B \tag{2.52b}$$

$$\alpha_2 = 0 \quad \text{or} \quad \underline{x}_1{}^2 = \underline{P}_B'{}^2 \tag{2.52c}$$

$$\alpha_3 = 0 \quad \text{or} \quad (\underline{x}_1 + 1/2\underline{P}_B')^2 = B \quad . \tag{2.52d}$$

The relationship between the real vectors $\underline{x}$, $\underline{P}_B$ and $\underline{P}_B'$ is shown in Fig.2.4 for the case $\alpha_1>0$, $\alpha_2>0$, $\alpha_3>0$. One sees that there are no singularities in this case, for if $\alpha_2>0$, $x_1 = P_B$, which is impossible since $\underline{x}_1$ is inside the triangle; the equations are inconsistent and there is no singularity due to a simultaneous pinch of all three

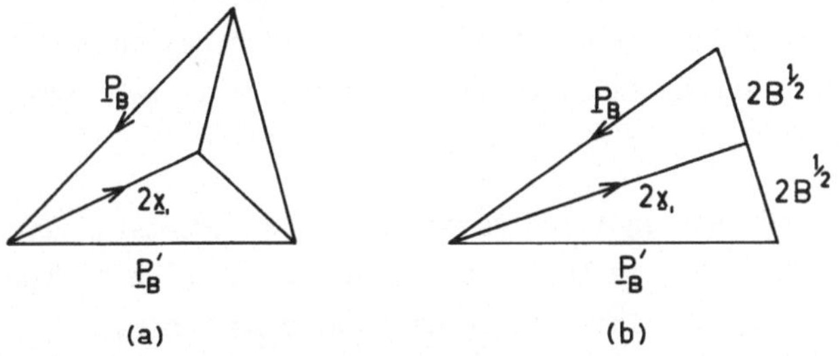

Fig.2.4a and b. Landau diagrams for the elastic amplitude $T_1(\underline{p}_B,\underline{p}_B;W)$

denominators in T_1. In fact all diagrams with $\alpha_2>0$ are eliminated for this reason. In the remaining case $\alpha_2 = 0$, $\alpha_1>0$ and $\alpha_3>0$, $\underline{x}_1$ is on the line $1/2(\underline{P}_B-\underline{P}_B')$ and the other constraints imply $|\underline{P}_B-\underline{P}_B'| = 4B^{1/2}$ so that a singularity occurs when

$$P_B{}^2 = 8B/(1-u), \quad u = \hat{n}\cdot\hat{n}' \quad . \tag{2.53}$$

In the general case STELBOVICS [2.18] has shown that there are no solutions to the Landau equations for T_n, $n\geq 2$.

In summary the only singularities of the on-shell elastic amplitude in the complex energy plane due to the vanishing of energy denominators are poles in the Born term $Z(\underline{p}_B,\underline{p}_B';W)$ at (cf. (2.46))

$$W = -4B(u+2)/(5+4u) \tag{2.54}$$

and the pinch singularity in T_1 at

$$W = - B(7 - u)/(1 - u) \quad . \tag{2.55}$$

These singularities play the same role as the left-hand cuts in potential scattering. Of course when singularities arising from the vertex functions are taken into account, the structure of the left-hand cut becomes much more complicated. In the circumstance that the inverse range parameter β is much larger than the binding energy parameter $B^{1/2}$, as for neutron-deuteron scattering, singularities (2.54) and (2.55) are the nearest to the physical region (W real, $W>-B$) and might be expected to give in a dispersion calculation the dominant contribution to the left-hand cuts. In addition to the left-hand cuts, there are unitarity cuts running along the positive, real energy axis. There is the usual square root branch point at the elastic threshold $W = -B$. There is also a logarithmic branch point at the threshold for break-up [2.29] whose location is given by (2.49) as $W = 0$.

2.3.3 Minimal Three-Body Scattering

If the limit $\beta\to\infty$ is taken, the possible singularities generated by the vertex functions disappear and a model with zero-range two-body interactions is obtained. In this limit the basic equation (2.5) reduces to an equation formulated by SKORNIAKOV and TER-MARTIROSIAN [2.30] in the 1950's preceding the work of Faddeev. From the results of the last subsection the on-shell amplitude of this equation has only two singularities along the negative real energy axis, contained in the first two terms of the multiple scattering.

Recently AMADO [2.26] has rederived this equation using general arguments of unitarity and analyticity, suggesting that this is the simplest form of three-body bound state scattering compatible with the general constraints of quantum mechanics.

The analogue of the equation of [2.30] in one dimension also provides an interesting example. Here the potential in coordinate space is

$$V(r,r') = - \lambda\delta(r - r')\delta(r) = - \lambda\delta(r)\delta(r')$$

and the two-body bound state energy is $-B = -\lambda^2/4$. The half off-shell amplitude for this model can be expressed in closed form (DODD [2.31])

$$X(p,p_B;W) = \frac{1}{2}\,\frac{[2(W - 3/4p^2)^{1/2} - i\lambda](3p_B + i\lambda)p}{(p - p_B)[(p + 1/2p_B)^2 + \lambda^2/4](p_B - i\lambda)(W - 3/4p^2)^{1/2}} \quad . \tag{2.56}$$

There are two possible on-shell amplitudes. For forward scattering $p\to p_B$ and

$$X(p_B,p_B;W) = 12p_B^2 \,/\, [(3p_B - i\lambda)(p_B - i\lambda)\,\lambda^2], \qquad \mathrm{Im}\{p_B\} > 0 \quad .$$

The Fredholm denominator vanishes at $p_B = i\lambda$, giving a three-body bound state with energy $W = -4B$. The singularity at $p_B = i\lambda/3$ agrees with (2.54) with $u = -1$. For backward scattering, $p \to -p_B$, both Z and T_1 have a pole at $p_B = i\lambda$ as required by (2.54) and (2.55), but the residues at these poles cancel, allowing $X(-p_B, p_B; W)$ to vanish.

2.3.4 The Break-up and Free-Scattering Amplitudes

We next turn to results on the analytic properties in the complex energy plane of the on-shell break-up and free-scattering amplitudes.

For the separate potential model, the analytic properties of these amplitudes follow from the properties of the off-shell elastic amplitude. The methods and types of argument are analogous to those already outlined for the elastic amplitude, and details will not be presented here. Again we shall be mainly concerned with the singularities arising from pinches between energy denominators. This class of singularities is independent of the form of the potentials. Consequently the results stated here also appear in the work of RUBIN et al. [2.17] on the more complicated case of local Yukawa interactions.

Break-up Amplitude: To define a continuation in complex energy W of the on-shell break-up amplitude $X_{0B}(\underline{p}_i', \underline{q}_i', \underline{p}_B; W)$ of (2.10), let

$$\underline{p}_i = k\underline{v}_i, \underline{q}_i = k\underline{u}_i \quad \text{where} \quad k = W^{1/2} \quad , \tag{2.57}$$

the on-shell condition (2.11) then becomes

$$3v_i^2 + 4u_i^2 = 4 \tag{2.58}$$

for the final state containing three free particles, and

$$3p_B^2 = 4(k^2 + B) \tag{2.59}$$

for the initial state containing a bound pair. From (2.10) the analytic properties of the break-up amplitude are determined by the off-shell elastic amplitude $X(\underline{p}_i', \underline{p}_B; W)$. It is convenient to split X into single, double and higher order scattering terms,

$$X(\underline{p}_i', \underline{p}_B; W) = Z(\underline{p}_i', \underline{p}_B; W) + T_1(\underline{p}_i', \underline{p}_B; W) + X'(\underline{p}_i, \underline{p}_B; W) \quad . \tag{2.60}$$

A region of analyticity for X' may be found by studying the terms in its Fredholm expansion. This region is extended by defining a rotated amplitude which is identical with X' in a region of overlap with X'. The rotated amplitude is therefore the analytic continuation of X'. The result is that X' is analytic in the entire upper half of the $k = W^{1/2}$ plane, with the exception of parts of the imaginary axis defined by

$$B^{1/2} < \mathrm{Im}\{k\} < (4B/3)^{1/2} \quad , \quad \mathrm{Im}\{k\} > 2B^{1/2} \quad \text{and} \quad k = 0 \quad .$$

The domain of analyticity of T_1 is also found to be the upper half of the k plane except for possible singularities on the imaginary axis in the ranges $\mathrm{Im}\{k\}>(4B/3)^{1/2}$, $B^{1/2}<\mathrm{Im}\{k\}<(4B/3^{1/2}$. The Born term $Z(\underline{p}_i,\underline{p}_B;W)$ has pole singularities which for $1<v_i^2<4/3$ may occur at complex energies. The break-up amplitude in the separable model has pole singularities from the terms $g(q_i')$ and $\tau(W-3/4p_i'^2)$ in (2.10) which lie on the imaginary axis with $\mathrm{Im}\{k\}>B^{1/2}$.

Thus with the exception of the Born term all the singularities of the break-up amplitude lie on the real energy axis.

Hermitian analyticity,

$$X^*(W) = X(W^*) \tag{2.61}$$

is a useful property in formulating dispersion relations, since the discontinuity of the amplitude X across the real axis is then equal to the imaginary part of X, $\mathrm{disc}\,\{X\} = X(E+i\varepsilon) - X(E-i\varepsilon) = 2i\mathrm{Im}\,\{X(\varepsilon+i\varepsilon)\}$. To ensure that the break-up amplitude has the property (2.61) it is necessary to *define* the analytic continuation of the single scattering term Z(W) in the lower half of the physical sheet by $Z(W^*) = Z^*(W)$, thereby introducing an additional cut on the real axis. The other terms are Hermitian analytic because each has a part of the real axis free of singularities and the Schwartz reflection principle applies.

The location of singularities generated by pinches between energy denominators may be determined by the Landau rules in the same manner as those of the elastic amplitude. Consider first the double scattering term $T_1(\underline{p}_i,\underline{p}_B;W)$; (2.52) become

$$\begin{aligned}
&2\underline{x}_1(\alpha_1+\alpha_2+\alpha_3) + \alpha_1\underline{P}_B + \alpha_3\underline{P}_i' = 0 \\
&\alpha_1 = 0 \quad \text{or} \quad (\underline{x}_1 + 1/2\underline{P}_B)^2 = B \\
&\alpha_2 = 0 \quad \text{or} \quad \underline{x}_1^2 = P_B^2 \\
&\alpha_3 = 0 \quad \text{or} \quad (\underline{x}_1 + 1/2\underline{P}_i') = Q_i'^2 \quad .
\end{aligned} \tag{2.62}$$

There are two different pinch singularities corresponding to the two Landau diagrams of Fig.2.5.

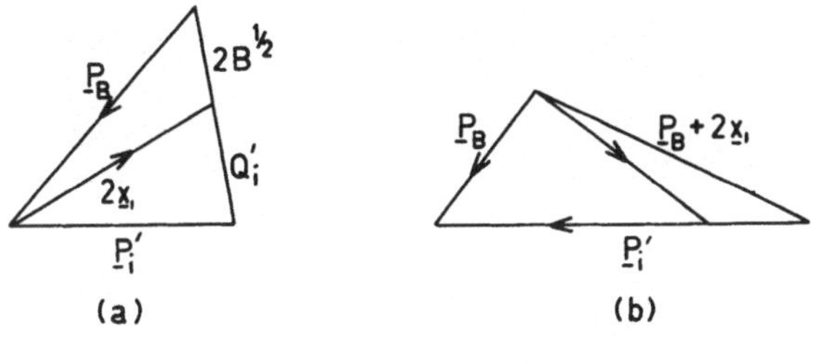

Fig.2.5a and b. Landau diagrams for the break-up amplitude $T_1(\underline{p}_i,\underline{p}_B;W)$

Fig.2.5a represents a pinch between the two Green's function denominators of T_1 which occurs when

$$|\underline{P}_B - \underline{P}_i'|/2 = B^{1/2} + Q_i' \quad .$$

This singularity is restricted to negative energies $E<-4B/3$ in the W plane. Fig.2.5b corresponds to a pinch between the Green's function in the final state and the bound state pole in the propagator τ, which occurs when

$$1/2P_i' = P_B + Q_i' \quad .$$

This singularity is located in the range $-4/3B<E<-B$ provided $v_i'>1$. For $v_i\leq 1$ it is not on the physical sheet.

Analysis of the higher order terms T_n, $n\geq 2$, shows that each one contains the second type of pinch singularity, restricted to $-4B/3<E<-B$, but not the first. *Free Particle Amplitude:* A continuation of the amplitude $X_{00}(\underline{p}_i',\underline{q}_i',\underline{p},\underline{q};W)$ of (2.12) is defined by constraining both initial and final momenta by the conditions (2.57) and (2.58). From (2.12) the analytic properties of the scattering amplitude for three free particles X_{00} are determined by the behavior of the off-shell amplitude $X(\underline{p}_i',\underline{p}_i;W)$. As before the sum of all the higher order terms ($n\geq 2$) may be continued everywhere in the upper half of the complex k plane, with the exception of the imaginary axis.

The term T_1 now has four singularities associated with pinches between energy-dependent denominators. The pinch for Fig.2.6a occurs when $|\underline{P}_i-\underline{P}_i'| = P_i' + P_i$ or equivalently when $2u_i = v_i+v_i'$. This singularity is the well-known triangle scattering singularity [2.8,17] due to the rescattering of three particles on the energy shell. (Note that the double scattering term in the off-shell elastic amplitudes corresponds to three collisions in the amplitude X_{00}).

Fig.2.6b and 2.6d both give the same type of singularity already discussed in connection with the break-up amplitude. We denote these singularities by $S_1(v_i')$ and $S_4(v_i)$. For $v_i>1$ and $v_i'>1$, these singularities are on the physical sheet, and restricted to $-4/3B<E<-B$. The branch cuts associated with these singularities is

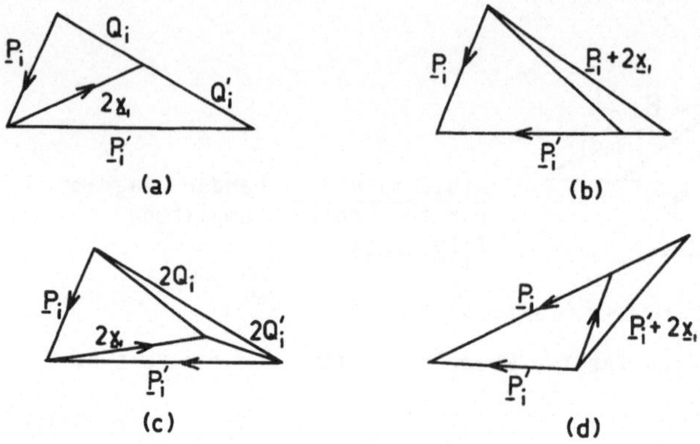

Fig.2.6a-d. Landau diagrams for the free-scattering amplitude $T_1(\underline{p}_i',\underline{p}_i;W)$

usually taken to run in the direction of the unitarity cuts, so that there is no overlap of left-hand and right-hand cuts. Thus the effect of this type of singularity is to extend the right-hand cut below the two-body threshold to an "anomalous" threshold.

Fig.2.6c is associated with a singularity $S_2(v_i,v_i',\hat{v}_i\cdot\hat{v}_i')$ which arises from the simultaneous pinch of all three energy denominators. If $v_i,v_i'>1$, S_2 lies on the physical shelf in the region $-4/3B<S_2<-B$.

With regard to the Hermitian analyticity of T_1, RUBIN et al.[2.17] have demonstrated that if $2u_i>v_i+v_i'$, T_1 is analytic in the first quadrant of the k-plane, but if one continues to the second quadrant, complex singularities are encountered. Therefore it is necessary to *define* T_1 in the second quadrant by $T_1(-k^*) = T_1(k)^*$, thereby introducing a branch cut along the negative real energy axis.

The higher order terms T_n, $n\geq2$ are Hermitian analytic, but each has the singularities $S_1(v_i')$ and $S_1(v_i)$. Finally we note that the full free-particle amplitude, (2.12), in addition to the connected term containing $X(\underline{p}_i',\underline{p}_i;W)$, has disconnected terms involving δ-functions which also make a contribution to the cut running along the negative energy axis.

2.3.5 Summary

It has been demonstrated by RUBIN et al. [2.17] for the case of local Yukawa interactions and by STELBOVICS [2.18] for the case of separable interactions that the nonrelativistic on-shell three-body amplitudes have an analytic structure suitable for the application of dispersion relations. With the exception of the Born term in the free-particle amplitude, all singularities in the complex energy plane are restricted to the real energy axis. The singularities are of four different types.

i) Unitarity cuts starting at the most tightly bound two-body state and running to the right along the positive energy axis. Physically these singularities are associated with the thresholds in energy at which various processes become possible, as discussed in Section 1.6.

ii) Left-hand cuts which are similar to the potential cuts familiar from two-particle potential scattering.

iii) Anomalous cuts in the break-up and free-particle amplitudes.

iv) Rescattering cuts which result from the requirements of Hermitian analyticity.

2.4 On-Shell Calculations

2.4.1 Dispersion Relations and the N/D Method

Partial wave amplitudes are obtained by integrating the fixed angle amplitudes together with appropriate compact functions over angular variables. The analytic properties in energy of the full amplitudes are therefore also shared by the partial wave amplitudes. From the considerations of the previous sections, it follows that dispersion relations may be formulated for the three-body partial wave amplitudes and solved using standard N/D techniques. Most nonrelativistic applications have been concerned with the three-nucleon system but before discussing the results of such calculations the simpler example of three spinless bosons will be used to show how the N/D method is applied.

The on-shell partial wave elastic amplitude X_ℓ for angular momentum ℓ is obtained from the amplitude X_{BB} of (2.9) by

$$z^\ell X_\ell(z) = \int_{-1}^{1} du\ X_{BB}(\underline{p}_B, \underline{p}_B', W) P_\ell(u) \quad , \tag{2.63}$$

where $u = \hat{\underline{p}}_B \cdot \hat{\underline{p}}_B'$ and the dimensionless energy variable z is defined by

$$zB = 3p_B^2/4 = 3p_B'^2/4 = W + B \quad . \tag{2.64}$$

The factor z^ℓ has been introduced to make the low-energy behavior of the amplitude explicit. As we have seen, all the singularities in energy of X_{BB} and hence $X_\ell(z)$ are confined to the real energy axis and are separated into two distinct types, left-hand or potential cuts generated by particle exchanges, and right-hand or unitarity cuts associated with the physical thresholds for scattering processes. In this example, the branch cut on the left closest to the physical region begins at z = -1/3, and arises from a particle in a bound pair exchanging with the free particle. The elastic scattering threshold occurs at z = 0, and the break-up threshold at z = 1. From Cauchy's theorem [2.32] and the Hermitian analytic property of the amplitudes, (2.61), the partial wave amplitude has the following representation:

$$X_\ell(z) = \frac{1}{\pi}\int_{-\infty}^{-1/3} ds\frac{\mathrm{Im}\{X_\ell(s+i0)\}}{s-z} + \frac{1}{\pi}\int_0^\infty ds\frac{\mathrm{Im}\{X_\ell(s+i0)\}}{s-z} , \tag{2.65}$$

where the variable s is used instead of z for real energies.

In the usual applications of dispersion relations the integral over the left-hand cuts

$$B_\ell(z) = \frac{1}{\pi}\int_{-\infty}^{-1/3} ds\,\frac{\mathrm{Im}\{X_\ell(s+i0)\}}{s-z} \tag{2.66}$$

is regarded as part of the input information and must be approximated in some way. Generally the assumption that the nearest singularities dominate the behavior for physical energies (s>0) is made.

The discontinuity across the right-hand cuts is determined by three-body unitarity ([2.22,33], and Sec.1.6),

$$\begin{aligned} 2i\mathrm{Im}\{X_{BB}(\underline{p}_B,\underline{p}_B';E+i0)\} &= X_{BB}(\underline{p}_B,\underline{p}_B';E+i0) - X_{BB}(\underline{p}_B,p_B';E-i0) \\ &= \int X_{BB}(\underline{p}_B,\underline{p}_B'',E-i0)\delta(E - 3/4p_B''^2 + B)X_{BB}(\underline{p}_B'',\underline{p}_B',E+i0)d^3p_B'' \\ &+ \frac{1}{3}\iint X_{B0}(\underline{p}_B,\underline{p}_1'',\underline{q}_1'';E-i0)\delta(E - 3/4p_1''^2 - q_1''^2)X_{0B}(\underline{p}_1'',\underline{q}_1'',\underline{p}_B';E+i0)d^3p_1''d^3q_1'' \quad . \end{aligned} \tag{2.67}$$

Upon projection (2.67) below the break-up threshold yields

$$\mathrm{Im}\{X_\ell(s+i0)\} = \rho_\ell(s)|X_\ell(s+i0)|^2 \qquad 0 < s < 1 \tag{2.68}$$

$$\text{with } \rho_\ell(s) = -s^{1/2+\ell} \quad ,$$

i.e., elastic unitarity holds.

Above the breaking threshold, a full treatment of unitarity couples the elastic and break-up amplitudes and the simplicity of the single-channel formulation is lost. In a phenomenological approach, (2.68) may be modified by replacing $\rho_\ell(s)$ by

$$\rho_\ell'(s) = \rho_\ell(s)(1+\lambda_\ell(s)) \qquad s > 0 \tag{2.69}$$

with $\lambda_\ell(s)$ vanishing for 0<s<1 but nonzero for s>1 to take account of inelasticity. The question of how the three-body contributions (i.e., the terms in (2.67) involving integrations over both $\underline{p}_1$ and $\underline{q}_1$) to unitarity are best treated is perhaps the main difficulty in the on-shell treatment of three-body problems. We will return to the problem of estimating the inelasticity $\lambda_\ell(s)$ later.

With (2.66) and (2.69), the dispersion relation (2.65) for X_ℓ becomes

$$X_\ell(z) = B_\ell(z) + \frac{1}{\pi} \int_0^\infty \frac{\rho'_\ell(s)|X_\ell(s + i0)|^2}{s - z} ds \quad . \tag{2.70}$$

In the N/D approach, X_ℓ is expressed in the form

$$X_\ell(z) = N_\ell(z) D_\ell^{-1}(z) \quad . \tag{2.71}$$

The numerator $N_\ell(z)$ contains all the left-hand singularities of $X_\ell(z')$ (excluding bound state poles) and $D_\ell(z)$ contains the right-hand singularities. Any bound state poles appear as zeros of $D_\ell(z)$.

From their definitions

$$\mathrm{Im}\{N_\ell(s+i0)\} = D_\ell(s) s^{-\ell} \mathrm{Im}\{B_\ell(s+i0)\} \qquad s < -1/3$$

and (2.72)

$$\mathrm{Im}\{D_\ell(s+i0)\} = -\rho'_\ell(s) N_\ell(s) \qquad s > 0$$

and equations for N_ℓ and D_ℓ follow by the usual arguments [2.1],

$$D_\ell(z) = 1 + \frac{(z - s_0)}{\pi} \int_0^\infty ds \frac{\rho'(s) N_\ell(s)}{(s - z)(s - s_0)} \tag{2.73}$$

and

$$N_\ell(z) = B_\ell(z) z^{-\ell} - \frac{1}{\pi} \int_0^\infty ds \rho'(s) \left[\frac{B(s)s^{-\ell}(s-s_0) - B(z)z^{-\ell}(z-s_0)}{(s - z)(s - s_0)}\right] \quad . \tag{2.74}$$

In the dispersion relation for D_ℓ, one subtraction has been made at $z = s_0$ and D normalized so that $D_\ell(s_0) = 1$. In the separable potential model the amplitude falls off sufficiently rapidly for large $|z| [X_\ell = O(z^{-\ell-1})]$ so that no subtraction in the integral for N is necessary. Unlike (2.70), (2.73) and (2.74) are linear and, provided the inelasticity has suitable asymptotic behavior at high energies, Fredholm. Hence, reduction to matrix equations by numerical quadratures is straightforward. Alternatively following PAGELS [2.34], the left-hand cuts may be replaced by a set of poles to obtain algebraic equations.

At first sight the on-shell approach has several advantages when compared with calculations using the off-shell Faddeev equations. The N/D equations are nonsingular, linear integral equations in contrast to the Faddeev equations which are singular at physical energies and must be modified by contour rotation [2.12] or other methods [2.35,36] in order to be tractable to numerical solution. Unlike the Faddeev equations, which must be solved anew for each energy of interest, the N/D equations yield the on-shell amplitude for a range of physical energies. Off-shell two-body information is not needed explicitly. Furthermore, the N/D equations are still one dimensional when local rather than separable potentials are considered, whereas the Faddeev equations are two dimensional integral equations, which are difficult to solve numerically. Generalization of the N/D method is also possible for few particle nuclear reactions involving several different two-body channels [2.37].

On the other hand, in applying the N/D method several uncertainties are introduced. The input information is constructed from the simplest particle exchange processes; the basic assumption that the singularities nearest to the physical region determine the amplitude with sufficient accuracy at low energies is questionable. As discussed in the next section, calculations seem to suggest that if off-shell behavior of the two-body amplitude is important in the Faddeev approach, then one cannot neglect the effects of distant singularities. Thus in some applications, the apparent advantage of requiring only on-shell information may be spurious. Unlike the calculations with the Faddeev equations, the contribution of break-up processes to unitarity has only been estimated in existing calculations.

It seems desirable to keep these limitations in mind when assessing the physical content of a particular calculation by comparison with experimental data. It is, however, possible to draw some conclusions about the reliability of the on-shell method by comparing the results with numerical solutions of the Faddeev equations for a sufficiently simple model where the exact model amplitudes are calculable.

2.4.2 Calculations for the Three-Nucleon System

In this section applications of the N/D method to nonrelativistic neutron-deuteron scattering will be reviewed.

If the coupling between spin and spatial degrees of freedom is ignored, there are two channels of interest, the doublet with spin $s = 1/2$ and the quartet with spin $s = 3/2$. In the quartet channel partial wave elastic amplitudes satisfy single-channel N/D equations similar to those of the last section. In the doublet channel there is mixing between the singlet and triplet two-nucleon states and the singularity structure of the left-hand cuts is more complicated.

An early calculation by BLANKENBECLER et al. [2.38] of low-energy forward neutron-deuteron scattering used a forward dispersion relation with the left-hand cut being approximated by the contribution from single proton exchange.

BARTON and PHILLIPS [2.39] solved the N/D equations for the s-wave quartet and doublet amplitudes, also using as input the nearest singularity to the physical region, the cut between s = - 1/3 and s = - 3 (cf. (2.54)), associated with nucleon exchange. Inelastic effects were ignored entirely, elastic unitarity (2.68) being assumed above the break-up threshold. In view of the simplicity of the calculation, agreement with the low-energy quartet scattering was excellent. In the doublet channel, no zero of D for the triton bound state emerged, and the results for s-wave scattering were poor. By repeating the calculation with an arbitrary subtraction in the equation for N at the two-body threshold, which was then fixed by the experimental value of the scattering length, a triton binding energy of 6 MeV and the well-known rapid variation at low energies of the doublet amplitude, observed experimentally, were obtained.

In a more extensive calculation, AVISHAI et al. [2.40] tackled the problem of inelasticity. They observed that in the separable potential or isobar model, the break-up amplitude was simply determined by the half-off-shell elastic amplitude (cf.(2.10)). In the quartet channel the inelasticity λ_ℓ of (2.69) was expressed in terms of known quantities and a function χ_ℓ, defined as the ratio of the half-off-shell and on-shell elastic amplitudes. The sensitivity of the N/D results to the inelasticity could then be tested by assuming various parameterization of χ_ℓ. A similar method of estimating the effect of break-up inelasticity on an elastic amplitude has been given in the context of deuteron-alpha scattering by FISHBANE and NOBLE [2.41].

In the isobar or separable potential model, the doublet amplitude for elastic n-d scattering is coupled to an amplitude involving the spin-zero, antibound singlet state with small positive energy of the two-nucleon system, which is not directly observable but which has a large effect on the scattering. A possible approach is to replace the antibound state by a genuine bound state with very small binding energy. A matrix N/D representation in [2.42] is then valid, but a further serious complication arises as the left- and right-hand cuts now overlap, requiring an analytic continuation of the N/D equations to be made [2.43]. AVISHAI et al. [2.40] therefore chose the single channel FRYE-WARNOCK equations [2.44] to describe the doublet scattering. In this formulation the inelasticity is a part of the input information and must be taken from experiment. Unfortunately the experimental data are not adequate at higher energies and SLOAN [2.45] has questioned whether the inelasticities used in [2.40] are correct. The left-hand cut was represented by the single nucleon and double nucleon exchange terms in the separable potential model, i.e., terms analogous to Z(W) and $T_1(W)$ of Section 2.2.

Again the results for the quartet channel are good. Apparently the quartet elastic amplitudes are dominated by the single nucleon exchange mechanism, and are insensitive to inelastic effects and the off-shell extrapolation in χ. The predictions for the doublet amplitudes show a strong dependence on the choice of vertex function and on the inelasticity, and are not very successful in reproducing the experimental data. By introducing a weak pole singularity some distance from the physical region (s = -60 MeV), AVISHAI et al. [2.40] were able to improve considerably the agreement of their calculations with experiment. These results seem to indicate that in the doublet channel the amplitudes are sensitive to distant singularities corresponding in the potential picture to details of the short-range nucleon-nucleon interaction. These effects may be simulated by either a subtraction or a distant pole.

The work of BOWER [2.46] also supports this conclusion. He attempts a calculation of n-d scattering using only on-shell information. The discontinuities over the left-hand cuts are found from the nonrelativistic limits of those Feynman diagrams whose singularities lie closest to the physical region. To avoid the problem of coupling to continuous break-up channels, the break-up inelasticity is parameterized by a simple step funtion. Again the doublet results are sensitive to the parameterization of distant singularities in the form of cutoffs and parameters for σ-meson exchange.

In summary although on-shell methods have met with some success, it is difficult to assess their reliability because of the interplay of both theoretical approximations and experimental uncertainties. STELBOVICS and DODD [2.47] have suggested that N/D solutions for neutron-deuteron scattering in the separable model can be tested by comparing them directly with numerical solutions of the Faddeev equations. Also in this case the analytic properties of the amplitudes and partial wave dispersion relations can be rigorously established [2.18].

In order to check the s-wave dispersion relations, the exact input information was computed from accurate numerical solutions of the Faddeev equations. On solving the N/D equations with this input, the exact amplitude was reproduced both in the quartet and doublet channels. The doublet and quartet s-wave dispersion relations were also solved using input functions constructed from the first to the fourth orders of the multiple scattering series obtained by iteration of the Faddeev equations, i.e., terms analogous to $T_n = \sum_{i=0}^{n} (ZG_0)^i Z (n = 0,1,2,3)$ of Subsection 2.2.2. The exact inelasticities, computed from the Faddeev equations, were regarded as part of the input information.

In the quartet channel, the phase shifts, complex above the break-up threshold, of the second, third and fourth order approximate inputs agree remarkably well with the exact model phase shifts, confirming that single and double nucleon exchange terms dominate the low-energy behavior of the quartet amplitude (see Fig.2.7).

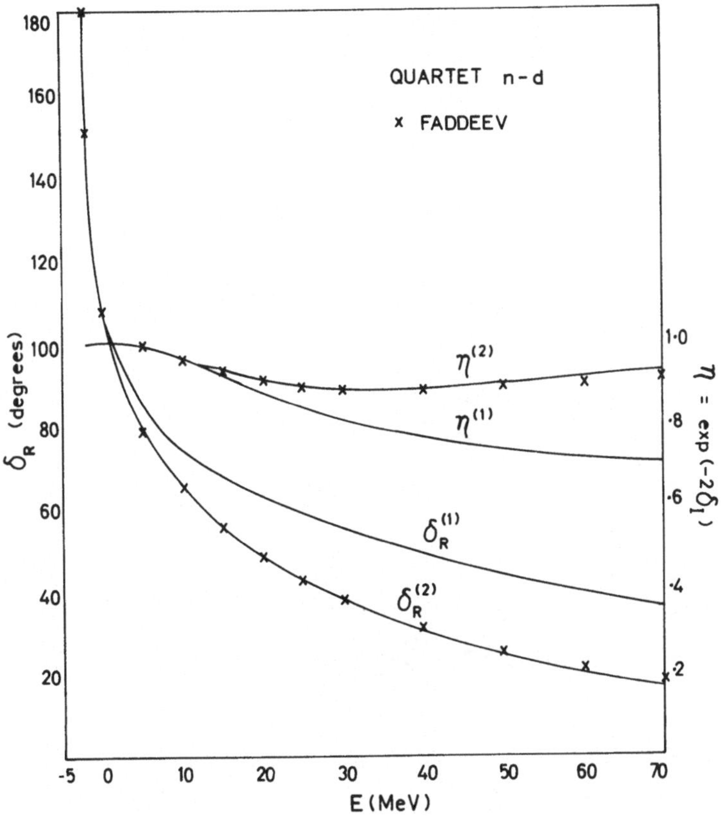

Fig.2.7. Real parts δ_R of the quartet s-wave phase shifts and absorption coefficients ($\eta = \exp(-2\delta_I)$) for neutron-deuteron scattering in the separable potential model. The crosses represent the exact model values obtained from the Faddeev equations. The solid lines indicate the approximate solutions calculated from the N/D equations, using as input first and second orders of the multiple scattering series

In the doublet the qualitative behavior of the approximate phase shifts agrees with the exact phase shifts, but even in the fourth order, the phase shifts still differ by as much as ten degrees (see Fig.2.8). In first order there is no triton pole and even in fourth order the triton pole is displaced by 1.6 MeV from the exact model value. It is also clear from the details of these calculations that while the solution of the N/D equations for the quartet channell is relatively insensitive to changes in the input information, the doublet solution is quite sensitive to small changes in input through both the driving terms and the inelasticities.

These properties are consistent with the physical picture proposed by BARTON and PHILLIPS [2.39]. They argued that because the interaction in the channel is repulsive the neutron does not approach the deuteron too closely so that only the longest

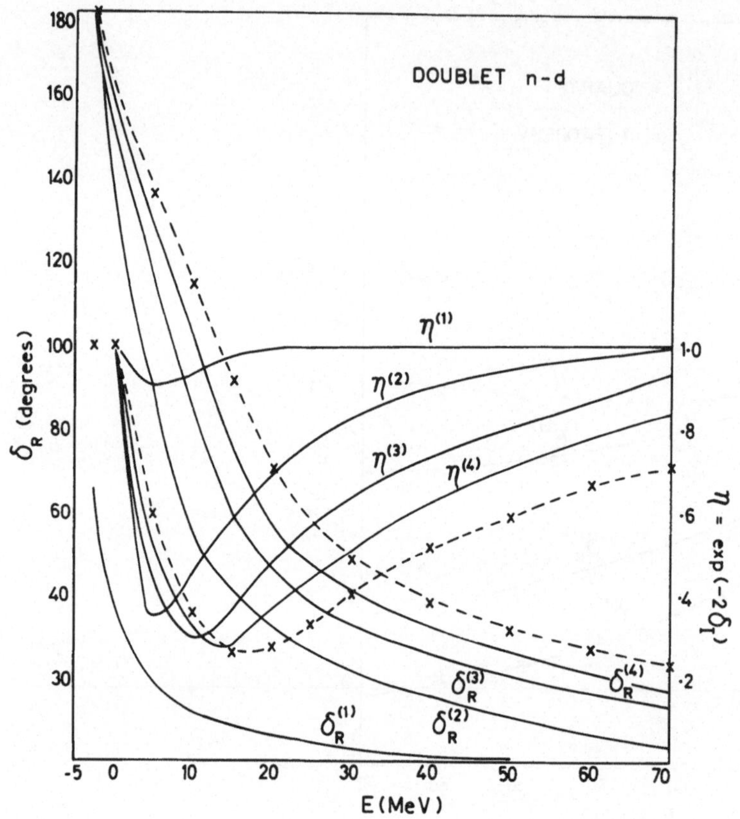

Fig.2.8. Real parts δ_R of the doublet s-wave phase shifts and absorption coefficients ($\eta = \exp(-2\delta_I)$) for neutron-deuteron scattering in the separable potential model. The broken curve represents the exact model solution obtained from the Faddeev equations. The solid lines represent approximate solutions calculated from the N/D equations, using as input various orders of the multiple scattering series

range forces, corresponding to nearby singularities, are important. On the other hand in the doublet channel the effective interaction is attractive so that short-range forces have a stronger effect. This implies that the more distant singularities in the higher oder multiple scattering terms are significant. We have also seen from Subsection 2.3.2 that in the limit of structureless vertex functions the Born and double exchange terms give the entire discontinuity on the left. Thus it may be concluded that the structure of the vertex function plays a significant role only in the doublet scattering.

Recently GREBEN, and KOK [2.48] have also compared N/D solutions with solutions of the Faddeev equations for n-d scattering in the separable potential model. In their work they extend the approximation for inelasticity given by AVISHAI et al. [2.40] to the doublet channel by replacing the antibound singlet deuteron state with a loosely bound singlet state. Thus there are two distinct channels for doublet scattering. In this approach the right-hand and left-hand cuts overlap, and the N/D equations must be modified [2.43].

They also find strong dependence of the doublet scattering on the form of the vertex function associated with the short-range part of the nucleon-nucleon interaction and sensitivity to the parameterization of the inelasticity.

2.4.3 Other Applications

RINAT and STINGL [2.37] have proposed the application of on-shell methods to low-energy reactions between light nuclei. Amplitudes for the scattering of any two clusters with a total number of N particles are assumed to satisfy coupled-channel N/D equations, preserving two-cluster unitarity. If channels where the N particles are broken into four or more clusters are ignored, the methods discussed in the last subsection for the three nucleon problem are applicable; channels with three clusters may be reduced to effective two-body channels of the interaction between any pair of clusters is assumed to be separable. The driving terms in the N/D equations are to be calculated from the lowest orders of cluster exchange amplitudes. RINAT and STINGL [2.49] have made a preliminary calculation for the N=5 system the channel containing an alpha and a neutron being coupled to the channel containing a triton plus deuteron. Three-cluster channels were ignored entirely so that the methods suggested for approximating three-particle unitarity have not been tested by this claculation. Also the problem of overlapping right- and left-hand cuts [2.43] in the amplitude for $t+d\to\alpha+n$ by single proton exchange was avoided. More complete calculations are, however, feasible.

KAC [2.50] has performed a similar calculation with two two-cluster channels, one containing a proton p and the ground state of ^{13}N, the other a proton and the first resonant state $^{13}N^*$ of ^{13}N. The coupling of these two channels through the N/D equations generates a J = 0+ three-body resonance level in ^{14}O, below the threshold for break-up into $^{13}N^*$ and p. This calculation also ignores three-cluster contributions to the unitarity equations (cf. (2.67) and (2.69)).

FISHBANE and NOBLE [2.41] have used an N/D formalism to estimate the effect of the deuteron break-up channel on low energy deuteron-alpha elastic scattering. They obtain a modified effective range expansion which includes the effect of inelasticity. In their work the left-hand cut is approximated by a pole with adjustable parameters; there is no attempt to calculate the discontinuity across the left-hand cut from a model.

A three-body system of considerable current interest is the pion-deuteron system. Calculations of elastic scattering and reaction processes have been attempted from both the Faddeev [2.51] and dispersion relation [2.52] methods. At present it is difficult to make a comparison of the two approaches as different models of the two-body interactions have been assumed. Further work should clarify their relationship.

2.5 Conclusions

The analytic structure in the total energy variable of nonrelativistic on-shell three-body amplitudes is well established and suitable for the application of dispersion relations.

The N/D solutions for the scattering amplitudes are approximate solutions of partial-wave dispersion relations. The success of on-shell methods of calculation of the amplitudes depends on the adequacy of the driving terms, calculated from the multiple scattering series or nonrelativistic limits of Feynman diagrams. If the nearest singularities to the physical region are not dominant, the method is unreliable. It is difficult to estimate the order of the driving terms necessary to produce sufficient accuracy. In calculations of neutron-deuteron scattering in the separable potential model, where the exact amplitudes are known, the convergence in the quartet channel is excellent. Here single neutron exchange is the dominant process, whereas convergence in the doublet channel is rather slow, distant singularities being important. Break-up channels must be included in an approximate way to take account of three-particle unitarity. The N/D solutions may be sensitive to the approximate parameterization of inelasticity. Only detailed calculations will show whether these problems are significant, and whether on-shell methods will prove to be a valuable complement, if not alternative, to the Faddeev equations.

From the theoretical viewpoint, much can be learned from studies of the relationship between the Faddeev and on-shell approaches. As previously discussed various methods of approximating the left-hand cut and inelasticity parameters may be tested by a comparison of the N/D solutions with solutions of the Faddeev equations. It seems reasonable to expect that other features of the N/D approach, such as three-body contributions to unitarity, anomalous thresholds, ghost states and subtractions will be clarified by such comparisons.

Appendix

In this Appendix explicit forms for the numerator N, and denominator D, of the resolvent kernel R of (2.17) are given (SMITHIES [2.21]).

If

$$K^n(\underline{p},\underline{p}';W) = \int K(\underline{p},\underline{p}_1;W)K(\underline{p}_1,\underline{p}_2;W)\ldots K(\underline{p}_{n-1},\underline{p}';W)d^3p_1\ldots d^3p_{n-1} \quad ,$$

let

$$\sigma_n = \mathrm{Tr}(K^n) = \int d^3pK^n(\underline{p},\underline{p};W) \quad \text{for} \quad n \geq 2 \quad .$$

Then

$$D(W) = 1 + \sum_{n=2}^{\infty} D_n(W)$$

with the determinant

$$D_n(W) = \frac{(-1)^n}{n!} \begin{vmatrix} 0 & n-1 & 0 & \ldots\ldots & 0 & 0 & 0 \\ \sigma_2 & 0 & n-2 & \ldots\ldots & 0 & 0 & 0 \\ \sigma_3 & \sigma_2 & 0 & \ldots\ldots & 0 & 0 & 0 \\ \ldots & \ldots & \ldots & \ldots & \ldots & \ldots & \ldots \\ \sigma_{n-1} & \sigma_{n-2} & \sigma_{n-3} & \ldots\ldots & \sigma_2 & 0 & 1 \\ \sigma_n & \sigma_{n-1} & \sigma_{n-2} & \ldots\ldots & \sigma_3 & \sigma_2 & 0 \end{vmatrix} \quad , \tag{2.75}$$

and

$$N(\underline{p},\underline{p}';W) = K(\underline{p},\underline{p}';W) + \sum_{n=1}^{\infty} N_n(\underline{p},\underline{p}';W)$$

with

$$N_n = \frac{(-1)^n}{n!} \begin{vmatrix} K & n & 0 & 0 & \ldots & 0 & 0 & 0 \\ K^2 & 0 & n-1 & 0 & \ldots & 0 & 0 & 0 \\ K^3 & \sigma_2 & 0 & n-2 & \ldots & 0 & 0 & 0 \\ \ldots & \ldots & \ldots & \ldots & \ldots & \ldots & \ldots & \ldots \\ K^n & \sigma_{n-1} & \sigma_{n-2} & & \ldots & \sigma_2 & 0 & 1 \\ K^{n+1} & \sigma_n & \sigma_{n-1} & & \ldots & \sigma_3 & 2 & 0 \end{vmatrix} \quad . \tag{2.76}$$

Thus the resolvent kernel is constructed from the multiple scattering terms K^n of the perturbation expansion.

Note on convergence: The series $\sum_{n=1}^{\infty} f_n(z)$ *converges absolutely* for a particular value of z, if for any $\varepsilon>0$ a number n exists such that $\left|\sum_{i=1}^{n+p}|f_i(z)|^2 - \sum_{i=1}^{n}|f_i(z)|^2\right|<\varepsilon$ for all positive p, the values of n depending on ε. If the values of n may be chosen independently of z in some region, then the series *converges uniformly* throughout the region. Absolute convergence implies the terms in the series for N and D may be manipulated like finite series. Uniformity of convergence in p and p' of N(p,p',W) ensures that the continuity properties in p and p' of the terms in N will be shared by N.

References

2.1 S.C. Frautschi: *Regge Poles and S-Matrix Theory* (W.A. Benjamin Inc., N.Y. 1963)

2.2 L.D. Faddeev: *Mathematical Aspects of the Three-Body Problem* (Daniel Davey and Co., Inc., New York 1965)

2.3 R.D. Amado, M.H. Rubin: Phys. Rev. Lett. 25, 194 (1970); R.D. Amado, D.F. Freeman, M.H. Rubin: Phys. Rev. D4, 1032 (1971)

2.4 R.T. Cahill: Phys. Rev. C9, 473 (1974)

2.5 R.D. Amado: Phys. Rev. C11, 719 (1975)

2.6 S.K. Adhikari, R.D. Amado: Phys. Rev. Lett. 27, 485 (1971)

2.7 R. Dashen, S. Ma, M.J. Bernstein: Phys. Rev. 187, 345 (1969)

2.8 R. Dashen, S. Ma: J. Math. Phys. 12, 689 (1971)

2.9 V.N. Efimov: Yad. Fiz. 12, 1080 (1970) (English Transl: Sov. J. Nucl. Phys. 12, 589 (1971)); Phys. Lett. 33B, 563 (1970)

2.10 R.D. Amado, J.V. Noble: Phys. Lett. 35B, 25 (1971); Phys. Rev. D5, 1992 (1972)

2.11 A.T. Stelbovics, L.R. Dodd: Phys. Lett. 39B, 450 (1972)

2.12 J.H. Hetherington, L.H. Schick: Phys. Rev. 137B, 935 (1965)

2.13 D.D. Brayshaw: Phys. Rev. 176, 1855 (1968)

2.14 S. Mandelstam: Phys. Rev. 140, B375 (1965)

2.15 R. Aaron, V. Teplitz: Phys. Rev. 167, 1284 (1968)

2.16 R.G. Newton: Nuov. Cim. 29, 399 (1963)

2.17 M. Rubin, R. Sugar, G. Tiktopoulos: Phys. Rev. 146, 1130 (1966); 159, 1348 (1967); 162, 1555 (1967)

2.18 A.T. Stelbovics: Ph.D. Thesis, University of Adelaide, 1975 (unpublished)

2.19 R. Aaron, R.D. Amado, Y.Y. Yam: Phys. Rev. 136, B650 (1964)

2.20 C.B. Lovelace: Phys. Rev. 135, B1225 (1964)

2.21 F. Smithies: *Integral Equations* (Cambridge University Press, 1958)

2.22 C.B. Lovelace: *Strong Interactions and High Energy Physics*, ed. by R.G. Moorhouse (Oliver and Boyd, London 1964)

2.23 L. Brown, D. Fivel, B. Lee, R. Sawyer: Ann. Phys. (N.Y.) 23, 167 (1963)

2.24 R.J. Eden, P.V. Landshoff, D.I. Olive, J.D. Polkinghorne: *The Analytic S-Matrix* (Cambridge University Press, 1966)

2.25 T. Bjorken, S. Drell: *Relativistic Quantum Fields* (McGraw-Hill, New York 1965)

2.26 R.D. Amado: Phys. Rev. Lett. 33, 323 (1974)

2.27 K.M. Watson, J. Nuttall: *Topics in Several Particle Dynamics* (Holden Day, Inc., San Fransisco 1967), Chap.5

2.28 R. Sugar, R. Blankenbecler: Phys. Rev. 136, 472 (1964)

2.29 P.R. Graves-Morris: Ann. Phys. (N.Y.) 41, 477 (1967)

2.30 G. Skorniakov, K. Ter-Martirosian: Sov. Phys. - JETP 31, 775 (1956)

2.31 L.R. Dodd: J. Math. Phys. 11, 207 (1970); Phys. Rev. D3, 2536 (1971)

2.32 E.T. Whittaker, G.N. Watson: *Modern Analysis*, 4th ed. (Cambridge University Press, 1927), Chap.5

2.33 E.O. Alt, P. Grassberger, W. Sandhas: Nucl. Phys. B2, 167 (1967)

2.34 H. Pagels: Phys. Rev. 140B, 1599 (1965)

2.35 N.M. Larson, J.H. Hetherington: Phys. Rev. C9, 699 (1974)

2.36 T.A. Tjon: Phys. Rev. D1, 2109 (1970)

2.37 A.S. Rinat, M. Stingl: Ann. Phys. (N.Y.) 65, 141 (1971)

2.38 R. Blankenbecler, M. Goldberger, F. Halpern: Nucl. Phys. 12, 629 (1959)

2.39 G. Barton, A.C. Phillips: Nucl. Phys. A132, 97 (1969)

2.40 Y. Avishai, W. Ebenhöh, A.S. Rinat-Reiner: Ann. Phys. (N.Y.) 55, 341 (1969); W. Ebenhöh, A.S. Rinat-Reiner: Phys. Lett. 29B, 638 (1969)

2.41 P.M. Fishbane, J.V. Noble: Phys. Rev. 160, 880 (1967); 171, 1150 (1968)

2.42 J.D. Bjorken: Phys. Rev. 4, 473 (1960)

2.43 J.M. Greben: Ph.D. Thesis, University of Groningen, 1974 (unpublished); J.M. Greben, L.P. Kok: Phys. Rev. C13, 489 (1976)

2.44 G. Frye, R.L. Warnock: Phys. Rev. 130, 478 (1963)

2.45 I. Sloan: Nucl. Phys. A168, 211 (1971)

2.46 R.M.J. Bower: Ann. Phys. (N.Y.) 73, 372 (1972)

2.47 A.T. Stelbovics, L.R. Dodd: Phys. Lett. 48B, 13 (1974)

2.48 J.M. Greben, L.P. Kok: Phys. Rev. C13, 1352 (1976)

2.49 A.S. Rinat, M. Stingl: Phys. Rev. C10, 1253 (1974)

2.50 M. Kac: Ann. Phys. (N.Y.) 81, 113 (1973)

2.51 I.R. Afnan, A.W. Thomas: Phys. Rev. C10, 109 (1974)

2.52 D. Schiff, J. Tranthanhvan: Nucl. Phys. 63, 671 (1967); B5, 1883 (1968)

3. Theory of Three-Body Final States

R. D. Amado*

With 3 Figures

The classic difficulty of the three-body problem is in describing the system when interactions among the pairs overlap. This overlap is present in any genuine three-body situation, but it is most clearly manifested in states containing three free particles. Except for certain atomic problems where genuine three-body collisions can occur, these states are realized either in three-body decay ($1 \to 3$ reactions such as $K \to 3\pi$) or in break-up or particle production reactions ($2 \to 3$ reactions such as $n+d \to n+n+p$ or $\pi+N \to \pi+\pi+N$). From a practical point of view these reactions are interesting because they provide information on the interaction of unstable particles or on reaction mechanisms. For theory, the problem is to provide a framework for analysis of these reactions that is easy to use, makes the physics clear, and embodies the full constraints of quantum mechanics. The purpose of this chapter is to develop those constraints and their implementation and to relate them to general three-body theory.

3.1 The Problem

The basic question is how the two-body interaction information is distributed over a three-body final state and how that affects reaction mechanisms. The hope is that this question of final state spectrum can be studied without having to solve the full three-body dynamics. The study of such problems is called final state interaction theory and is not restricted to the three-body problem. Many of the methods and problems we encounter in the three-body case are more simply and familiarly revealed in two-body final state theory [3.1]. Hence we shall often turn to the two-body case for clarification and emphasis.

A general three-body state is a complicated quantity that requires nine scalar variables for the specification of its orbital motion, but if we work in the center of mass, specify the total three-body energy E and angular momentum J and if we also

* Supported in part by the National Science Foundation.

specify the angular momentum of a given pair, there remains only one free continuous variable, which may be taken as the relative energy of that pair in its own center of mass, or the energy of the third particle. These two are related by energy and momentum conservation since $E = \frac{p_1^2}{2m_1}+\frac{p_2^2}{2m_2}+\frac{p_3^2}{2m_3}$ but since $\underline{p}_1+\underline{p}_2+\underline{p}_3 = 0$

$$E = \frac{p_i^2}{2\mu_i} + \frac{q_i^2}{2\mu_{jk}} \equiv \frac{p_i^2}{2\mu_i} + \sigma_i \tag{3.1}$$

any ijk cyclic.

The problem of three-body final state interaction theory is then to determine the dependence of the amplitude on this remaining free variable. Of particular interest is the case in which each pair in the final state has a strong interaction in some state. How is that strong interaction distributed over the three-body final state? It might be thought that that question is clearly answered by the Faddeev decomposition of (1.106). Here the three-body amplitude is decomposed into three terms depending on which pair interacts last. Since each term carries the half on-shell two-body t-matrix of the pair, it would seem to carry the corresponding final state interaction information. This deceptively simple way of apparently partitioning the final state interaction information has been one of the attractions of the Faddeev decomposition over the many other formally different but completely equivalent rearrangements that solve the disconnectedness problem.

However the Faddeev amplitudes are the solution of a set of coupled equations. There is no reason to assume that the superficial partitioning of final state information into three terms means those terms are incoherent. In fact, it was discovered in early calculations [3.2] that the final state interaction information of a given pair is far more clearly developed in the full three-body amplitude than in the Faddeev term ending with that pair's interaction. This surprising result was subsequently shown to follow from an inportant coherence among the Faddeev amplitudes, imposed not by the detailed dynamics of a particular problem but by the general constraints of quantum mechanics [3.3]. The primary subject of this chapter is developing that coherence, largely from the constraints of unitarity and analyticity, and then showing that implementation of these conditions leads nearly all the way back to the separable interaction Faddeev equations.

3.2 Coherence

Let us see how this coherence arises in a simple case. Consider a 2→3 amplitude decomposed in the Faddeev form

$$T = T^1 + T^2 + T^3 \quad , \tag{3.2}$$

where T^i is the sum of all terms ending with the half-shell τ_i.[1] Let us focus on T^3 as an example. T^3 can be expressed in terms of T^2 and T^1 by

$$T^3 = B\tau_3 + (T^1 + T^2)G_0\tau_3 \tag{3.3}$$

where $B\tau_3$ represents the Born or potential terms. In terms of (3.3), (3.2) becomes

$$T = B\tau_3 + T^1 + T^2 + (T^1 + T^2)G_0\tau_3 \tag{3.4}$$

or equivalently

$$T = B\tau_3 + (T^1 + T^2)(1 + G_0\tau_3) \quad . \tag{3.5}$$

The factor $1+G_0\tau_3$ carries the essential coherence between the second and third terms of (3.4). It is in fact just the 1-2 scattering wave function (cf. (1.26). To see the importance of this combination, consider the equation for τ_3 itself,

$$\begin{aligned} \tau_3 &= V_3 + V_3G_0\tau_3 \\ &= V_3(1 + G_0\tau_3) \quad . \end{aligned} \tag{3.6}$$

Clearly the 1-2 scattering information is manifested in the τ_3 on the left. For example, if there is a resonance the appropriate rapid variation of magnitude and phase will be in this τ_3. But since V_3 on the right is presumably real and slowly varying, the rapid variation of τ_3 can only occur if the $V_3G_0\tau_3$ term has a slowly varying real part that just cancels the V_3 on the right. The half shell τ_3 in the $V_3G_0\tau_3$ terms has the scattering phase, hence the extra phase needed to accomplish the cancellation comes from G_0, or strictly from its imaginary part. Thus the coherence between the Faddeev terms comes about because T^3 has a "slowly varying" part that is just what is needed to "cancel" the "slowly varying" parts T^1 and T^2, and give the entire three-body amplitude the dependence of τ_3, not $G_0\tau_3$. Since the coherence is brought about by the imaginary part of G_0, which imaginary part corresponds to on-shell propagation, this coherence must be required by unitarity (recall that unitarity gives the relationship among amplitudes that can be connected by on-shell propagation). We therefore now turn to a much fuller and more careful discussion of the coherence constraints using unitarity as our principal tool.

---

[1] Throughout this chapter the time sequence in equations is left - right corresponding to initial - final.

3.3 The Two-Body Case

In order to clarify the physics, let us begin by examining the more familiar problem of final state interactions in two-body reactions and the implications of unitarity [3.1]. The problem here is to find how final state rescattering affects the final state spectrum in a two-body reaction in which a weak initial state channel leads to a strong one. Classic examples are photo-pion production (γ+N→π+N) in particle physics, and photodisintegration of the deuteron (γ+d→n+p) in nuclear physics. Let t_{WS} be the amplitude for the process in an appropriate eigenchannel (fixed J, etc.), where W and S stand for weak and strong. Assume further that S is the only strong channel energetically allowed. If we apply unitarity to t_{WS} we obtain only to first order in the W channel

$$\mathrm{Im}\{t_{WS}\} = -\,k_S t_{WS} t^*_{SS} - k_W t_{WW} t^*_{WS} \tag{3.7}$$

where we may drop the second term since we work only to first order in the weak process, and where k_S is the relative momentum in the strong channel, and $k_S{}^2$ is proportional to the kinetic energy in that channel. (We use nonrelativistic language in this discussion, but the argument is easily made with relativistic kinematics with no essential change in the conclusions). The strong interaction elastic t-matrix t_{SS} is normalized so that (this is different from the Chap.1 normalization)

$$t_{SS} = \frac{-\exp\,(i\delta)\,\sin\,\delta}{k_S} \tag{3.8}$$

in terms of the strong interaction phase shift δ. It should be recalled that we are working in a particular state of angular momentum, etc. Since $\mathrm{Im}\{t_{WS}\}$ is real, (3.7) and (3.8) may be combined to give

$$t_{WS} = M_{WS}\,\exp\,(i\delta) \quad , \tag{3.9}$$

where M_{WS} is any real quantity. Thus through unitarity the strong rescatterings give the weak to strong amplitude the strong scattering phase. But for most applications we are interested not only in the phase of t_{WS}, but in its energy dependence, and (3.9) is of little help. We need to know more about M_{WS}. We might argue that after the weak process acts to produce the final state, i.e., the photon breaks up the deuteron, the strong force rescatters the particles many times, just as it does in t_{SS}, and that therefore we should choose

$$M_{WS} = N_{WS}\,\sin\delta/k_S \quad , \tag{3.10}$$

where N_{WS} is a slowly varying factor that carries information on the break-up process itself. This choice is often called Watson's theorem or better Watson's approximation, since only (3.9) is a theorem. But it is also possible for the final state to form without rescatterings, while (3.10) has the embarrassing feature of going to zero as the strong interaction is switched off. Perhaps we should add B_{WS}, the first Born approximation for the W-S process. Doing so will violate the constraint (3.9) but yet $t_{WS} \to B_{WS}$ as $t_{SS} \to 0$. To resolve this puzzle let us turn to a full dynamical theory of the process in which we also know the strong potential V_{SS}. We would then write the exact expansion

$$\begin{aligned} t_{WS} &= B_{WS} + B_{WS}G_0V_{SS} + B_{WS}G_0V_{SS}G_0V_{SS} + \dots \\ &= B_{WS} + B_{WS}G_0t_{SS} \end{aligned} \tag{3.11}$$

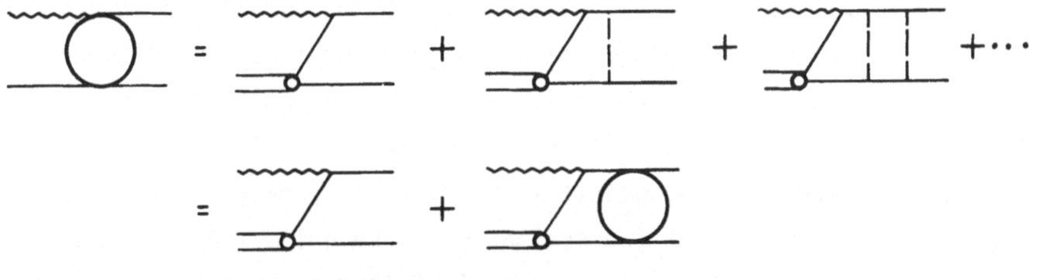

Fig.3.1. Diagrammatic representation of (3.11)

which equation is represented diagrammatically in Fig.3.1. The t_{SS} that appears is the half on-shell strong t-matrix and itself has the phase δ. How then can the form (3.11), which is a complete and exact theory of the process, agree with (3.9) which is a general constraint, when the first term in (3.11) is the purely real quantity B? The answer is that the propagator, G_0, in the BG_0t term is also complex and its imaginary part combines with the B term in an essential way that is independent of the dynamics. It is just this feature that made the Faddeev amplitudes coherent. Let us examine it in more detail here. We write

$$G_0(E) = (E - H_0 + i\varepsilon)^{-1} = P(E - W_0)^{-1} - i\pi\delta(E - H_0) \tag{3.12}$$

where P stands for principal part. The delta function part puts t_{SS} on shell, so that we get for (3.11) after some algebra

$$t_{WS} = B_{WS} + B_{WS}P(E - H_0)^{-1}t_{SS} + iB_{WS}\exp(i\delta)\sin\delta \tag{3.13}$$

The second term has only the phase δ, since the principal part integral is purely real. The first and third terms can be combined to give

$$B_{WS}[1 + i\exp(i\delta)\sin\delta] = B_{WS}\exp(i\delta)\cos\delta \tag{3.14}$$

Hence both terms have the phase δ and satisfy (3.9), but more importantly we see that it is the on-shell part of the propagator in the rescattering term and the term without rescattering which have an important coherence. This is precisely how the coherence arises in the Faddeev case (3.2), or in the Lippmann-Schwinger equation (3.6). The moral then is that terms which end with the multiple scattering of a given pair do not necessarily carry only the final state scattering information of that pair but, by the coherence mechanism we outlined above, also contain slowly varying pieces made to cancel terms that do not carry the rescattering so that the pair scattering information (e.g., the phase) is distributed over the entire amplitude.

It is precisely the problem of constructing three-body final state amplitudes that build in that coherence that we must consider. But let us first complete the discussion of the two-body case by seeing how we construct the amplitudes there. We need to find a general form that satisfies (3.7). Of course (3.11) is such a form, but it requires knowledge of the potential or off-shell t-matrix. We would like to satisfy (3.7) with knowledge of the scattering phase only. The unitarity relation (3.7) tells us more than (3.9); it also implies that t_{WS} has a square-root branch cut as a function of the center of mass energy, and $\mathrm{Im}\{t_{WS}\}$ is the discontinuity across that cut. We could try to exploit this fact and write

$$t_{WS} = A'_{WS} + ik_S B'_{WS} \tag{3.15}$$

with A' and B' both real. Since $\mathrm{Im}\{t_{WS}\} = k_S B'_{WS}$, we can combine (3.7), (3.8) and (3.15) to solve for B'_{WS} in terms of A'_{WS}. We get for (3.15) after some algebra

$$t_{WS} = \frac{A'_{WS} \exp(i\delta)}{\cos\delta} \quad . \tag{3.16}$$

Like (3.9) this is a correct representation of t_{WS}, but is useful only if we can assume that A'_{WS} is slowly varying and is at least independent of the strong interactions. For example, we might want to take A'_{WS} to be the Born term B_{WS} for the weak process. We certainly cannot do so near a resonance in this channel since on resonance, $\delta = \pi/2$ and $\cos\delta = 0$. The difficulty is that t_{WS} is a real analytic function of $E_S = k_S^2$ the center of mass kinetic energy, but its real and imaginary parts, A' and B', are not. There is no reason to assume that they are simple or slowly varying or independent of dynamics. To implement (3.7) we must include the analyticity information it implies. To do this, we write a dispersion relation (Cauchy's theorem). Since $t_{WS}(E_S)$ has a branch cut running from $E_S = 0$ to $E_S = \infty$, and since $\mathrm{Im}\{t_{WS}(E_S)\}$ is its discontinuity across that cut, we get

$$t_{WS}(E_S) = C_{WS}(E_S) + \frac{1}{\pi} \int_0^{\infty} \frac{\mathrm{Im}\{t_{WS}(E')\}}{E' - E_S} dE' \quad , \tag{3.17}$$

where C is the part of t that has no right-hand unitarity cut.

If we take it as real and slowly varying, we do no violence to analyticity, unitarity or the physics. We can, for example, take it to be the Born term, $C_{WS}=B_{WS}$. In fact the form (3.11), which is "exact" if one knows the potential, will also give t_{WS} left-hand cuts in E, but these are not required by unitarity or its corresponding analyticity. Hence (3.17) with a simple (e.g., constant) C is the minimal form for t_{WS} consistent with unitarity and analyticity. If we substitute (3.7) into (3.17), we get a singular integral equation for t_{WS}. Equations of this form have been studied extensively by MUSKHELISHVILI [3.4] and OMNES [3.5] and their general solution is known. In the simple case that C is a constant and no subtractions are required, the solution is

$$t_{WS}(E) = C \exp[\Delta(E)] = C/D(E) \quad , \tag{3.18}$$

where

$$\Delta(E) = \frac{1}{\pi} \int \frac{\delta(E')dE'}{E' - E} \quad , \tag{3.19}$$

and where D(E) is the usual denominator or D function of scattering theory. For the special case of a separable potential, D is also easily constructed in terms of the potential, or the potential easily obtained from D, as discussed in Chapter 1. But (3.18) and (3.19) are much more general than separable potentials. They assure not only that t_{WS} has the phase δ, but also give the variation of its magnitude. Eq. (3.18) is therefore the minimal embodiment of the unitarity constraint, where the constraint is properly viewed as giving information not just on the imaginary part of t_{WS} but also about its analytic structure.

3.4 The Three-Body Case

Let us now turn to the problem of applying our lessons to the three-body case [3.6]. As in the two-body case, we begin with unitarity, this time for an amplitude $T_{2,3}$ for going from some initial two-body state to a three-body state. Assuming only the two- and three-body channels are open, unitarity for $T_{2,3}$ can be written (as discussed in Sec.1.6)

$$\mathrm{Im}\{T_{2,3}\} = -\pi\sum_{2'} T_{2,2'}\delta(E - E_{2'})T^{\dagger}_{2',3} - \pi\sum_{3'} T_{2,3'}\delta(E - E_{3'})T^{\dagger}_{3',3} \quad , \tag{3.20}$$

where $T_{2,2'}$ and $T_{3',3}$ are the elastic two-body and three-body amplitudes, respectively. $T_{3',3}$ can be decomposed into a connected part $T_{3',3\ \mathrm{con}}$, and a sum of disconnected parts $T_{3',3\ \mathrm{discon}}$, which represent one particle going by while the other two scatter. These disconnected pieces of $T_{3',3}$ are a correct and necessary consequence of the definition of the S and T matrix. Hence the above can be written

$$\begin{aligned}\mathrm{Im}\{T_{2,3}\} = &-\pi\sum_{2'} T_{2,2'}\delta(E - E_{2'})T^{\dagger}_{2',3}\\ &-\pi\sum_{3'} T_{2,3'}\delta(E - E_{3'})T^{\dagger}_{3',3\ \mathrm{con}}\\ &-\pi\sum_{3'} T_{2,3'}\delta(E - E_{3'})T^{\dagger}_{3',3\ \mathrm{discon}} \quad .\end{aligned} \tag{3.21}$$

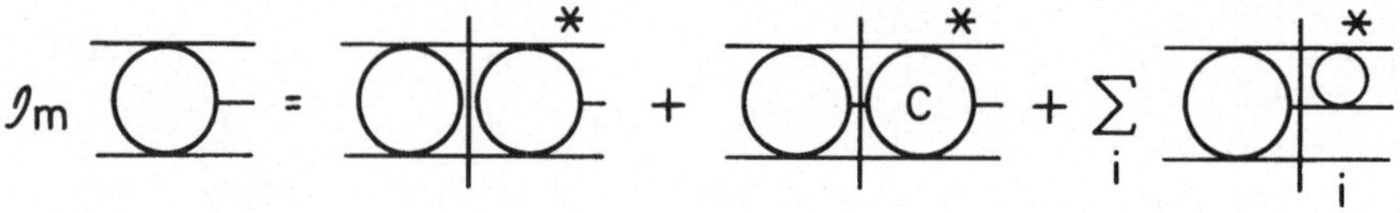

Fig.3.2. Diagrammatic representation of (3.21)

This equation is represented schematically in Fig.3.2. We are interested in exploiting (3.21) to obtain the dependence of $T_{2,3}$ on pair sub-energies, $q_i^2/2\mu_{jk} \equiv \sigma_i$, for fixed total energy E. We are particularly interested in any singularities $T_{2,3}$ might have in the sub-energy, since a singularity represents rapid dependence. As is well known, each term in unitarity represents a singularity at the threshold of that term [3.7]. For example in the case of coupled two-body channels there is a square-root branch point in the amplitude as a function of the total energy at the opening of each coupled channel. Unitarity signals this by acquiring a new terms as the threshold of the channel is crossed. Strictly speaking, each term in unitarity contributes the discontinuity across the singularity beginning at the threshold and in the variable carrying that threshold. We are interested in singularities in the pair sub-energies, since the final state interaction problem is to find how the two-body interaction information is distributed over the final state phase space for fixed total energy. This information should be most sharply focused in the pair sub-energy dependence of the amplitudes. The $T_{2,2'}T^{\dagger}_{2',3}$ term has a threshold in E, the total energy, at $E = E_{2,0}$, the minimum two-body energy. Similarly the $T_{2,3'}T_{3',3\ con}$ term has a threshold in E at $E = E_{3,0}$ the minimum three-body energy. So, apparently does the $T_{2,3}T_{3',3\ discon}$ term from (3.21), but as is clear from Fig.3.2 and as will become clear in our development, the δ function in $T_{3',3\ discon}$ coming from the fly-by particle will give it a threshold in the interacting pair's sub-energy. Hence this is the term we need keep to study sub-energy singularities. Keeping this term alone, we no longer have $\mathrm{Im}\{T_{2,3}\}$, but only the discontinuity of $T_{2,3}$ across the sub-energy singularity. We call it $\mathrm{Disc}\{T_{2,3}\}$, but must keep in mind that it is only the discontinuity across the sub-energy threshold that is given by

$$\mathrm{Disc}\{T_{2,3}\} = -\pi \sum_{3'} T_{2,3'}\delta(E - E_{3'})T_{3',3\ discon} \quad . \qquad (3.22)$$

(Note that we here define $\mathrm{Disc}\{f(\sigma)\} = [f(\sigma + i\varepsilon) - f(\sigma - i\varepsilon)]/2i$).

We now flesh out (3.22) for three final particles of mass, m_1, m_2, m_3 (for simplicity we neglect internal degrees of freedom such as spin, isospin, etc.). For $T_{3,3'\ discon}$ we have

$$\langle \underline{p}_1, \underline{p}_2, \underline{p}_3 | T_{3,3'\ discon} | \underline{p}_1', \underline{p}_2', \underline{p}_3' \rangle = \sum_i \delta(\underline{p}_i - \underline{p}_i') \langle \underline{q}_i | \tau | \underline{q}_i' \rangle \quad , \qquad (3.23)$$

in terms of the two-body t-matrix. For $T_{2,3}$ we take a particular form suggested by the Faddeev equations and called the sequential decay model in nuclear physics, and the isobar model in particle physics. We separate $T_{2,3}$ into a sum of three terms depending on which pair interacts last. Each ends with the half on-shell t-matrix of that pair. We assume that t-matrix is dominated by a single partial wave, and the interaction in that wave by a particularly strong state. It is simplest to think of it as a resonance. The three-body final state is then formed in two steps. First the initial state is transformed to a state of resonance plus spectator particle and then the resonant pair propagates and decays. The first process is represented by the quasi-two-body amplitude for forming the spectator and the resonance considered as a particle. The propagation of the pair is carried by the half-shell t-matrix, but half-shell and on-shell t-matrix differ only by a real factor that has no physical region singularities. Hence in the usual isobar or sequential decay phenomenology the propagation and decay factor is taken to be the on-shell pair t-matrix (with certain kinematic corrections). In fact, there are a number of reasons for preferring to represent this factor by the two-body D function (again with kinematic correction) of the pair instead. This function can be constructed directly from the two-body scattering data, as we discussed above, Assuming for simplicity that each pair interaction is dominated by a single partial wave, we write

$$\langle \underline{k}|T_{2,3}|\underline{p}_1,\underline{p}_2,\underline{p}_3\rangle = \delta(\underline{p}_1+\underline{p}_2+\underline{p}_3) \times \sum_{\substack{i,j,k\\ \text{cycic}\\ m}} \langle \underline{k}|f_i(E)|\underline{p}_i \ell_i m\rangle q_i^{\ell_i} Y_{\ell_i,m_i}(\hat{q}_i) D_{\ell_i}^{-1}(\sigma_i) \quad (3.24)$$

where $\hat{q}$ is a unit vector. Strictly f_i is defined by (3.24), but it is clearly the two-body amplitude for going from the initial state of relative momenta $\underline{k}$, to a state of spectator particle i with momentum $\underline{p}_i$ and a correlated jk pair with angular momentum quantum number $\ell_i m$. The factor q^{ℓ} is needed to get the correct threshold dependence. If the form (3.24) is to be used far from threshold, or if Coulomb forces are important, this factor should be replaced by the appropriate penetrability factor. To use (3.23) and (3.24) in unitarity, we first decompose the two-body t-matrix in (3.23) according to

$$\langle \underline{q}_i|\tau|\underline{q}_i'\rangle = \sum_m Y^*_{\ell_i,m}(\hat{q}_i)\tau_{\ell_i}(\sigma_i)Y_{\ell_i,m}(\hat{q}_i') \quad (3.25)$$

recalling that we are assuming only one pair partial wave is important. Our convention for unitarity gives

$$\text{Im}\{\tau_{\ell_i}(\varepsilon)\} = -\pi\mu_{jk}q_i|\tau_{\ell_i}(\varepsilon)|^2 \quad (3.26)$$

with q_1 on-shell

$$q_i^{\,2} = 2\mu_{jk}\varepsilon \quad . \quad (3.27)$$

We write

$$\tau_{\ell_i} = N_{\ell_i}(\varepsilon)/D_{\ell_i}(\varepsilon) \quad , \quad (3.28)$$

where D has only right-hand cuts and N only left-hand cuts.

If we now use (3.24) and (3.25) in unitarity (3.22) and equate coefficients of $Y_{\ell_i,m}(\hat{q}_i)$ on both sides, we get

$$\begin{aligned}&\delta(\underline{p}_1+\underline{p}_2+\underline{p}_3)\ \text{Disc}\ \{\langle \underline{k}|f_i(E)|\underline{p}_i,\ell_i m\rangle\}q_i^{\ell_i}D_{\ell_i}^{-1}(\sigma_i)\\ &= -\pi\sum_{\substack{t,m'\\ r,s}}\int d^3p_1'd^3p_2'd^3p_3'\delta(\underline{p}_1'+\underline{p}_2'+\underline{p}_3')\delta(\underline{p}_i-\underline{p}_i')\\ &\quad \delta(\underline{p}_j+\underline{p}_k-\underline{p}_j'-\underline{p}_k')\langle \underline{k}|f_t(E)|\underline{p}_t',\ell_t m'\rangle q_t'^{\ell_t}D_{\ell_t}^{-1}(\sigma_t') \quad (3.29)\\ &\quad \times Y_{\ell_t m'}(\hat{q}_t')Y^*_{\ell_i m}(\hat{q}_i)\tau^*_{\ell_i}(\sigma_i)\\ &\quad \times \delta(E-p_t^2/2\mu_t-q_t^2/2\mu_{rs})\end{aligned}$$

where we have used the fact that the various δ functions force $|\underline{q}_i| = |\underline{q}_i'|$. On the left in (3.29) we use

$$\begin{aligned} \text{Disc}\,\{f/D\} &= (\text{Disc}\{f\})/D^* + f\,\text{Disc}\{1/D\} \\ &= (\text{Disc}\{f\})/D^* + f\,\text{Im}\{1/D\} \end{aligned} \tag{3.30}$$

which follows from the definition of Disc.

$$\begin{aligned} \text{Disc}\,\{f/D\} &= (f^+/D^+ - f^-/D^-)/2i \\ &= (f^+/D^+ - f^+/D^- + f^+/D^- - f^-/D^-)/2i \\ &= f\,\text{Disc}\,\{1/D\} + (\text{Disc}\{f\})/D^- \quad . \end{aligned} \tag{3.31}$$

We also used $\text{Disc}\{1/D\} = \text{Im}\{1/D\}$ and $D^- = D^*$ because the only discontinuity of 1/D comes from its one threshold and that gives its imaginary part. There are two types of terms on the right-hand side of (3.29). In the first t = i, rs = jk. In this term the $Y_{\ell m}$ integrals are easily done by orthonormality since the arguments are the same. One then finds that the $f\text{Im}\{1/D\}$ term on the left exactly cancels with the $f1/D\tau^*$ term on the right by two-body unitarity (3.26). One is now left with the $\text{Disc}\{f\}/D^*$ term on the left and the terms on the right where $t \neq i$. Cancelling the $1/D^*$ in both of these one finally gets

$$\begin{aligned} &\text{Disc}\{\langle\underline{k}|f_i(E)|\underline{p}_i,\ell_i m\rangle\} \cdot \\ &- \pi \frac{N_{\ell_i}(\sigma_i)}{q_i^{\ell_i}} \int d^3q_i'\,\delta(\sigma_i - \sigma_i')\,Y^*_{\ell_i,m}(\hat{q}_i) \\ &\times \sum_{m'} \{\langle\underline{k}|f_j(E)|\underline{p}_j',\ell_j m'\rangle q_j'^{\ell_j} Y_{\ell_j m'}(\hat{q}_j') D^{-1}_{\ell_j}(\sigma_j') \\ &+ \langle\underline{k}|f_k(E)|\underline{p}_k',\ell_k m'\rangle q_k'^{\ell_k} Y_{\ell_k m'}(\hat{q}_k') D^{-1}_{\ell_k}(\sigma_k')\} \quad . \end{aligned} \tag{3.32}$$

Thus we see that the quasi-two-body amplitude $\langle\underline{k}|f_i(E)|\underline{p}_i,\ell_i m\rangle$ has a singularity in the j-k pair subenergy. We shall see that it is a $(\sigma_i)^{1/2}$ singulartiy. The strength of the singularity depends on the "non-i" parts of the amplitude, that is on f_j/D_j, and f_k/D_k, and hence it is important when the pair interactions are strong in overlapping parts of the final state phase space.

3.5 Spinless Boson Example

In order to explore the content and implications of (3.32) in greater detail, let us make even more simplifications. Consider the case of three spinless bosons of mass m in the final state and assume further that their dominant interaction is in s-waves. There is only one function f in that case, and dropping unnecessary labels, we get for (3.32)

$$\mathrm{Disc}\{\langle \underline{k}|f(E)|\underline{p}\rangle\} = -\frac{1}{2}N\left(E-\frac{3p^2}{4m}\right)\int d^3p'\langle \underline{k}|f(E)|p'\rangle$$
$$\times\ \delta\left[E-\frac{p^2}{2m}-\frac{p'^2}{2m}-\frac{(\underline{p}+\underline{p}')^2}{2m}\right]\times D^{-1}\left(E-\frac{3}{4m}p'^2\right)\quad . \tag{3.33}$$

We make a partial wave reduction of f according to

$$\langle \underline{k}|f(E)|\underline{p}\rangle = \sum_{\ell,m} Y_{\ell,m}(\hat{k})\langle k|f_\ell(E)|p\rangle Y^*_{\ell,m}(\hat{p}) \tag{3.34}$$

where now ℓ labels the total three-body angular momentum.

Eq. (3.33) becomes

$$\mathrm{Disc}\{\langle k|f_\ell(E)|p\rangle\} = -\pi N\left(E-\frac{3p^2}{4m}\right)\int p'^2dp'\langle k|f_\ell(E)|p'\rangle$$
$$\times\int dzP_\ell(z)\delta\left(E-\frac{p^2}{m}-\frac{p'^2}{m}-\frac{pp'z}{m}\right)D^{-1}\left(E-\frac{3}{4m}p'^2\right) \tag{3.35}$$

in terms of the Legendre polynomial $P_\ell(z)$. In obtaining (3.33) - (3.35) we have made explicit use of the fact that $\mu_i = 2m/3$, $\mu_{jk} = m/2$ and $Y_{00} = (4\pi)^{-1/2}$. The δ function in (3.35) can be used to do the z integral, but since $|z|\leq 1$, there are restrictions on p' for fixed E and p. These are most easily expressed by changing variables to $p^2/2m = x$ and $p'^2/2m = y$. Let us call $\langle k|f_\ell(E)|p\rangle = f_\ell(E,x)$. Eq. (3.35) then becomes

$$\mathrm{Disc}\{f_\ell(E,x)\} = \frac{-\pi^2}{(2mx)^{1/2}}N\left(E-\frac{3}{2}x\right)\int_{y_-}^{y_+}dyf_\ell(E,y)$$
$$\times\ D^{-1}\left(E-\frac{3}{2}y\right)\times P_\ell\left[\left(\frac{E}{2}-x-y\right)/(xy)^{\frac{1}{2}}\right] \tag{3.36}$$

where

$$y_I = \frac{1}{2}\{E-x\pm[2x(E-\frac{3}{2}x)]^{1/2}\}\quad . \tag{3.37}$$

These limits in y are just those allowed by phase space. Note that E - (3/2)x is the sub- energy of the pair associated with the spectator of energy x. One sees clearly that (3.36) gives f a singular part at a given x in terms of an integral over f at different values. For example from (3.37) we see that, for x near zero, y values near (1/2)E contribute, while for x near (2/3)E, y near (1/6)E contributes.

Since as x approaches (2/3) E the integration region shrinks to zero, Disc$\{f_\ell(E,x)\}$ goes like $[E - (2/3)x]^{1/2}$ in this region. This is just the square-root branch point we have stressed. $E - (3/2)x = 0$ is precisely the two-body threshold of the pair associated with the "x" spectator. The integration region also shrinks to zero as $x \to 0$ like $x^{1/2}$, but because of the $x^{-1/2}$ in front, $f_2(E,x)$ is finite at $x = 0$, and has no singularity there. In fact, the branch cut associated with (3.36) runs from $E - (3/2)x = 0$ to $E - (3/2)x = \infty$ or from $x = (2/3)$ E to $x = -\infty$, just as we would expect for a two-body scattering cut, rather than from $x = (2/3)E$ to $x = 0$, as one might naively expect from phase-space considerations only. In the physical region $[0 \le x \le (2/3)E]$, (3.33) or (3.36) may be used to calculate the discontinuity across the cut; for negative x these expressions do not apply and some form of analytic continuation is required if one wants the discontinuity in that region.

3.6 Implementing Unitarity and Analyticity

We have seen that unitarity alone forces the quasi-two-body amplitude f to have a part with a singular dependence on the sub-energy of the correlated pair and a part is proportional to the reaction strength in the other isobar groupings. We would like to understand that result in terms of the Faddeev formulation and also to implement it in a phenomenology. That is to write down some form for f, or constraint on f, that assures that (3.32) is satisfied, just as using the phase-shift parameterization ensures that two-body unitarity is maintained. Let us begin with relating (3.32) to the Faddeev or multiple scattering formulation. Consider some particular term in the multiple scattering series for a 2→3 reaction, represented graphically in Fig.3.3,

Fig.3.3. Diagrammatic representation of (3.38)

$$BG_0\tau_1 \cdots\cdots \tau_1 G_0 \tau_3 G_0 \tau_1 G_0 \tau_3 \qquad (3.38)$$

where B is the driving term or Born term for the process. This is a contribution to the Faddeev amplitude T^3. Note that in Fig.3.3 the label p_3 disappears to the left of the dotted line, so that any dependence that (3.38) gets on p_3 must arise from parts to the right of that line. There are no right-hand cut singularities at all that come from p_1 and p_2 since all such singularities must come from the vanishing of propagators. There are propagators in the last τ_3 which can vanish (when

$E-(p_3^2/2\mu_3) = 0$) but these just give rise to the normal threshold singularity of τ_3. These are already accounted for in our sequential decay or isobar or Faddeev formalism by the $1/D_3$ or half-shell τ_3 factor. But p_3 also occurs in the G_0 just before this last τ_3. That propagator too can vanish, again at $E = p_3^2/2\mu_3$. But all of (3.88) before this last τ_3 is a part of the isobar amplitude f, hence the singularity associated with G_0 is in f. Of course this is the singularity given by (3.32). It is easy to see that the δ-function in (3.32), or particularly in (3.33), has just the argument to be the imaginary part of that last G_0. To account for the rest of (3.32) note that Fig.3.3 or (3.38) is just one of the many terms in the multiple scattering series that ends with τ_3. The only requirement for such a term is that the last scattering before the $G_0\tau_3$ factor *not* be a scattering of 1 and 2.

The sum of all ways of going from the initial state to three particles that do not end in a 1-2 scattering is just the sum of the other Faddeev, or sequential decay, or isobar terms. The parts we called "non-i". That is just what (3.32) says. There is a special feature of (3.38) that we can exploit for phenomenology. The p_3 or sub-energy singularity comes from the last rescattering. The rest of the terms of the multiple scattering series are going to build up the correct total E singularity and hence scale of the amplitude. (Note that in unitarity the total E singularities enter in a nonlinear way and hence are sensitive to scale while the sub-energy singularities enter only linearly and hence are not). If we are not concerned with the scale of the amplitude, but only its sub-energy dependence, we might be able to use nearly any simple model that has the correct singularity structure to get it. In fact, it has been shown [3.8] that the first multiple scattering terms can, in many cases, give a very good description of the sub-energy dependence, even in cases where the multiple scattering series itself diverges badly. Of course in such cases the scale of the amplitude is completely wrong. We shall return to another manifestation of this point later.

The singularity we have been discussing, associated with the on-shell part of the last G_0, is just the singularity responsible for the coherence we discussed earlier. One can see from (3.22) that the disconnected terms in unitarity require that the two-body phase be in some sense distributed over the entire amplitude, not just in one isobar term. That is the role of the coherence. In fact, if the three pair sub-energies were independent, that would suggest that a product form for the T_{2-3} amplitude is better than a sum. Except for very special static models, they are not independent and hence the product form will not do. Thus we are faced with the problem that we know, particularly when pair interactions overlap strongly, that the quasi-two-body amplitudes in the sequential decay or isobar approximation are neither constant nor incoherent and that that coherence is important in ensuring the distribution of two-body information over the entire three-body amplitude, but we do not yet have a simple phenomenology for implementing that coherence. We turn now to that problem.

Unitarity says that the quasi-two-body amplitude f_j as a function of the pair sub-energy σ_j has a branch cut. Hence we can write,

$$f_j(\sigma_j) = \mathrm{Disp}\{f_j(\sigma_j)\} + i\,\mathrm{Disc}\{f_j(\sigma_j)\} \qquad (3.39)$$

where

$$\mathrm{Disc}\{f_j(\sigma_j)\} = \frac{1}{2i}\,[f_j(\sigma_j + i\varepsilon) - f_j(\sigma_j - i\varepsilon)] \qquad (3.40)$$

and the dispersion part (Disp{f}) (analogous to the real part for a real analytic function) is

$$\mathrm{Disp}\{f_j(\sigma_j)\} = \frac{1}{2}\,[f_j(\sigma_j + i\varepsilon) + f_j(\sigma_j - i\varepsilon)] \quad . \qquad (3.41)$$

Unitarity (3.32) then reads (symbolically)

$$\mathrm{Disc}\{f_j(\sigma_j)\} = \sum_{k\neq j} \int K(\sigma_j,\sigma_k) f_k(\sigma_k) \qquad (3.42)$$

where K is a finite integral kernel made up of phase space, delta functions, and two-body scattering information. If we substitute (3.39) in (3.42) we get a set of linear, Fredholm integral equations for $\mathrm{Disc}\{f_j\}$ with inhomogeneous parts that depend on $\mathrm{Disp}\{f_j\}$. They are Fredholm because the kernel K is finite and the delta function in it ensures that the integration region is finite (phase space). For any choice of $\mathrm{Disp}\{f_j\}$, the solution of these equations will automatically satisfy unitarity. (Unfortunately no closed-form solution of these equations is known even for the simplest example). However $\mathrm{Disp}\{f_j\}$ is not in fact arbitrary. As (3.40) and (3.41) show, $\mathrm{Disp}\{f_j\}$ and $\mathrm{Disc}\{f_j\}$ are the dispersive and absorptive parts (corresponding to real and imaginary parts) of an analytic function of σ. It is not enough to satisfy unitarity at a fixed σ, but we must also preserve those analytic properties. A simple choice of Disp{f} could lead to the troubles encountered in (3.16) with a simple choice of A'_{WS} for implementing two-body unitarity, that is spurious singularities can be introduced because we use unitarity without analyticity.

In the three-body case this can be particularly serious in the case of resonant two-body interactions. If threshold final state interactions are important as they are in n-d break-up, we are interested in the behavior of $f(\sigma)$ for small σ. That is where the square-root singularity comes and it is entirely in Disc{f}, the assumption then that Disp{f} is simple and analytic is then consistent and causes no difficulty. In fact, for those cases it has been shown [3.9] in a number of model calculations that the $\sigma^{1/2}$ singularity dominates the dependence of f and can be obtained from (3.42), with Disp{f} taken as constant, although as mentioned above much more is needed to get the magnitude of f as well. In cases of important resonant final state interactions the situation is different. It turns out that the resonant pair final state interactions produce logarithmic singularities in $f(\sigma)$, but these are in the unphysical amplitude $f(\sigma-i\varepsilon)$. If we construct Disp{f} and Disc{f} consistently via (3.40) and (3.41) the $f(\sigma-i\varepsilon)$ terms will cancel and the logarithmic singularities will not appear in the physical $[f(\sigma+i\varepsilon)]$ amplitude. But if we use a simple guess for Disp{f(σ)} and then use (3.42) to generate Disc{f(σ)}, these logarithmic singularities will eppear in Disc{f(σ)} from the $f(\sigma-i\varepsilon)$ term of (3.41) but will not be compensated by a corresponding term in (3.40) [3.10]. Hence (3.42) cannot be used in that case with arbitrary Disp{f} to generate a useful Disc{f}. We need a scheme that simultaneously generates Disp{f} and Disc{f}, that satisfies unitarity and analyticity at the same time.

The way to do this, as in the two-body case, is to write a dispersion relation,

$$f_j(\sigma) = R_j(\sigma) + \frac{1}{\pi}\int_0^\infty \frac{d\sigma' \,\mathrm{Disc}\{f_j(\sigma')\}}{\sigma' - \sigma} \quad , \tag{3.43}$$

where R_j is some driving term that does not have the σ cut required by unitarity. Since (3.32) relates Disc$\{f_j\}$ back to f_j, the set (3.43) for all three j with (3.32) form a set of coupled, linear integral equations for the f_j with the R_j as inhomogeneous terms. Since the cut in (3.43) runs from $\sigma = 0$ to ∞, but unitarity only gives us the discontinuity in phase space from $\sigma = 0$ to some maximum σ, we need to use some form of analytic continuation to realize (3.43). Furthermore we see from (3.1) that σ_i, E and p_i^2 are not all independent and we must decide what we will keep fixed while we integrate over σ. Any choice of continuation or of variable will satisfy sub-energy unitarity and analyticity, but some may produce strange E analyticity which must be compensated by special choice of R. In our discussion below we will make those arbitrary choices that allow R to be chosen as simply as possible and that give actual integral equations for f not just integral constraints.

3.7 Three Identical Bosons Again

In order to see what we are doing, let us use the three-body example of (3.33), and simplify it even further by making a zero range approximation for the two-body force so that we can replace $N[E-(3/4)(p^2/m)]$ by N_0 a constant.

It is clear from the structure of (3.33) that we run into far fewer problems with continuation outside the phase space if we choose to disperse in E for fixed p, rather than in σ or p^2 for fixed E. We then get, putting (3.33) in (3.43),

$$\langle \underline{k}|f(E)|\underline{p}\rangle = \langle \underline{k}|R(E)|\underline{p}\rangle - \frac{\pi}{2} N_0 \int \left[\frac{\langle \underline{k}|f(E_0)|\underline{p}'\rangle}{D(E_0 - \frac{3p'^2}{4m})} + F(\underline{k},\underline{p},\underline{p}',E) \right] \times \frac{d^3p'}{E - \frac{p'^2}{2m} - \frac{p^2}{2m} - \frac{(\underline{p}+\underline{p}')^2}{2m}} \tag{3.44}$$

where $E_0 = p^2/2m+p'^2/2m+(\underline{p}+\underline{p}')^2/2m$, and where F is any function such that $F(\underline{k},\underline{p},\underline{p}',E_0) = 0$ since all that unitarity (3.33) tells us is what the residue at the $E = E_0$ pole of f must be and (3.44) will have that residue if F satisfies this condition. There are many impediments to using (3.44), not the least of which is that it contains an arbitrary unknown function F and requires our knowledge of f at $E = E_0$. Of course (3.44) satisfies sub-energy analyticity and unitarity for any choice of F, but one choice makes the problem particularly simple. That one is

$$F(\underline{k},\underline{p},\underline{p}',E) = \langle \underline{k}|f(E)|\underline{p}'\rangle D^{-1}(E - \frac{3}{4}\frac{p'^2}{m}) - \langle \underline{k}|f(E_0)|\underline{p}'\rangle D^{-1}(E_0 - \frac{3}{4}\frac{p'^2}{m}) \quad . \tag{3.45}$$

Eq. (3.44) then becomes

$$\langle \underline{k}|f(E)|\underline{p}\rangle = \langle \underline{k}|R(E)|\underline{p}\rangle - \pi\frac{N_0}{2}\int d^3p'\langle \underline{k}|f(E)|\underline{p}'\rangle \left\{ D\left[E - \frac{3p'^2}{4m}\right]\left[E - \frac{p^2}{2m} - \frac{p'^2}{2m} - \frac{(\underline{p}+\underline{p}')^2}{2m}\right]\right\}^{-1} \tag{3.46}$$

This is just the usual Faddeev equation with s-wave separable potentials, but of zero range [3.11]. It has the unfortunate liability that the kernel does not go to zero sufficiently rapidly so that it is not a Fredholm integral equation [3.12]. We shall return to repair this fault. Meanwhile let us remark that we have derived this equation with no mention of the Schrödinger equation or the like, but only with unitarity and analyticity and a few well-placed arbitrary choices. It is gratifying that we can do this, but it is also disheartening for phenomenology. Recall that the dispersion relation (3.43) represents the minimal expression of the analyticity and unitarity required by a correct quantum mechanical treatment of final state interactions. Without these conditions the coherence of various pieces of the final state amplitude will be lost, and/or spurious singularities may be introduced in the final state. Now we see that in order to build these conditions in we have to do something very close to solving the full problem. If we do not have identical bosons, s-wave interactions, etc., but still use the sequential decay ansatz of only one or a few states dominating for each pair, we will get instead of (3.46), a sum of coupled integral equations, but again the structure will be very similar to the set obtained with the corresponding separable interactions.

We now turn to the problem of finite range. For interaction in pair states of $\ell \neq 0$, we have factors in unitarity of q^ℓ. These are to represent threshold behavior and are valid only near threshold. The measure of near is provided by recalling that it is not q^ℓ that should occur but $(qR)^\ell$ where R is some range parameter. Hence the threshold question and the finite range question are related. But as is clear from (3.43), in implementing unitarity with a dispersion relation we integrate over all momenta and hence need the "threshold" factors far from threshold. In that case we should make the replacement in (3.32)

$$q^\ell \to v_\ell(q) \quad . \tag{3.47}$$

If we further assume

$$v_\ell^2(q) = N_\ell(q^2/2m) \tag{3.48}$$

unitarity (3.32) takes on a particularly simple form. Of course (3.47) and (3.48) are the final "arbitrary" choices on the path to making our realization of unitarity and analyticity correspond to separable interactions, but we have found no other way of realizing these constraints and incorporating the finite range of forces that also satisfies total energy analyticity within the context of the isobar formalism. To keep the algebra simple in employing (3.47) in analyticity, we go back to the three-boson case and the assumption of $\ell = 0$ only, but assume the v's are now due to some finite range effect. We then get for (3.44)

$$\begin{aligned}\langle \underline{k}|f(E)|\underline{p}\rangle &= \langle \underline{k}|R(E)|\underline{p}\rangle \\ &- \frac{\pi}{2}\int d^3p'[E - p^2/2m - p'^2/2m - (\underline{p} + \underline{p}')^2/2m]^{-1}\Big\{v[(2mE_0 - 3p^2/2)^{1/2}] \\ &\times v[(2mE_0 - \tfrac{3}{2}p'^2)^{1/2}]D^{-1}(E_0 - 3p'^2/4m)\langle \underline{k}|f(E_0)|\underline{p}'\rangle - F(\underline{k},\underline{p},p',E)\Big\} \quad .\end{aligned} \tag{3.49}$$

We note that

$$2mE_0 - \frac{3}{2}p^2 = 2(\underline{p}' + \frac{\underline{p}}{2})^2$$

and

$$2mE_0 - \frac{3}{2}p'^2 = 2(\underline{p} + \frac{\underline{p}'}{2})^2 \quad . \tag{3.50}$$

Hence the natural choice for F is

$$F(\underline{k},\underline{p},\underline{p}',E) = v\{[2(\underline{p} + \underline{p}'/2)^2]^{1/2}\}\, v\{[2(\underline{p}' + \underline{p}/2)^2]^{1/2}\} \tag{3.51}$$
$$\times\, [\langle\underline{k}|f(E)|\underline{p}'\rangle\, D^{-1}(E - 3p'^2/4m) - \langle\underline{k}|f(E_0)|\underline{p}'\rangle D^{-1}(E_0 - 3p'^2/4m)]$$

giving for (3.49)

$$\langle\underline{k}|f(E)|\underline{p}\rangle = \langle\underline{k}|R(E)|\underline{p}\rangle$$

$$- \frac{\pi}{2}\int d^3p'\, \frac{\langle\underline{k}|f(E)|\underline{p}'\rangle v\{[2(\underline{p} + \underline{p}'/2)^2]^{\prime 2}\} v\{[2(\underline{p}' + \underline{p}/2)^2]^{\prime 2}\}}{[E - p^2/2m - p'^2/2m - (\underline{p} + \underline{p}')^2/2m]D(E - 3p'^2/4m)}\ , \tag{3.52}$$

which is now precisely the s-wave separable interaction equation. Implementation of more complex unitarity equations with higher partial wave interactions, distinguishable particles, etc., would lead to the corresponding coupled separable interaction equations - so long as we make the simplifying assumptions (3.47) and (3.48).

3.8 Conclusions

We have seen that a proper accounting of the fundamental principles of quantum mechanics forces amplitudes normally taken as independent and constant in phenomenology to be neither. We have also seen that full implementation of the quantum mechanical constraints of unitarity and analyticity leads back to the separable interaction formalism. It may be that if one only wants unitarity and analyticity over the three-body phase space and is not concerned with the absolute cross section or its dependence on total center of mass energy, less complicated approaches than full solution of the separable potential equations are possible. Many such are presently being tried, both theoretically and for phenomenology. There is no difficulty with extending our discussion to the relativistic problem, and some of these studies are in particle physics [3.13]. But until simpler schemes are proven, we must continue to be surprised and somewhat disappointed to find that attempts to develop a phenomenology consistent with the principles of quantum mechanics for overlapping final state interactions lead us all the way back to full dynamical solution of the problem.

References

3.1 M.L. Goldberger, K.M. Watson: *Collision Theory* (John Wiley and Sons, New York 2964)

3.2 R. Aaron, R.D. Amado: Phys. Rev. 150, 857 (1966)

3.3 R.D. Amado, Phys. Rev. 158, 1414 (1967)

3.4 N.I. Muskhelishvili: *Singular Integral Equations* (P. Noordhoff, N.V. Groningen, Holland 1953)

3.5 R. Omnes: Nuovo Cimento 8, 316 (1958)

3.6 This material is discussed in more detail in R.D. Amado: Phys. Rev. C11, 719 (1975)

3.7 R.J. Eden, P.V. Landshoff, D.I. Olive, J.C. Polkinghorne: *The Analytic S-Matrix* (Cambridge U. Press, Cambridge, England 1966)

3.8 S.K. Adhikari, R.D. Amado, Phys. Rev. D9, 1467 (1974)

3.9 R.D. Amado, T. Takahashi: Phys. Rev. C12, 1134 (1975)

3.10 I.J.R. Aitchison, R.J.A. Golding: Physics Letters 59B, 288 (1975)

3.11 This result was first obtained by I.J.R. Aitchison: Phys. Rev. B137, 1070 (1965)

3.12 R.D. Amado: Phys. Rev. 132, 485 (1963)

3.13 R. Aaron, R.D. Amado: Phys. Rev. D13, 2581 (1976)

4. The Boundary Condition Method

D. D. Brayshaw

With 2 Figures

4.1 Fundamentals

The goal of potential theory is to construct a realistic description of the two-particle interaction. In practice such potentials are largely empirical, with parameters adjusted to reproduce observables (phase shifts, bound state properties) via solutions of the two-body Schrödinger equation. However, except under artificially restrictive mathematical assumptions (locality), it is impossible to uniquely determine potentials from such information. Thus a class of "realistic" models may be constructed which provide comparable fits to the two-particle data. These potentials are distinguished by specific dynamical assumptions which can be probed only by applying them to a new class of phenomena.

In this circumstance one may adopt opposing views regarding the effectiveness of potential theory. To the extent that crucial (to the application at hand) aspects of the potential are *uniquely* determined, the potential becomes a powerful tool for predicting new observables, and to that extent succeeds in providing a physical interpretation which transcends the particular data employed in its construction. On the other hand, it obscures the distinction between these elements and the untested dynamical ingredients by inextricably linking them in the fitting parameters. Thus, while alternative potentials may be judged on the basis of their effectiveness in describing a certain (new) set of experiments, there is an unavoidable ambiguity in ascertaining which aspects are actually being tested. This property of the potential description tends to hinder efforts to further understand the interaction mechanism.

These questions become of practical importance at the three-particle level. Thus, if empirically determined potentials accurately characterize the physics, one would hope to predict at least the gross features of three-body systems. This can provide a useful guide, particularly if experimental informations is incomplete. On the other hand, three-body observables may in general depend on aspects of the potentials which are inaccessible in direct two-particle experiments. This is clear, at least formally, from the explicit appearance of off-shell two-particle amplitudes in the

Faddeev equations. Providing that true three-body forces can be neglected (or reliably estimated), this suggests that a comparison of theoretical predictions with experiment could lead to new dynamical information. Whether or not this potentiality can be realized is a complicated question, and the tendency has been to proceed in a spirit of naive optimism. Unfortunately, in view of the effort and expense required to pursue such information via three-body calculations and experiments, this question is crucial. In order to answer it one must achieve an effective separation of the on-shell and off-shell input, and ascertain which aspects of this input are actually relevant to the accessible phenomena. For the reasons noted above, one must go outside the framework of potential theory in order to accomplish this objective.

An instructive example is provided by early theoretical calculations of the triton ground state properties [4.1]. On the one hand, simple phenomenological NN forces were shown to produce quite reasonable results, and considerable sensitivity was exhibited to such effects as the momentum-dependence of the separable interaction, the presence of short-range repulsion, tensor vs. central forces, etc. However, at this level of sophistication the corresponding models can also be distinguished by the *two*-particle data, and hence one cannot accurately characterize this as "off-shell" sensitivity. In fact, as more sophisticated models were employed this apparent sensitivity tended to vanish, and it now appears that the differences exhibited by competing "realistic" potentials are comparatively minor [4.2,3]. Moreover, only the triton binding energy (E_T) and the minimum in the ^{3}He charge form factor are truly sensitive, and the interpretation of the latter is dependent on unknown mesonic corrections [4.4,5]. Furthermore, if one enlarges the data base to include low energy n-d scattering, it has been known for sometime that the doublet scattering length ^2a is closely correlated with E_T, whereas ^4a is quite insensitive to the details of the forces [4.6].

Two conclusions may be drawn from this example. The first is that a reliable estimate of off-shell effects can only be obtained by comparing models which are virtually indistinguishable at the two-particle level (preferably phase equivalent). The second is that only a few three-body observables are likely to prove *independently* sensitive to such effects, and hence it is essential to isolate such parameters and correlate them with the off-shell degrees of freedom. These factors motivated the development of the boundary condition formalism (BCF) as a practical alternative to three-particle potential (Faddeev) theory [4.7]. By placing the emphasis on three-body *observables*, the BCF explicitly exhibits both the model-independent constraints implied by the two-particle data and the independent parameters which characterize the off-shell structure. This is achieved by a basic reformulation of the three-body problem.

4.2 Two-Particle Description

The basic idea of the approach is to characterize a three-body state in such a way as to completely isolate the off-shell degrees of freedom. Its philosophy can best be illustrated by first examining a two-particle problem (for simplicity we consider identical spinless particles interacting only in relative s-waves). Suppose that two such particles are separated by a distance x, and described by a wave function $\psi_>(x)$ in the region $x>r_b$. In ordinary potential theory we may also calculate an interior wave function $\psi_<(x)$, which is valid for $x<r_b$. In such cases both ψ and its derivative (ψ') are continuous, and $\psi_>$ is constrained by the matching boundary condition (BC)

$$(\psi_>'/\psi_>)_{x=r_b} = (\psi_<'/\psi_<)_{x=r_b} \equiv \lambda^{\text{int}}(\kappa^2) \quad , \tag{4.1}$$

where κ is the on-shell value of the momentum in the CM ($\kappa=(mE)^{1/2}$). However, if we regard r_b as a radius characteristic of the highly nonlocal short-range behavior, a potential description is at best an idealization for $x<r_b$, and a hard core is often used to simulate the net empirical effect (strong repulsion) of this region. In this sense (4.1) is more general than the potential description since, with a suitable interpretaion of the parameters, it may also describe the latter possibility. Two choices often made are $\lambda^{\text{int}}\to\infty$ (hard core), and λ^{int} = constant (the boundary condition model). Since scattering experiments measure only the asymptotic form ($\psi_>(x), x\to\infty$), it is clear that the entire complexity of the interior region can be summarized by the function $\lambda^{\text{int}}(\kappa^2)$; i.e., a single number at a given energy.

If we were given the half-on-shell t-matrix $t_0(p,\kappa,:E)$, we could calculate $\psi_>(x)$ via the relation (1.26)

$$\begin{aligned}\psi(x) &= (\pi/2)^{1/2}\langle x|1 + G_0 t|\kappa\rangle \\ &= j_0(\kappa x) - m\int_0^\infty dp p^2 j_0(px)(p^2 - \kappa^2 - i\varepsilon)^{-1} t_0(p,\kappa;E) \quad ,\end{aligned} \tag{4.2}$$

by evaluating the integral for $x>r_b$ (for simplicity we drop normalization factors like $(2/\pi)^{1/2}$ for the wave functions employed in this chapter). Here $j_0(z) = \sin z/z$ (our conventions for the spherical Bessel functions $j_\ell(z)$, $h_\ell(z)$ are consistent with those of MORSE and FESHBACH [4.8]). In general (cf. (1.43))

$$t_0(p,\kappa;E) = f_0(p,\kappa)t_0(\kappa) \quad , \tag{4.3}$$

where $t_0(\kappa) \equiv t_0(\kappa,\kappa;E)$ is the on-shell amplitude, and $f_0(p,\kappa)$ is the Kowalski-Noyes half-on-shell extension function [4.9,10]. The latter is real-valued, and satisfies $f_0(\kappa,\kappa) = 1$. We may then write

$$\psi_>(x) = \psi^{ext}(x) + \psi^{pot}(x) \quad , \tag{4.4}$$

where

$$\begin{aligned}\psi^{ext}(x) &= j_0(\kappa x) + i \sin \delta_0 \exp (i\delta_0) h_0(\kappa x) \\ &= j_0(\kappa x) - (m\pi/2) t_0(\kappa) \exp (i\kappa x)/x \quad ,\end{aligned} \tag{4.5}$$

and

$$\begin{aligned}\psi^{pot}(x) &= \chi_0(x,\kappa) t_0(\kappa) \quad , \\ \chi_0(x,\kappa) &= m \int_0^\infty dp p^2 j_0(px)(p^2 - \kappa^2)^{-1}[1 - f_0(p,\kappa)] \quad .\end{aligned} \tag{4.6}$$

We note that the integral defining $\chi_0(x,\kappa)$ is nonsingular, and hence χ_0 is real-valued. If one imagines performing the integral by closing an infinte contour in the upper half-plane, it is rather easy to deduce some general properties of χ_0. Thus

$$\chi_0(x,\kappa) = (m/2ix) \int_{-\infty}^{\infty} dp p \exp (ipx) (p^2 - \kappa^2)^{-1}[1 - f_0(p,\kappa)] \quad , \tag{4.6'}$$

where we have used the reflection property under $p\to(-p)$ to eliminate the negative-frequency exponential. The integral can now be performed by the method of residues given the analytic properties of $f_0(p,\kappa)$. If, for example, the interaction vanishes for $x>R$, it may be shown that $f_0(p,\kappa)$ is an *entire* function of p increasing like $\exp (-ipR)$ as $p\to i\infty$ [4.11]. For $x>R$ the contribution at infinity vanishes, and so does the integral since there are no singularities. Thus $\chi_0(x,\kappa) \equiv 0$ in the region required ($x>r_b$) if $r_b>R$. Similarly, if the potential is meson-theoretic (and hence represented as a superposition of Yukawa potentials), $f_0(p,\kappa)$ is analytic except for cuts with branch points p_n, where $|\mathrm{Im}\{p_n\}|\geq\mu$ (the mass of the lightest exchanged particle) [4.12]. In this case $\chi_0(x,\kappa)$ may be evaluated as a sum of terms, each proportional to $\exp (ip_n x)$, and hence is proportional to $\exp (-\mu x)$. We thus infer that $\psi_>(x)$ approaches $\psi^{ext}(x)$ asymptotically, modulo exponentially declining pieces characteristic of meson exchange.

Comparing (4.1), (4.5) and (4.6), we deduce that a knowledge of $\lambda^{int}(\kappa^2)$, $\chi_0(r_b,\kappa)$, $\chi_0'(r_b,\kappa)$ is sufficient to determine $t_0(\kappa)$, and hence all scattering observables (at the two-particle level). This is a good deal less information than is contained in $f_0(p,\kappa)$ and illustrates the manner in which the asymptotic properties probe only certain aspects of the full dynamical input (it should be clear that an analogous situation will pertain at the three-body level). Equivalently, we can introduce the (real) function

$$B_0(\kappa) = \lambda^{int}(\kappa^2)\chi_0(r_b,\kappa) - \chi_0'(r_b,\kappa) \quad , \tag{4.7}$$

and write (4.1) as

$$\psi^{ext\prime}(r_b) - \lambda^{int}(\kappa^2)\psi^{ext}(r_b) = B_0(\kappa)t_0(\kappa) \quad , \tag{4.8}$$

from which $t_0(\kappa)$ can be determined by substituting (4.5). The content of (4.8) is that both interior and exterior effects can be represented as effective BC's on ψ^{ext}, thus summarizing the measureable consequences of the dynamics.

It is convenient in what follows to restate (4.8) in the form

$$\left(\psi^{ext\prime}/\psi^{ext}\right)_{r=r_b} = \lambda(\kappa^2) \quad , \tag{4.9}$$

which, upon substitution of (4.5) for ψ^{ext}, has the explicit solution $t_0(\kappa) = N(\kappa)/D(\kappa)$;

$$\begin{aligned} N(\kappa) &= f(\kappa^2)j_0(r_b\kappa) - \cos r_b\kappa \quad , \\ D(\kappa) &= (\pi m/2r_b)\left[f(\kappa^2) - ir_b\kappa\right] \exp\,(ir_b\kappa) \quad . \end{aligned} \tag{4.10}$$

Here we have employed the (real) function

$$f(\kappa^2) \equiv r_b\lambda(\kappa^2) + 1 \quad . \tag{4.11}$$

In order to connect $\lambda(\kappa^2)$ with the quantity $B_0(\kappa)$ defined above, we first note that (4.5) and (4.10) imply that

$$\psi^{ext}(r_b) = (\pi m/2r_b)D^{-1}(\kappa) \quad . \tag{4.12}$$

A comparison of (4.8) and (4.9) then yields the result

$$\begin{aligned} \lambda(\kappa^2) &= \lambda^{int}(\kappa^2) + \lambda^{eff}(\kappa^2) \quad , \\ \lambda^{eff}(\kappa^2) &= (2r_b/\pi m)B_0(\kappa)N(\kappa) \quad . \end{aligned} \tag{4.13}$$

Although $N(\kappa)$ itself depends (linearly) on $\lambda^{eff}(\kappa^2)$, it is clear that one may solve (4.13) to display $\lambda^{eff}(\kappa^2)$ as an explicit function of $\lambda^{int}(\kappa^2)$, $B_0(\kappa)$ and known functions of $r_b\kappa$. Thus, insofar as two-particle scattering is concerned, the net consequence of the interaction may be represented as a single real-valued function $\lambda(\kappa^2)$(or $f(\kappa^2)$). This, of course, is no surprise, since it is well known that the phase shift (δ_0) plays an equivalent role. The point here is that the boundary condition representation is easily generalized to systems of three or more particles; this is *not* the case for the phase shift.

In what follows it will be convenient to decompose $f(\kappa^2)$ in the form

$$f(\kappa^2) = f^c + r_b\Delta(\kappa^2) \quad , \tag{4.14}$$

where $\Delta(\kappa^2)$ is taken to approach zero as $|\kappa|\to\infty$. In fact, for suitable choices of r_b characteristic of the particular system, this is consistent with empirical experience in many cases (including the NN system with $r_b = 0.7$ fm $\simeq (2m_\pi)^{-1}$). However, we will only require this description for $-\infty<\kappa^2<\kappa^2_{max}$, and the behavior as $\kappa\to i\infty$ may be chosen at our convenience. In the physical region $f(\kappa^2)$ may be taken directly from the empirical values for δ_0, using the relation

$$f(\kappa^2) = r_b\kappa \cot(\delta_0 + r_b\kappa) \quad , \tag{4.15}$$

although it is more convenient (and usually quite adequate) to parametrize $\Delta(\kappa^2)$ in the form

$$\Delta(\kappa^2) = \sum_j \frac{r_j}{\kappa^2 - \beta_j^2} \quad . \tag{4.16}$$

In practice, at most two such terms are required to produce good quality fits to the elastic amplitude for NN, πN, or $\pi\pi$ scattering. Below it will be convenient to employ the notation N^c, D^c to represent the functions N, D defined in (4.10) with $f(\kappa^2)$ replaced by f^c. In view of the above remarks, N^c and D^c correspond to the limiting values as $\kappa\to i\infty$. Introducing $\lambda^c \equiv (f^c-1)/r_b$, we also note that (4.8) can be restated as

$$\psi^{ext\prime}(r_b) - \lambda^c\psi^{ext}(r_b) = B(\kappa)t_0(\kappa) \quad , \tag{4.17}$$

providing that $B(\kappa)$ is given by

$$B(\kappa) = (\pi m/2r_b)\Delta(\kappa^2)N^{-1}(\kappa) \quad . \tag{4.18}$$

Below we shall introduce a precise analogue of (4.17) for a three-body system.

4.3 Boundary Conditions for Three-Particle Scattering

We now consider the three-body problem, our goal being to define analogous BC's on the asymptotic three-particle wave function in order to isolate those parameters which actually control the scattering observables. We also impose an additional requirement; namely, that we clearly distinguish between parameters which are in principle known to us from two-particle experiments (e.g., determined by the phase shifts) and the remaining degrees of freedom. The latter include both properties of the pairwise interactions which are "off-shell" at the two-body level, and true three-body forces. (Although we shall not discuss this question here, one consequence of this approach is a demonstration that such effects cannot be distinguished empirically, even if a complete set of three-particle scattering data were at our disposal [4.13]). We therefore require the most general representation consistent with three-particle unitarity and a given set of phase shifts. From the discussion above, we already anticipate that the free parameters will correspond to only a subset of the information buried in the potentials.

For definiteness, we consider the projection of the incoming plane wave onto an s-wave with respect to the relative motion of particles 2 and 3, and denote this by $|\phi_{1;c}>$. It is convenient to employ

$$\underline{y}_i = (m_i/\mu_i)\underline{r}_i \tag{4.19}$$

which is the displacement of particle i from jk CM. We may then write

$$\begin{aligned}\phi_{1;c}(\underline{r}_1\underline{r}_2\underline{r}_3) &= <\underline{r}_1\underline{r}_2\underline{r}_3|\phi_{1;c}> \\ &= j_0(y_1p_1^0)j_0(r_{23}q_1^0)\end{aligned} \tag{4.20}$$

for a state of total angular momentum L = 0. The three-particle wave function is given by (cf. (1.107))

$$\begin{aligned}\psi_{1;c}(\underline{r}_1\underline{r}_2\underline{r}_3) &= <\underline{r}_1\underline{r}_2\underline{r}_3|\left(1 + G_0^{(+)}(E)\sum_j T^j\right)|\phi_{1;c}> \\ &= \psi^1(r_{23},y_1) + \psi^2(r_{31},y_2) + \psi^3(r_{12},y_3) \quad .\end{aligned} \tag{4.21}$$

Thus

$$\begin{aligned}\psi^k(r_{ij},y_k) &= \delta_{k1}j_0(y_1p_1^0)j_0(r_{23}q_1^0) + \psi^{k;out}(r_{ij},y_k) \quad , \\ \psi^{k;out}(r_{ij},y_k) &= \int_0^\infty dp_kp_k^2j_0(y_kp_k)\int_0^\infty dq_kq_k^2j_0(r_{ij}q_k) \\ &\quad \times T^k(q_k,p_k)\left(E - q_k^2/2\mu_{ij} - p_k^2/2\mu_k + i\varepsilon\right)^{-1} \quad ,\end{aligned} \tag{4.22}$$

where we have suppressed the dependence of T^k on E and the initial variables p_1^0,q_1^0. On the other hand, the Faddeev equations imply that $T^k(q_k,p_k)$ is of the form

$$T^k(q_k,p_k) = \int_0^\infty dq_k'q_k'^2t_k\left(q_k,q_k';E - p_k^2/2\mu_k\right)F_k(q_k',p_k) \quad , \tag{4.23}$$

and hence its behavior as a function of q_k is controlled by that of the corresponding off-shell t-matrix. Introducing the on-shell value of momentum κ_k in channel k ($q_k = \kappa_k$)

$$\begin{aligned}\kappa_k &\equiv \kappa_k(E,p_k)\\ &= \left[2\mu_{ij}(E - p_k^2/2\mu_k)\right]^{1/2}\end{aligned} \tag{4.24}$$

we can rewrite (4.22) as

$$\begin{aligned}\psi^{k;out}(r_{ij},y_k) = -\,2\mu_{ij}\int_0^\infty dp_k p_k^2 j_0(y_k p_k)\int_0^\infty dq_k q_k^2 j_0(r_{ij}q_k)\\ \times\, T^k(q_k,p_k)\left(q_k^2 - \kappa_k^2 - i\varepsilon\right)^{-1}\quad ,\end{aligned} \tag{4.25}$$

where it is understood that κ_k is positive imaginary for $p_k^2 > 2\mu_k E$. If we again imagine performing the inner integration by contours, a simple generalization of the arguments given above in connection with (4.2-6) yields the result

$$\psi^{k;out}(r_{ij},y_k) = \psi^{k;ext}(r_{ij},y_k) + O[\exp(-\mu r_{ij})] \quad , \tag{4.26a}$$

$$\psi^{k;ext}(r_{ij},y_k) = -(\pi\mu_{ij}/r_{ij})\int_0^\infty dp_k p_k^2 j_0(y_k p_k)\,\exp(i\kappa_k r_{ij})T^k(\kappa_k,p_k) \quad . \tag{4.26b}$$

Here we have assumed that the lightest meson exchanged by particles i and j has mass μ. If t_k corresponds to an interaction which vanishes identically for $r_{ij} > R$, then $\psi^{k;out} = \psi^{k;ext}$ for $r_{ij} > R$ (the asymptotic form is exact in that region). As in the two-body problem, we observe that the asymptotic behavior is specified by the *on-shell* scattering amplitude (T^k with $q_k = \kappa_k(E,p_k)$).

We therefore deduce that the three-particle analogue of (4.5) is given by

$$\psi^{ext}_{1;c}(\underline{r}_1\underline{r}_2\underline{r}_3) = P^{ext}(r_{12},r_{23},r_{31})\left[\phi_{1;c}(\underline{r}_1\underline{r}_2\underline{r}_3) + \sum_k \psi^{k;ext}(r_{ij},y_k)\right] \quad , \tag{4.27}$$

where

$$P^{ext}(r_{12},r_{23},r_{31}) = \theta\left(r_{12} - r_b^{(3)}\right)\theta\left(r_{23} - r_b^{(1)}\right)\theta\left(r_{31} - r_b^{(2)}\right) \tag{4.28}$$

defines a projection onto the region exterior to *all three* boundary radii ($r_b^{(k)}$). We next seek BC's analogous to (4.17) in order to determine the unknown functions $T^k(\kappa_k,p_k)$. In doing so we will be guided by an obvious (but crucial) general principle: *three-particle scattering must reduce to two-particle scattering as a special case when one particle is taken to infinte distance.* It is instructive to note how this requirement is embodied in our expressions above for the asymptotic wave function. Suppose that $r_{23}>r_b^{(1)}$ and consider the limit as r_1 (or y_1) is taken to infinity. Clearly all the r_{ij} and y_k will also become infinite in that limit (except r_{23}), and

$$\psi_{1;c}(\underline{r}_1\underline{r}_2\underline{r}_3) \underset{y_1\to\infty}{\longrightarrow} \phi_{1;c}(r_1r_2r_3) + \psi^{1;out}(r_{23},y_1)$$
$$+ \psi^{2;ext}(r_{31},y_2) + \psi^{3;ext}(r_{12},y_3) \quad , \tag{4.29}$$

plus negligible exponentially damped terms. It is easy to see that in this limit the dominant contribution arises from the single (2-3) scattering term, since

$$T^1(q_1,p_1) = \delta(p_1 - p_1^0)p_1^{-2}t_1(q_1,\kappa_1^0;E - p_1^{02}/2\mu_1) + \cdots \quad , \tag{4.30}$$

where the remaining terms are bounded integrable functions of p_1. Recalling (4.2), the contribution from this term combines with $\phi_{1;c}$ to yield

$$\psi_{1;c}(\underline{r}_1\underline{r}_2\underline{r}_3) \underset{y_1\to\infty}{\longrightarrow} \tilde{\psi}_{1;c}(r_{23})j_0(y_1p_1^0) + O(y_1^{-2}) \quad , \tag{4.31}$$

the latter estimate following from (4.25) and (4.26) and the Riemann-Lesbeque lemma. The form of (4.31) corresponds to treating particle 1 as a noninteracting spectator, as is required by the physical situation in this limit.

As a consequence, we observe that

$$\lim_{y_1\to\infty}\left[\left(\partial\psi_{1;c}^{ext}/\partial r_{23}\right) - \lambda_1^c\psi_{1;c}^{ext}\right]_{r_{23}=r_b^{(1)}} = B_1(\kappa_1^0)t_1(\kappa_1^0)j_0(y_1p_1^0) \quad , \tag{4.32}$$

where we have used (4.17), adding a channel index to the quantities λ^c, $B(\kappa)$ introduced in our discussion of the two-body BC. This suggests the following simple generalization of (4.17);

$$\Psi_k^{ext\prime}(r_b^{(k)},y_k) - \lambda_k^c\Psi_k^{ext}(r_b^{(k)},y_k) = \sum_j \int_0^\infty dp_j p_j^2 \bar{B}_{kj}(y_k,p_j)T^j(\kappa_j,p_j) \quad , \tag{4.33}$$

where we have introduced $\Psi_k^{ext}(r_{ij},y_k)$ to denote $\psi_{1;c}^{ext}(\underline{r}_1\underline{r}_2\underline{r}_3)$ projected onto s-waves in the variables $\underline{r}_{ij}$ and $\underline{y}_k$ (i.e., averaged over $\hat{r}_{ij}$ and $\hat{y}_k$), and used a prime to denote the partial derivative with respect to r_{ij}. We note that (4.33) is actually *three* separate BC's corresponding to the possible values of k (above we chose arbitrarily to project the initial plane wave onto s-waves in channel 1). The only constraint on $\bar{B}$ which is necessary to guarantee that two-particle scattering is recovered as a special case is that

$$\bar{B}_{kj}(y_k,p_j) \underset{y_k\to\infty}{\longrightarrow} \delta_{kj}B_j(\kappa_j)i_0(y_jp_j) \quad , \tag{4.34}$$

in view of (4.30) and (4.32). With somewhat more effort one can show that $\overline{B}$ is real-valued (as in the two-body case) and approaches the limit stated in (4.34) modulo terms proportional to exp $(-\mu y_k)$; we shall return to this point in Section 4.6.

In evaluating (4.33), it is important to keep in mind that the partial-wave projection is complicated by the factor P^{ext}, which depends on $\hat{r}_i \cdot \hat{r}_j$ (or $\hat{r}_{ij} \cdot \hat{y}_k$). In general, for $r_{ij} = r_b^{(k)}$, there will be displacements b_k, y_k^0 such that

$$
\begin{aligned}
P_k^{ext}\left(r_b^{(k)}, y_k\right) &= 1 \qquad \text{if } y_k > b_k \quad , \\
&= 0 \qquad \text{if } y_k < y_k^0 \quad ,
\end{aligned}
\tag{4.35}
$$

where $P_k^{ext}(r_{ij}, y_k)$ represents the s-wave projection of P^{ext} (in the sense of Ψ_k^{ext}). The precise values of b_k, y_k^0 depend on the mass ratios and the $r_b^{(k)}$ (y_k^0 may be zero). For identical particles and common radius r_b one finds $y_k^0 = \sqrt{3}\, r_b/2$, $b_k = 3r_b/2$. Physically, b_k corresponds to a distance of particle k from the ij CM such that $r_{jk} > r_b^{(i)}$ and $r_{ki} > r_b^{(j)}$, regardless of the relative directions of $\underline{r}_{ij}$ and $\underline{r}_k$. Conversely, $y_k < y_k^0$ implies that all three particles are sufficiently close together that at least one pair is within their corresponding radius (irrespective of $\hat{r}_{ij} \cdot \hat{y}_k$). There is thus an intermediate region of finite size $(y_k^0 < y_k < b_k)$ where $\Psi_k^{ext}(r_b^{(k)}, y_k)$ has support and P^{ext} is nontrivial. In the special case under consideration one finds

$$
\begin{aligned}
P_k^{ext}(r_b, y_k) &= 0, & y_k &\le \sqrt{3} r_b/2; \\
&= \left(y_k^2 - 3r_b^2/4\right)/r_b y_k, & \sqrt{3} r_b/2 \le y_k &\le 3r_b/2; \\
&= 1, & y_k &\ge 3r_b/2 \quad .
\end{aligned}
\tag{4.36}
$$

4.4 The BCF Integral Equation

Given (4.17) and the simple representation for $\psi^{ext}(x)$ stated in (4.5), it is trivial to solve for $t_{0}(\kappa)$ in terms of r_b, λ^c and $B(\kappa)$. Similarly, (4.26), (4.27) and (4.33) constrain the T_0^k by linking them to the quantities $r_b^{(k)}$, λ_k^c and $\overline{B}_{kj}$. We next show that these constraints are equivalent to a one-dimensional integral equation.

Let us first consider the contribution of $\psi^{k;ext}$ to (4.33). It is straight-forward (using (4.26b) with $\mu_{ij} = m/2$) to obtain

$$P_k^{ext}(r_b,y_k)\left[\psi^{k;ext'}(r_b,y_k) - \lambda_k^c\psi^{k;ext}(r_b,y_k)\right]$$
$$= P_k^{ext}(r_b,y_k)r_b^{-1}\int_0^\infty dp_k p_k^2 j_0(y_k p_k)D_k^c(\kappa_k)T^k(\kappa_k,p_k) \quad , \tag{4.37}$$

where we have identified $D^c(\kappa_k)$ from (4.10). Note that P^{ext} is *not* differentiated entiated with respect to r_{ij}; as in the two-body case, the exterior wave function is differentiated and *then* the limit to the boundary radius is taken ($r_{ij}\to r_b+$). The presence of P^{ext} is merely a bookkeeping device to ensure that we do not employ our exterior representation in a region in which it is invalid (interior to any core). If we further define

$$\overline{X}_k(p_k) = - D_k^c(\kappa_k)T^k(\kappa_k,p_k) \quad , \tag{4.38}$$

we see that the integral in (4.37) is essentially the Fourier transform of $\overline{X}_k$ (i.e., $\overline{X}_k(y_k)$). It thus becomes clear that (4.33) can be stated in the form

$$\theta(y_k - y_k^0)\overline{X}_k(y_k) = r_b\left[P_k^{ext}(r_b,y_k)\right]^{-1}\theta(y_k - y_k^0)G(y_k) \quad , \tag{4.39}$$

where $G(y_k)$ is known up to integrals containing $\overline{X}_k$, and hence (4.39) has the structure of a one-dimensional integral equation for $\overline{X}_k$.

In order to make this more explicit, it is convenient to introduce the notation

$$_k\langle y_k| = \lim_{r_{ij}\to r_b}(4\pi)^{-2}\int d\hat{r}_{ij}d\hat{y}_k\langle \underline{r}_{ij}\underline{y}_k| \tag{4.40}$$

to represent the projection of a state described by the two independent vectors $\underline{r}_{ij},\underline{y}_k$ onto s-waves in the k-th channel (evaluated at the common BC radius r_b). That is,

$$\psi_k^{ext}(r_b,y_k) = {}_k\langle y_k|P^{ext}\left(|\phi_{1;c}\rangle + \sum_j |\psi^{j;ext}\rangle\right) \quad . \tag{4.41}$$

The contribution of $\psi^{k;ext}$ to this is (using (4.26b) and (4.38))

$$_k\langle y_k|P^{ext}|\psi^{k;ext}\rangle = \pi\mu_{ij}r_b^{-1}P_k^{ext}(r_b,y_k)\int_0^\infty dp_k p_k^2 j_0(y_k p_k)$$
$$\times \exp(ir_b\kappa_k)\overline{X}_k(p_k)/D_k^c(p_k) \quad . \tag{4.42}$$

In order to derive the contributions from $j\neq k$ without further complicating our notation, we will assume for definiteness that ijk is cyclic, and employ the relations

$$\underline{r}_{ki} = -(\mu_{ij}/m_i)\underline{r}_{ij} + \underline{y}_k \quad ,$$
$$\underline{y}_j = -(\mu_{ki}/\mu_k)\underline{r}_{ij} - (\mu_{ki}/m_i)\underline{y}_k \quad , \tag{4.43}$$

to show that

$$_k\langle y_k|P^{ext}|\psi^{j;ext}\rangle = (\pi\mu_{ki}/2)\int_{-1}^{1} dx_k \theta(r_k^+ - r_b)\theta(r_k^- - r_b)r_{ki}^{-1}$$
$$\times \int_0^\infty dp_j p_j^2 j_0(y_j p_j) \exp(i\kappa_j r_{ki})\overline{X}_j(p_j)/D_j^c(\kappa_j) \quad . \tag{4.44}$$

Here we have introduced $x_k = \hat{y}_k \cdot \hat{r}_{ij}$, and defined (for equal masses)

$$r_k^\pm = \left(r_b^2/4 \pm r_b y_k x_k + y_k^2\right)^{1/2} \quad . \tag{4.45}$$

If we further define

$$\overline{x}_k \equiv \overline{x}_k(y_k)$$
$$= P_k^{ext}(r_b, y_k) \tag{4.46}$$

(see (4.36)), (4.44) can be written in the form

$$_k\langle y_k|P^{ext}|\psi^{j;ext}\rangle = (\pi\mu_{ki}/2)\int_{-\overline{x}_k}^{\overline{x}_k} dx_k \int_0^\infty dp_j p_j^2 j_0(y_j p_j)$$
$$\times r_{ki}^{-1} \exp(i\kappa_j r_{ki})\overline{X}_j(p_j)/D_j^c(\kappa_j) \quad . \tag{4.47}$$

In (4.44) and (4.47) it is understood that $y_j = |\underline{y}_j|$ and $r_{ki} = |\underline{r}_{ki}|$ are to be evaluated in terms of r_b, y_k and x_k via (4.43).

The contribution of $\psi^{j;ext}$ $(j\neq k)$ to (4.33) can now be expressed as

$$\left(\partial/\partial r_b - \lambda_k^c\right)_k\langle y_k|P^{ext}|\psi^{j;ext}\rangle \quad , \tag{4.48}$$

provided that one takes care not to differentiate the θ-function part (i.e., one differentiates y_j and r_{ki} in (4.47)). The result of this operation is given by the following expression:

$$(\pi\mu_{ki}/2r_b)\int_0^\infty dp_j p_j^2 \hat{N}_{kj}(y_k, p_j)\overline{X}_j(p_j)/D_j^c(\kappa_j) \quad , \tag{4.49a}$$

where (for ijk cyclic)

$$\hat{N}_{kj}(y_k,p_j) = -r_b\int_{-\bar{x}_k}^{\bar{x}_k} dx_k I_0(y_k,x_k,p_j)r_{ki}^{-1}\exp(i\kappa_j r_{ki}) \quad ,$$

$$I_0(y_k,x_k,p_j) = \left[\lambda_k^c + (\mu_{ij}/m_i r_{ki}^2)(\mu_{ij}r_b/m_i - y_k x_k)(1 - i\kappa_j r_{ki})\right]j_0(y_j p_j) \quad (4.49b)$$

$$+ (\mu_{ki}^2/\mu_k m_i y_j)(m_i r_b/\mu_k + y_k x_k)p_j j_1(y_j p_j) \quad .$$

Here we have used $j_1(z) = -dj_0(z)/dz$ [4.8]. In this special case (identical particles) one can easily show that the contribution from kji cyclic is identical to the above. It can similarly be shown that

$$\begin{aligned} {}_k\langle y_k|P^{ext}|\phi_{1;c}\rangle &= P_k^{ext}(r_b,y_k)j_0(y_k p_1^0)j_0(r_b\kappa_1^0), \quad \text{if } k = 1 \quad ; \\ &= 1/2\int_{-\bar{x}_k}^{\bar{x}_k} dx_k j_0(y_1 p_1^0)j_0(r_{23}\kappa_1^0), \quad \text{if } k \neq 1 \quad ; \end{aligned} \quad (4.50)$$

where y_1 and r_{23} are to be computed via (4.43) with ijk = 312. The remaining contribution to (4.33) can thus be expressed in terms of the function

$$\begin{aligned} \Omega_k(y_k) &\equiv r_b\bar{x}_k^{-1}(\partial/\partial r_b - \lambda_k^c)\,{}_k\langle y_k|P^{ext}|\phi_{1;c}\rangle \quad . \\ &= -N_1^c(\kappa_1^0)j_0(y_k p_1^0)\theta(y_k - y_k^0) \quad , \qquad k = 1 \quad ; \\ &= -r_b/2\bar{x}_k\int_{-\bar{x}_k}^{\bar{x}_k} dx_k\,\tilde{I}_0(y_k,x_k,p_1^0) \quad , \qquad k \neq 1 \quad ; \end{aligned} \quad (4.51)$$

where

$$\begin{aligned} \tilde{I}_0(y_k,x_k,p_1^0) &= (\mu_{23}^2/\mu_2 m_3 y_1)(m_3 r_b/\mu_2 + y_k x_k)p_1^0 j_1(y_1 p_1^0)j_0(r_{23}\kappa_1^0) \\ &+ \left[\lambda_k^c j_0(r_{23}\kappa_1^0) + (\mu_{31}/m_3 r_{23})(\mu_{31}r_b/m_3 - y_k x_k)\kappa_1^0 j_1(r_{23}\kappa_1^0)\right]j_0(y_1 p_1^0) \quad . \end{aligned} \quad (4.52)$$

Although this is strictly true only for k=2, one can easily show that the expression for k=3 corresponds simply to the reflection $x_k \to -x_k$ (for identical particles). In (4.49b) and (4.52) we have refrained from inserting the specific values for the mass ratios (e.g., μ_{ij}=m/2) in order to emphasize the origin of the various numerical factors.

We may now combine (4.37), (4.48) and (4.51) to state explicitly the relation implied by (4.33), and sketched in (4.39). Thus

$$\overline{X}_k(y_k) = \Omega_k(y_k) + \theta(y_k^0 - y_k)\overline{X}_k(y_k)$$

$$+ \sum_j \int_0^\infty dp_j p_j^2 N_{kj}(y_k,p_j)\overline{X}_j(p_j)/D_j^c(\kappa_j) \quad . \tag{4.53}$$

where

$$N_{kj}(y_k,p_j) = N_{kj}^{(0)}(y_k,p_j) + r_b B_{kj}(y_k,p_j) \quad ,$$

$$N_{kj}^{(0)}(y_k,p_j) = (1 - \delta_{kj})(\pi\mu_{ki}/2\overline{x}_k)\hat{N}_{kj}(y_k,p_j) \quad , \tag{4.54}$$

and we have defined B_{kj} such that

$$\overline{B}_{kj}(y_k,p_j) = P_k^{ext}(r_b,y_k)B_{kj}(y_k,p_j) \quad . \tag{4.55}$$

This involves no loss in generality since $P_k^{ext}(r_b,y_k)$(or $\overline{x}_k$) is nonvanishing for $y_k > y_k^0$. By definition, the first and third terms on the right-hand side of (4.53) vanish if $y_k < y_k^0$; it thus reduces to an identity in that region. Inasmuch as $\theta(y_k^0-y_k)\overline{X}_k(y_k)$ may be arbitrary, (4.53) cannot uniquely determine $\overline{X}_k(y_k)$, and hence the BC stated in (4.33) is insufficient for our purpose. We must therefore supplement (4.33) by an auxiliary BC which will constrain $\overline{X}_k(y_k)$ in that finite domain. The form of this constraint is most easily derived in the momentum representation, and we postpone this question momentarily.

As it stands, (4.53) is not an integral equation, although it is easy to obtain one by inserting

$$\overline{X}_j(p_j) = 2/\pi \int_0^\infty dy_j y_j^2 j_0(y_j p_j)\overline{X}_j(y_j) \quad . \tag{4.56}$$

(Due to the fact that we have used unnormalized plane-wave states, the (one-dimensional) Fourier transform of a coordinate-space function such as $\overline{X}_j(y_j)$ is defined with a factor $(2/\pi)$). However, it is preferable (as in the Faddeev equations) to work in momentum-space, since Green's function singularities are relatively easy to handle in comparison with coordinate-space asymptotics. For formal manipulations we introduce a state $|\overline{X}_k\rangle$ such that

$$\overline{X}_k(y_k) = {}_k\langle y_k|\overline{X}_k\rangle \quad , \tag{4.57}$$

and use the notation ${}_k\langle p_k|$ to denote the conjugate momentum projection, e.g.,

$${}_k\langle p_k|\overline{X}_k\rangle = \overline{X}_k(p_k) \quad . \tag{4.58}$$

It is also useful to define operaters θ_0, θ_b corresponding to the step functions $\theta(y_k^0-y_k)$, $\theta(b_k-y_k)$, respectively. Using relations found on p.1574 of [4.8], it is easy to obtain

$$_k\langle p_k|\theta_b|p_k'\rangle_k \equiv \theta_b(p_k,p_k')$$

$$= \left(2b_k^2/\pi\right)\left[p_k j_1(b_k p_k) j_0(b_k p_k') - p_k' j_1(b_k p_k') j_0(b_k p_k)\right]\left(p_k^2 - p_k'^2\right)^{-1} \quad . \tag{4.59}$$

We may now express (4.53) as an integral equation for $\overline{X}_k(p_k)$, namely

$$\overline{X}_k(p_k') = \Omega_k(p_k') + {}_k\langle p_k'|\theta_0|\overline{X}_k\rangle + \sum_j \int_0^\infty dp_j p_j^2 N_{kj}(p_k',p_j)\left[D_j^c(\kappa_j)\right]^{-1} \overline{X}_j(p_j) \quad . \tag{4.60}$$

Here

$$N_{kj}(p_k',p_j) = 2/\pi \int_{y_k}^{\infty} dy_k y_k^2 j_0(y_k p_k') N_{kj}(y_k,p_j) \quad , \tag{4.61}$$

with a similar expression for $\Omega_k(p_k')$ in terms of $\Omega_k(y_k)$.

4.5 Practical Evaluation of the Kernel

Although formally correct, (4.61) is somewhat impractical since it involves an infinite integral over the oscillating function $j_0(yp_k')$; this form also obscures its relation to the Faddeev kernel. We therefore digress briefly to show that the portion $b_k<y_k<\infty$ can be evaluated analytically with the aid of a few manipulations. We first introduce the Fourier transform of $N_k^c(q)$ by observing that (recall (4.10))

$$N_k^c(q) = \int_0^\infty drr^2 j_0(rq) N_k^c(r) \tag{4.62a}$$

if

$$N_k^c(r) = f_k^c \delta(r - r_b)/r_b^2 + \delta'(r - r_b)/r_b \quad . \tag{4.62b}$$

Employing (4.37) and (4.38) in (4.33), and noting that the projection $(1-\theta_b)$ sets $P^{ext} = 1$ as a consequence of (4.35), we deduce that the contribution of $|\psi^{j;ext}\rangle$ to ${}_k\langle p_k'|(1-\theta_b)|\overline{X}_k\rangle$ is given by ${}_k\langle p_k'|(1-\theta_b)|G_{kj}\rangle$, where (for ijk cyclic)

$$_k\langle p_k'|G_{kj}\rangle = 2r_b/\pi \int_0^\infty dy_k y_k^2 j_0(y_k p_k')\left(\partial/\partial r_b - \lambda_k^c\right)_k\langle y_k|\psi^{j;ext}\rangle \tag{4.63a}$$

$$= -(8\pi^3)^{-1} \int d\underline{y}_k d\underline{r}_{ij} \exp(-i\underline{y}_k \cdot \underline{p}_k') N_k^c(r_{ij}) \langle \underline{r}_{ij}\underline{y}_k|\psi^{j;ext}\rangle \quad . \tag{4.63b}$$

In going from (4.63a) to (4.63b) we have used (4.62b) and the properties of the delta function, as well as the fact that integrating $\langle \underline{r}_{ij}\underline{y}_k|\psi^{j;ext}\rangle$ over $d\hat{r}_{ij}$ removes its dependence on $\hat{y}_k$ (for s-wave forces). In evaluating (4.63b) it is helpful to employ the expression

$$\langle \underline{r}_{ij}\underline{y}_k|\psi^{j;ext}\rangle = (\mu_{ki}/8\pi^2)\int d\underline{p}_j d\underline{q}_j \exp(i\underline{y}_j\cdot\underline{p}_j + i\underline{r}_{ki}\cdot\underline{q}_j) \times \left(q_j^2 - \kappa_j^2 - i\varepsilon\right)^{-1}\overline{X}_j(p_j)/D_j^c(\kappa_j) \quad , \tag{4.64}$$

in which $\underline{y}_j$, $\underline{r}_{ki}$ are to be expressed in terms of $\underline{y}_k$, $\underline{r}_{ij}$ by the relations given in (4.43); note that (4.64) reduces to (4.26b). Substituting

$$N_k^c(r_{ij}) = (2\pi^2)^{-1}\int d\underline{q}_k' \exp(-i\underline{r}_{ij}\cdot\underline{q}_k')N_k^c(q_k') \quad , \tag{4.65}$$

which follows from (4.62a), the integrals over $\underline{y}_k$, $\underline{r}_{ij}$ can be performed, yielding

$$(2\pi)^6\delta(\underline{p}'_k - \underline{P}_{kj})\delta(\underline{q}_k' - \underline{Q}_{kj}) \quad ,$$

$$\underline{P}_{kj} = -\mu_{ki}\underline{p}_j/m_i + \underline{q}_j \quad , \tag{4.66}$$

$$\underline{Q}_{kj} = -\mu_{ki}\underline{p}_j/\mu_k - \mu_{ij}\underline{q}_j/m_i \quad .$$

The delta functions are then used to perform the $\underline{q}_k'$, $\underline{q}_j$ integrations, and one obtains

$${}_k\langle p_k'|G_{kj}\rangle = -\mu_{ki}\int_{-1}^{1} dz \int_0^\infty dp_j p_j^2 N_k^c(Q_{kj})\left(Q_{jk}^2 - \kappa_j^2 - i\varepsilon\right)^{-1}\overline{X}_j(p_j)/D_j^c(\kappa_j) \quad . \tag{4.67}$$

Here $z = \hat{p}_k'\cdot\hat{p}_j$, and

$$\underline{Q}_{kj} = -\underline{p}_j - \mu_{ij}\underline{p}_k'/m_i \quad ,$$
$$\underline{Q}_{jk} = \underline{p}_k' + \mu_{ki}\underline{p}_j/m_i \quad . \tag{4.68}$$

We next form

$${}_k\langle p_k'|\theta_b|G_{kj}\rangle = \int_0^\infty dp_j p_j^2 G_{kj}(p_k',p_j)\overline{X}_j(p_j)/D_j^c(\kappa_j) \quad , \tag{4.69}$$

where

$$G_{kj}(p_k'',p_j) = -\mu_{ki}\int_0^\infty dp_k' p_k'^2 \theta_b(p_k'',p_k')\int_{-1}^{1} dz N_k^c(Q_{kj})\left(Q_{jk}^2 - \kappa_j^2 - i\varepsilon\right)^{-1} \quad . \tag{4.70a}$$

The p_k' integration in (4.70a) can be performed by the method of contours, first writing

$$G_{kj}(p_k'',p_j) = -\mu_{ki}/2 \int_{-1}^{1} dz \int_{-\infty+i\varepsilon}^{\infty-i\varepsilon} dp_k' p_k'^2 \theta_b(p_k'',p_k') N_k^c(Q_{kj})\left(Q_{jk}^2 - \kappa_j^2\right)^{-1} \quad . \tag{4.70b}$$

Here we have used the symmetry of the integrand under $p_k' \to (-p_k')$, $z \to (-z)$, to write $\int_0^\infty dp_k' = 1/2 \int_{-\infty}^{\infty} dp'_k$ and distorted the latter contour so that it is antisymmetric about (and passes through) the origin, while passing *under* the singularity at $Q_{jk}^2 = \kappa_j^2$ in the right half-plane (Re $p_k' \geq 0$) in accord with the $i\varepsilon$ instruction in (4.70a). This choice guarantees that for each point p_k' on the contour, there is a corresponding point $(-p_k')$. We next decompose $\theta_b(p_k'',p_k')$ into its positive- and negative-frequency parts, $\theta_b = \theta_b^{(+)} + \theta_b^{(-)}$, using (4.59) to verify that

$$\theta_b^{(+)}(p_k'',p_k') = (-\,i/\pi p_k') f_0(p_k'',b_k,p_k')\left(p_k'^2 - p_k''^2\right)^{-1} \quad ,$$

$$f_0(x,r,y) = [\cos rx - iry\, j_0(rx)] \exp(iry) \quad , \tag{4.71}$$

$$\theta_b^{(-)}(p_k'',p_k') = \theta_b^{(+)}(p_k'',-p_k') \quad ,$$

(note that $f_0(x,r,x) = 1$). By taking advantage of the reflection symmetry under $p_k' \to (-p_k')$ in (4.70b), we can replace $\theta_b(p_k'',p_k')$ in that expression by twice its positive-frequency part $(2\theta_b^{(+)})$.

The p_k' integration can now be performed analytically by closing the contour in the upper half-plane; the exponentially damped behavior of $\theta_b^{(+)}$ is just adequate to cancel the exponential increase of $N_k^c(Q_{kj})$ as Im $\{p_k'\} \to \infty$. We observe that the integrand has poles at $p_k' = p_k''$, and at $Q_{jk} = \kappa_j$ within the closed contour. Using (4.68), the latter is equivalent to $p_k' = \overline{P}_{kj}$,

$$\overline{P}_{kj} = -\mu_{ki} z p_j/m_i + R_{kj} \quad ,$$
$$R_{kj} = \left[\kappa_j^2 - \mu_{ki}^2(1-z^2)p_j^2/m_i^2\right]^{1/2} \quad . \tag{4.72}$$

We note that for $E<0$ or large p_j the quantity R_{kj} is positive imaginary. We thus obtain

$$G_{kj}(p_k',p_j) = -\mu_{ki} \int_{-1}^{1} dz\Big[N_k^c(Q_{kj})\left(Q_{jk}^2 - \kappa_j^2 - i\varepsilon\right)$$
$$-\overline{P}_{kj} R_{kj}^{-1} f_0(p_k',b_k,\overline{P}_{kj}) N_k^c(\overline{K}_{kj})\left(p_k'^2 - \overline{P}_{kj}^2 - i\varepsilon\right)^{-1}\Big] \quad , \tag{4.73}$$

with Q_{jk}, Q_{kj} as defined in (4.68), and $\overline{K}_{kj} \equiv \kappa_k(E,\overline{P}_{kj})$. Physically, $\overline{P}_{kj}$ is the momentum-conserving value of p_k for an on-shell state specified by p_j and z, and $\overline{K}_{kj}$ is the corresponding on-shell value of q_k. Upon substitution into (4.69), the first term in the bracket gives simply ${}_k\langle p_k'|G_{kj}\rangle$, and we therefore obtain

$$
\begin{aligned}
{}_k\langle p_k'|(1-\theta_b)|G_{kj}\rangle &= \int_0^\infty dp_j p_j^2 \overline{N}_{kj}(p_k',p_j)\overline{X}_j(p_j)/D_j^c(\kappa_j) \quad , \\
\overline{N}_{kj}(p_k',p_j) &= -\mu_{ki}\int_{-1}^{1} dz\overline{P}_{kj}R_{kj}^{-1}f_0(p_k',b_k,\overline{P}_{kj})N_k^c(\overline{K}_{kj})\left(p_k'^2 - \overline{P}_{kj}^2 - i\varepsilon\right)^{-1} \quad .
\end{aligned} \tag{4.74}
$$

Finally, recalling (4.53), (4.60) and (4.61), we infer that

$$
N_{kj}^{(0)}(p_k',p_j) = \overline{N}_{kj}(p_k',p_j) + (2/\pi)\int_{y_k^0}^{b_k} dy_k y_k^2 j_0(y_k p_k') N_{kj}^{(0)}(y_k,p_j) \quad . \tag{4.75}
$$

The long-range ($y_k > b_k$) contribution to the minimal kernel ($\overline{B}=0$) is thus $\overline{N}_{kj}(p_k',p_j)/D_j^c(\kappa_j)$, which has a structure quite similar to the Faddeev kernel for a separable t-matrix. In particular, we note the Green's function singularity with a residue proportional to the *on-shell* amplitude (N^c/D^c in the simplest case). A more precise comparison is afforded by considering the coefficient of $\overline{X}_j(p_j)$ in (4.67), which has exactly the form of the Faddeev kernel if one makes the replacement

$$
N_k^c(Q_{kj}) \to N_k^c(Q_{kj})N_j^c(Q_{jk})/N_j^c(\kappa_j) \quad ; \tag{4.76}
$$

this corresponds to the separable amplitude

$$
\begin{aligned}
t_j^s(q',q;E) &= g_j(q',\kappa_j)t_j^c(\kappa_j)g_j(q,\kappa_j) \quad , \\
g_j(q,\kappa_j) &= N_j^c(q)/N_j^c(\kappa_j) \quad .
\end{aligned} \tag{4.77}
$$

We note that making this substitution would not affect the result derived from (4.67), i.e., (4.74). Thus the long-range projection of the minimal BCF kernel is equal to that of the Faddeev kernel for a somewhat unorthodox t-matrix (t_j^s does not correspond to an energy-independent separable potential).

4.6 The Interior Region

We now return to (4.60) and consider the problem posed by the θ_0 projection of $\overline{X}_k$. Providing that $\overline{B}$ (or B) is chosen to satisfy (4.34), the two-particle data impose no constraint on this region, and we are free to choose whatever form is most convenient. In doing so we are guided by two major requirements: 1) that the additional constraint uniquely specify $\overline{X}_k$ via an integral equation with a compact kernel, and 2) that it be sufficiently general to span the full range of possibilities permitted by unitarity. It is also desirable to achieve a real parametrization comparable to the two-particle phase shift, reasonable analyticity properties consistent with our knowledge of the Faddeev amplitudes, and (hopefully) simplicity of form. For the purpose of this discussion we will assume that the required constraint may be expressed as

$$ {}_k\langle p_k'|\theta_0|\overline{X}_k\rangle = \sum_j \int_0^\infty dp_j p_j^2 \overline{C}_{kj}(p_k',p_j)\overline{X}_j(p_j)/D_j^c(\kappa_j) \quad , \tag{4.78} $$

i.e., that there is no contribution to the driving term (Ω_k). It may in fact be shown that all apparent alternatives can be reduced to this form for a suitable choice of $\overline{C}$ (consistent with the general properties specified below); see [4.13].

Although any compact $\overline{C}$ will uniquely determine $\overline{X}_k$ via (4.60), it turns out that its imaginary part, Im $\{\overline{C}\}$, is entirely determined by the requirements of three-particle unitarity. To see this we will need the following statement of the unitarity relation in terms of the channel amplitudes (cf., Sec.1.6)

$$ \begin{aligned} &\mathrm{Im}\left\{T^{k_f k_i}(\kappa_{k_f},p_{k_f};\kappa_{k_i}p_{k_i})\right\} \\ &= -\pi\sum_j \int_0^{(2\mu_j E)^{1/2}} dp_j p_j^2 T_+^{k_f j}(\kappa_{k_f},p_{k_f};\kappa_j p_j)\mu_{ki}\kappa_j T_-^{jk_i}(\kappa_j,p_j;\kappa_{k_i}p_{k_i}) \quad , \end{aligned} \tag{4.79} $$

where we have expanded our previous notation in order to include the dependence on the variables of the initial state. Thus

$$ T^k(\kappa_k,p_k) \equiv T^{k1}(\kappa_k p_k;\kappa_1^0 p_1^0) \quad . \tag{4.80} $$

The notation $\pm$ corresponds to $E\pm i\varepsilon$, and $T_-=T_+^*$. Correspondingly, we observe that $\Omega_k(p_k)$ is a particular matrix element of an operator Ω_{kj}, i.e.,

$$ \Omega_{k1}(p_k) = {}_k\langle p_k|\Omega_k|\kappa_1^0 p_1^0\rangle_1 \quad , \tag{4.81} $$

where we have introduced the notation

$$ |q_k p_k\rangle_k \equiv (4\pi)^{-2}\int d\hat{q}_k d\hat{p}_k |\underline{q}_k\underline{p}_k\rangle \tag{4.82} $$

to represent the s-wave projection of the momentum-state conjugate to $|\underline{r}_{ij}\underline{y}_k\rangle$. Equivalently, $|\phi_{1;c}\rangle$ is a particular state of this type evaluated at the on-shell value $q_1^0=\kappa_1^0$. The expanded version of (4.60) thus reads

$$ \begin{aligned} \overline{X}_{k_f k_i}(p_{k_f};\kappa_{k_i}p_{k_i}) &= {}_{k_f}\langle p_{k_f}|\Omega_{k_f k_i}|\kappa_{k_i}p_{k_i}\rangle_{k_i} \\ &+ \sum_j \int_0^\infty dp_j p_j^2 N_{k_f j}^{tot}(p_{k_f},p_j)\overline{X}_{jk_i}(p_j;\kappa_{k_i}p_{k_i})/D_j^c(\kappa_j) \end{aligned} \tag{4.83} $$

where

$$N^{tot}_{kj}(p'_k,p_j) = N_{kj}(p'_k,p_j) + \overline{C}_{kj}(p'_k,p_j) \quad . \tag{4.84}$$

The connection of $\overline{X}$ to T^k is of course given by (4.38),

$$T^{kj}(\kappa'_k p'_k;\kappa_j p_j) = -\overline{X}_{kj}(p'_k;\kappa_j p_j)/D^c_k(\kappa'_k) \quad . \tag{4.85}$$

In order to exploit the unitarity relation, we first note that

$$\mathrm{Im}\{D^c_k(\kappa_k)\} = \pi\mu_{ij}\kappa_k N^c_k(\kappa_k) \quad , \tag{4.86}$$

and hence

$$\mathrm{Im}\{T^{kj}(\kappa'_k p'_k;\kappa_j p_j)\} = \Big[\pi\mu_{ij}\kappa'_k t^c_k(\kappa'_k)\overline{X}^+_{kj}(p'_k;\kappa_j p_j) - \mathrm{Im}\{\overline{X}_{kj}(p'_k;\kappa_j p_j)\}\Big]/D^{c-}_k(\kappa'_k) \quad . \tag{4.87}$$

To compute $\mathrm{Im}\{\overline{X}\}$ we use the relations

$$\begin{aligned} \mathrm{Im}\{V\} &= (2i)^{-1}\,\mathrm{disc}\,\{V\} = (2i)^{-1}(V^+ - V^-) \quad , \\ \mathrm{disc}\,\{(AB)\} &= A^+(\mathrm{disc}\,\{B\}) + (\mathrm{disc}\,\{A\})B^- \quad , \end{aligned} \tag{4.88}$$

writing (4.83) schematically in the form

$$\begin{aligned} \overline{X} &= \overline{\Omega} + K\overline{X} \\ &= Z\overline{\Omega} \quad , \end{aligned} \tag{4.89}$$

where $Z = (1-K)^{-1}$. Employing the identity

$$A^{-1} - B^{-1} = A^{-1}(B - A)B^{-1} \quad , \tag{4.90}$$

we deduce that

$$\begin{aligned} \mathrm{disc}\,\{\overline{X}\} &= Z^+\,(\mathrm{disc}\,\{K\})Z^-\overline{\Omega} \\ &= Z^+\,(\mathrm{disc}\,\{K\})\overline{X}^- \quad , \end{aligned} \tag{4.91}$$

since $\mathrm{disc}\,\{\overline{\Omega}\} = 0$ ($\overline{\Omega}$ is real). The problem thus reduces to evaluating $\mathrm{disc}\,\{K\}$; this will result in the desired constraint on $\mathrm{disc}\,\{\overline{C}\}$.

In order to compute the imaginary part of the kernel we first recall (4.51), which implies that

$$
{}_k\langle p_k'|\Omega_{kj}|\kappa_j p_j\rangle_j = (2r_b/\pi) \int_{y_k}^{\infty} dy_k y_k^2 j_0(y_k p_k') \bar{x}_k^{-1} \times \left(\partial/\partial r_b - \lambda_k^c\right)_k\langle y_k|P^{ext}|\kappa_j p_j\rangle_j \quad . \tag{4.92}
$$

This expression can be related to Im $\{K\}$ by using the following device. We introduce an artificial off-shell extension

$$
\tilde{T}^{kj}(q_k' p_k';\kappa_j p_j) = (j_0(aq_k')/j_0(a\kappa_k'))T^{kj}(\kappa_k' p_k';\kappa_j p_j) \quad , \tag{4.93}
$$

with $0<a\ll r_b$. The content of (4.26b) can then be expressed as the relation

$$
|\psi^{k;ext}\rangle = G_0^{(+)}\tilde{T}^{k1}|\phi_{1;c}\rangle \quad , \tag{4.94}
$$

and the analogue of (4.63a) is that

$$
{}_k\langle p_k'|(1-\theta_0)|\overline{X}_k\rangle = (2r_b/\pi) \int_{y_k}^{\infty} dy_k y_k^2 j_0(y_k p_k') \bar{x}_k^{-1} \times \left(\partial/\partial r_b - \lambda_k^c\right)_k\langle y_k|P^{ext}|\psi^{j;ext}\rangle + \cdots \quad . \tag{4.95}
$$

Together, (4.94) and (4.95) imply that

$$
\begin{aligned}
&{}_k\langle p_k'|(1-\theta_0)K^{(0)}|p_j\rangle_j D_j^c(\kappa_j) \\
&= 2\mu_{ki} \int_0^{\infty} dq_j q_j^2 \; {}_k\langle p_k'|\Omega_{kj}|q_j p_j\rangle_j j_0(aq_j)\left[j_0(a\kappa_j)(q_j^2 - \kappa_j^2 - i\varepsilon)\right]^{-1} \quad ,
\end{aligned} \tag{4.96}
$$

where $(1-\theta_0)K^{(0)}$ indicates that portion of the kernel which arises from $N_{kj}^{(0)}$ (see (4.75)). Using

$$
\text{disc}\,\{(q^2 - \kappa^2)^{-1}\} = \pi i\delta(q - \kappa)/q \quad , \tag{4.97}
$$

it is simple to obtain the result

$$
\text{Im}\{N_{kj}^{(0)}(p_k',p_j)\} = \pi\mu_{ki}\kappa_{jk}\langle p_k'|\Omega_{kj}|\kappa_j p_j\rangle_j \quad . \tag{4.98}
$$

Combining (4.54) and (4.84), we have

$$
\begin{aligned}
\text{Im}\{N_{kj}^{tot}(p_k',p_j)\} = (1 - \delta_{kj})\, \text{Im}\{N_{kj}^{(0)}(p_k',p_j)\} + r_b\, \text{Im}\{B_{kj}(p_k',p_j)\} \\
+ \text{Im}\{\overline{C}_{kj}(p_k',p_j)\} \quad .
\end{aligned} \tag{4.99}
$$

Finally, recalling (4.87), it is clear that

$$\mathrm{Im}\{K_{kj}(p_k',p_j)\} = \Big[-\pi\mu_{ki}\kappa_j t_j^c(\kappa_j) N_{kj}^{tot^+}(p_k',p_j) + \mathrm{Im}\{N_{kj}^{tot}(p_k',p_j)\}\Big]/D_j^{c-}(\kappa_j) \quad . \tag{4.100}$$

This result can be expressed in the form

$$\mathrm{Im}\{K_{kj}(p_k',p_j)\} = \pi\mu_{ki}\kappa_j\Big[\delta(p_k' - p_j)p_j^{-2}\delta_{kj} - K_{kj}^+(p_k',p_j)\Big]t_j^{c-}(\kappa_j) + \pi\mu_{ki}\kappa_{jk}\langle p_k'|\Omega_{kj}|\kappa_j p_j\rangle_j/D_j^{c-}(\kappa_j) + \Delta_{kj}(p_k',p_j)/D_j^{c-}(\kappa_j) \quad , \tag{4.101}$$

where

$$\Delta_{kj}(p_k',p_j) = r_b\,\mathrm{Im}\{B_{kj}(p_k',p_j)\} + \mathrm{Im}\{\overline{C}_{kj}(p_k',p_j)\} - \pi\mu_{ki}\kappa_j\theta_0(p_k',p_j)N_j^c(\kappa_j)\delta_{kj} \quad . \tag{4.102}$$

Here we have written the factor $\theta(y_k-y_k^0)$ occurring in Ω_{kk}, (4.51), as $1-\theta(y_k^0-y_k)$ before performing the Fourier transform; the first part produces the $\delta(p_k'-p_j)$ term in the square bracket.

Direct substitution of (4.101) into the relation symbolized by (4.91) shows that the first two terms are precisely what is necessary to prove (4.79), and hence unitarity requires that $\Delta_{kj} \equiv 0$. Inasmuch as B and $\overline{C}$ are projections onto orthogonal subspaces, i.e.,

$$B = (1 - \theta_0)B \quad , \qquad \overline{C} = \theta_0\overline{C} \quad , \tag{4.103}$$

this is equivalent to the conditions

$$\mathrm{Im}\{B_{kj}(p_k',p_j)\} = 0 \quad , \qquad \mathrm{Im}\{\overline{C}_{kj}(p_k',p_j)\} = \pi\mu_{ki}\kappa_j\theta_0(p_k',p_j)N_j^c(\kappa_j)\delta_{kj} \quad . \tag{4.104}$$

Thus, as anticipated in the discussion concerning (4.33), the function B_{kj} (and hence $\overline{B}_{kj}$ in view of (4.55)) is real-valued. It is clear that B_{kj} summarizes off-shell properties which are long range (in the three-body sense). Similarly, $\overline{C}_{kj}$, which represents the short-range effects, is determined up to an arbitrary real function. In order to put these input functions on the same footing, it is convenient to introduce a minimal form $\overline{C}^0$ which satisfies (4.104). Perhaps the simplest choice consistent with analyticity and the need for a compact kernel is

$$\overline{C}^0_{kj}(p'_k,p_j) = \delta_{kj}\theta_0(p'_k,p_j)\left[D^c_j(\kappa_j) - D^c_j(\bar{\kappa}_j)\right] \quad , \tag{4.105}$$

where $\bar{\kappa}_j \equiv \kappa_j(\bar{E},p_j)$, and we have recalled (4.86). The quantity $\bar{E}$ is an energy parameter which must be negative in order for the second term to be real, but is otherwise arbitrary. The subtraction is introduced to improve the convergence as $p_j \to \infty$, since $\kappa_j \to \bar{\kappa}_j + O(p_j^{-1})$ in that limit. This form also results in a common factor of $\exp(ir_b\kappa_j)$ for large p_j, which is necessary to compensate the exponential decrease of $D^c_j(\kappa_j)$ in forming this contribution to the kernel $(\overline{C}^0/D^c_j)$.

4.7 Summary of the Formal Work

We are now in a position to summarize the results of our derivation and cast the BCF equation into its final form. We first write

$$B_{kj}(y_k,p_j) = \delta_{kj}B_j(\kappa_j)j_0(y_jp_j) + r_b^{-1}\hat{B}_{kj}(y_k,p_j)D^c_j(\bar{\kappa}_j) \quad , \tag{4.106}$$

which satisfies (4.34) if $\hat{B}_{kj}(y_k,p_j)$ is suitably damped as $y_k \to \infty$. We have chosen to extract the (nonvanishing) factor $D^c_j(\bar{\kappa}_j)$ in order to exhibit the required factor $\exp(ir_b\kappa_j)$ explicitly; there is no loss in generality in so doing. With this convention suitable convergence will be obtained if $\hat{B}_{kj}(p'_k,p_j)$ is any *square-integrable* (L_2) real operator. Similarly, we take

$$\overline{C}_{kj}(p'_k,p_j) = \overline{C}^{(0)}_{kj}(p'_k,p_j) + \hat{C}_{kj}(p'_k,p_j)D^c_j(\bar{\kappa}_j) \quad . \tag{4.107}$$

Elsewhere it has been shown that real L_2 functions $\hat{B}$, $\hat{C}$ can always be chosen so as to exactly reproduce the (on-shell) Faddeev amplitudes, and hence the exponential factor built into $D^c_j(\bar{\kappa}_j)$ must of necessity arise (a proof is given in [4.13]). The complete statement of (4.60) becomes

$$\left[1 - r_bB_k(\kappa'_k)/D^c_k(\kappa'_k)\right]\overline{X}_k(p'_k) = \Omega_k(p'_k) + \sum_j \int_0^\infty dp_jp_j^2\overline{K}_{kj}(p'_k,p_j)\overline{X}_j(p_j) \quad , \tag{4.108}$$

where

$$\begin{aligned}
\overline{K}_{kj}(p'_k,p_j) &= K^{(0)}_{kj}(p'_k,p_j) + A_{kj}(p'_k,p_j)D^c_j(\bar{\kappa}_j)/D^c_j(\kappa_j) \quad , \\
A_{kj}(p'_k,p_j) &= \hat{B}_{kj}(p'_k,p_j) + \hat{C}_{kj}(p'_k,p_j) \quad , \\
K^{(0)}_{kj}(p'_k,p_j) &= \left[N^{(0)}_{kj}(p'_k,p_j) + \overline{C}^{(0)}_{kj}(p'_k,p_j)\right]/D^c_j(\kappa_j) \quad .
\end{aligned} \tag{4.109}$$

Our equation can now be brought into its final form by invoking the relation

$$1 - r_b B_k(\kappa_k)/D_k^c(\kappa_k) = t_k^c(\kappa_k)/t_k(\kappa_k) \equiv \rho_k^{-1}(\kappa_k) \quad , \tag{4.110}$$

which can be inferred from (4.17) and shown explicitly by combining (4.10) and (4.18). We define

$$X_j(p_j) \equiv \rho_j^{-1}(\kappa_j)\overline{X}_j(p_j) \quad , \tag{4.111}$$

and obtain

$$X_k(p_k') = \Omega_k(p_k') + \sum_j \int_0^\infty dp_j p_j^2 K_{kj}(p_k',p_j)X_j(p_j) \quad , \tag{4.112}$$

with

$$K_{kj}(p_k',p_j) = \overline{K}_{kj}(p_k',p_j)\rho_j(\kappa_j) \quad . \tag{4.113}$$

We note that the distinction between K and $\overline{K}$ corresponds to the denominator $D_j^c(\kappa_j)$ being replaced by

$$\tilde{D}_j(\kappa_k) = D_j(\kappa_j)N_j^c(\kappa_j)/N_j(\kappa_j) \quad . \tag{4.114}$$

One can easily verify that (4.112) has the expected singularity structure with poles arising in the kernel at the positions of two-particle bound states (zeros of $D_j(\kappa_j)$, and the usual (three-particle) Green's function singularity contained in $N_{kj}^{(0)}(p_k',p_j)$ (cf. (4.74)).

It is clear that (4.112) satisfies both the specific mathematical properties set forth at the beginning of Section 4.6, and the general requirements discussed at the beginning of Section 4.3; i.e., we have a one-dimensional integral equation of the Fredholm type whose solutions generate a unitary three-particle amplitude, and the input for this equation is cleanly separated into terms specified by the two-particle data (e.g., Ω_k, $K_{kj}^{(0)}$, $\tilde{D}_j$) and a totally unrestricted real operator (A_{kj}). The latter specifies that portion of the dynamical information which is actually probed in three-body scattering experiments and which plays the role of a "phase shift" for the three-particle system. Given any combination of two- and three-body potentials, it has been rigorously shown that a corresponding A operator can be uniquely constructed [4.13], but it is much more efficient to establish the characteristic properties of A and to parametrize it directly. By varying such parameters, one can rapidly generate the complete class of possible physical amplitudes consistent with a prescribed set of two-particle data. For this purpose the fact that the kernel is *linear* in A has important practical consequences, as we show below.

In concluding this section, we note that the assumptions introduced to simplify some aspects of the derivation (identical particles, s-wave forces) are in no sense crucial, and in fact there is no real difficulty in including the spin and isospin degrees of freedom as well as generalizing to arbitrary angular momentum (specific formulas of a more general nature are presented in [4.7]). What is more interesting is that generalizations of *qualitative* nature can also be defined in a very straightforward way (more particles, inclusion of Coulomb forces, relativistic covariance); we shall return to this point in Section 4.9.

4.8 An Example

In the preceding sections we have derived a one-dimensional integral equation, (4.112), as an alternative to the Faddeev equation. In doing so we have nowhere assumed separable t-matrices, and hence we have gained the practical (numerical) advantages commonly associated with separable potentials at no cost in generality. We have paid a price, however, in introducing a new type of operator (A in (4.109)) to represent the off-shell information usually contained in the potentials. Although this decomposition has many advantages, one loses the easy familiarity gained over a long period of time with potential theory and must adjust conceptually to the new representation. The most obvious way to bridge this gap is to study the behavior of A for various potentials, and hence we devote this section to some simple but representative examples.

Although general formulas have been derived for constructing the A operator equivalent to any combination of two- and three-body potentials [4.13], we shall simplify the discussion by assuming identical particles of mass m scattering in s-waves, with the separable t-matrix

$$t_0(q',q;E) = f_0(q',\kappa)t^c(\kappa)f_0(q,\kappa) \quad . \tag{4.115}$$

We thus take the on-shell amplitude to be the simple BC model stated in (4.10) with $f(\kappa^2) = f^c$ = constant, but allow the Kowalski-Noyes function $f_0(q,\kappa)$ (cf. (4.3)) to vary independently (note that we drop channel indices in this section in accord with our assumption of identical particles). In this case the function $\rho(\kappa)$ defined in (4.110) is unity, and the relationship between our function X(p) and the Faddeev channel function $\tau(q,p)$ is given by

$$\tau(q,p) = - f_0(q,\kappa)t^c(\kappa)X(p)/N^c(\kappa) \quad , \tag{4.116}$$

using (4.38) and (4.111) with $T^1 \equiv \tau$. If we consider the Faddeev equation corresponding to the above choice of $t_0(q',q;E)$, we infer that X(p) satisfies a one-dimensional equation with the kernel (c.f. (1.166))

$$K^f(p',p) = N^f(p',p)/D^c(\kappa) \quad ,$$
$$N^f(p',p) = -\, mN^c(\kappa') \int_{-1}^{1} dz f_0(Q',\kappa') f_0(Q,\kappa)(p'^2 + pp'z + p^2 - mE - i\varepsilon)^{-1} \quad , \tag{4.117}$$

where

$$\kappa = \kappa(E,p) = (mE - 3p^2/4)^{1/2} \quad ,$$
$$Q = (p'^2 + pp'z + p^2/4)^{1/2} \quad , \tag{4.118}$$
$$Q' = (p^2 + pp'z + p'^2/4)^{1/2} \quad .$$

On the other hand, we know that X(p) satisfies (4.112), and we might thus expect that

$$A(p',p)D^c(\bar{\kappa}) = N^f(p',p) - N^{(0)}(p',p) - \bar{C}^{(0)}(p',p) \quad , \tag{4.119}$$

in view of (4.109). The trouble with this argument is that the driving term of the Faddeev equation is not identical with that of the BCF equation ($\Omega(p')$), and hence the correspondence is actually more complicated. However, the $(1-\theta_b)$ projections of the driving terms *are* identical, as may be verified from (4.51), and we may thus use the corresponding projection (4.119) to deduce that

$$\hat{B}(y,p)D^c(\bar{\kappa}) = \int_0^\infty dp'p'^2 j_0(yp')\left[N^f(p',p) - \bar{N}(p',p)\right] \tag{4.120}$$

for $y>b = 3r_b/2$. Here we have used (4.75) with $\bar{N}\equiv\bar{N}_{12}+\bar{N}_{13}$. By comparing (4.74) and (4.117), one can easily show that the singular terms in N^f and $\bar{N}$ exactly cancel, and thus $\hat{B}$ is real-valued, as required.

In order to obtain some feeling for the functional behavior of $\hat{B}(y,p)$, we consider an exceptionally simple choice of $f_0(q,\kappa)$ which allows the integral in (4.120) to be evaluated analytically. We thus take

$$f_0(q,\kappa) = g(q)/g(\kappa) \quad ,$$
$$g(q) = N^c(q)(q^2 + \mu^2)^{-1} \tag{4.121}$$

and employ the method of contours previously utilized in connection with (4.70a). Due to the cancellation noted above, the only singularities enclosed in the upper half-plane are poles (for fixed z) corresponding to $Q'=i\mu$ and $Q=i\mu$. These are equivalent to $p'=P_1$ and $p'=P_2$, respectively, where

$$P_1 = -\, 2pz + iR_1 \quad ,$$
$$R_1 = 2\left[(1 - z^2)p^2 + \mu^2\right]^{1/2} \quad , \tag{4.122}$$

$$P_2 = -pz/2 + iR_2 \quad , \qquad (4.122)$$
$$R_2 = \left[(1 - z^2)p^2/4 + \mu^2\right]^{1/2} \quad .$$

The result may be expressed as

$$\hat{B}(y,p) = -(m\pi/iy)N^c(i\mu)\left[N^c(\kappa)D^c(\bar{\kappa})\right]^{-1} \int_{-1}^{1} dzU(y,p,z) \quad , \qquad (4.123)$$

with

$$\begin{aligned} U(y,p,z) = &-(P_2/2R_2)N^c(Q_2)\exp(iyP_2) \\ &+ (8/3)\Big[(P_1/R_1)N^c(Q_1)\exp(iyP_1)\left(p^2-P_1^2\right)^{-1} \\ &- (P_2/4R_2)N^c(Q_2)\exp(iyP_2)\left(p^2 - P_2^2\right)^{-1}\Big](\kappa^2 + \mu^2) \quad . \end{aligned} \qquad (4.124)$$

Here we have used

$$Q_1 = i\left[\mu^2 + 3\left(p^2 - P_1^2\right)/4\right]^{1/2} \quad , \qquad (4.125)$$
$$Q_2 = i\left[\mu^2 + 3\left(p^2 - P_2^2\right)/4\right]^{1/2} \quad .$$

Although it is clear that $\hat{B}(y,p)$ is rather complicated even for a trivial model, the important point is that it possesses certain simple general features which are apparent in (4.123). In particular, for real p, we note that $\mathrm{Im}\{P_1\} \geq 2\mu$, $\mathrm{Im}\{P_2\} \geq \mu$, and hence $\hat{B}(y,p) \propto \exp(-\mu y)$. Thus, for a two-body interaction which falls off like $\exp(-\mu x)$ at large x, we expect a corresponding decline of $\hat{B}(y,p)$ at large y *with the same range*. Equivalently, the function $\hat{B}(p',p)$ has branch points at $p'=P_1$, $p'=P_2$ $(z=\pm 1)$, but is analytic in the strip $|\mathrm{Im}\{p'\}| \leq \mu$ surrounding the real axis. This has the practical consequence that, although the structure of $\hat{B}(p',p)$ in the complex plane is very complicated, its behavior in the vicinity of the real axis is quite smooth for the short-range forces characteristic of nuclear physics. It may thus be represented by a far simpler structure in many applications, e.g.,

$$\hat{B}'(p',p) = \alpha(p'^2 + \bar{\mu}^2)^{-1}(p^2 + \bar{\mu}^2)^{-1} \qquad (4.126)$$

for suitable choices of α and $\bar{\mu}$ (comparable to μ), or by a superposition of such terms.

The form of (4.123) also illustrates two additional features which are general in character. The first is the appearance of an explicit factor $\exp(ir_b\kappa)$ to compensate the exponential term in $D^c(\bar{\kappa})$ as $p \to \infty$ (in this case in $N^c(\kappa)$); this is in accord with the discussion given earlier in connection with (4.106). Although this may appear to have arisen arbitrarily as a consequence of the specific choice for

$g(q)$ made in (4.121), one may verify from (4.115) that $t_0(q',q;E)$ will grow exponentially as $E\to-\infty$ unless $f_0(q,\kappa)$ has such a factor. (If we had chosen an on-shell amplitude different from $t^c(\kappa)$ the general arguments of [4.13] would apply). The second feature concerns the energy-dependence of $\hat{B}(y,p)$, which appears only via the factor $(\kappa^2 + \mu^2)$ in (4.124) and in the overall factor $[N^c(\kappa)]^{-1}$ in (4.123). Inasmuch as the BC represents the short-range behavior, we have by assumption that $r_b<\mu^{-1}$, and hence we would expect significant energy dependence in $\hat{B}(y,p)$ only if E is comparable to μ^2/m. Since the dominant part of the nuclear force has a range $\mu^{-1}\lesssim 1$fm, we infer that E should be at least 45 MeV in the trinucleon problem (T_L=70 MeV in N-d scattering) for this to occur. We would thus anticipate that the A operator is only weakly dependent on E, which has been confirmed by numerical studies [4.14].

Although the evaluation of $\hat{B}$ is quite tedious even in simple cases and $\hat{C}$ is even more complicated, one can gain a good qualitative understanding of their behavior by means of the following procedure. Let us represent them in the form

$$\hat{B}(y,p) = B(y)(p^2 + \mu^2)^{-1} \quad ,$$
$$\hat{C}(y,p) = C(y)(p^2 + \mu^2)^{-1} \quad , \qquad (4.127)$$

although this is strictly possible only as $p\to 0$. The advantage of this (separable) approximation is that, given a solution of the Faddeev equation for $X(p)$, we can substitute the result (numerically) into (4.112) and solve explicitly for $B(y)$ and $C(y)$. By construction we will the have $\hat{B}$, $\hat{C}$ operators which yield the same scattering amplitudes as does the Faddeev theory, and we can check the effectiveness of the approximation by confirming that $\mathrm{Im}\{\hat{B}\} = \mathrm{Im}\{\hat{C}\}\approx 0$. In particular, if we consider the class of models defined by (4.115), the corresponding calculation is quite trivial.

We thus choose m to be the (average) nucleon mass, and take r_b = 1.095 fm, f^c = -0.253, which ensures that $t^c(\kappa)$ behaves like the $^3S_1$ amplitude in the effective range region (including a "deuteron" pole at -2.2 MeV). Ignoring spin, we solve for the elastic amplitude for scattering the third particle from this bound state (analogue of N-d scattering), and subsequently compute $B(y)$, $C(y)$ as indicated above. We choose $f_0(q,\kappa)$ to have the form given in (4.121) with $g(q) = v(q)(q^2+\mu^2)^{-1}$; we take μ = 2 fm^{-1} and the following three choices for $v(q)$:

$$\begin{aligned} &a) \quad v(q) = N^c(q) \quad , \\ &b) \quad v(q) = j_0(r_b q) \quad , \\ &c) \quad v(q) = 1 \quad . \end{aligned} \qquad (4.128)$$

(In case c) we avoid the exponential problem by employing relativistic kinematics as described in [4.15]; this is roughly equivalent to cutting off the p-integrations

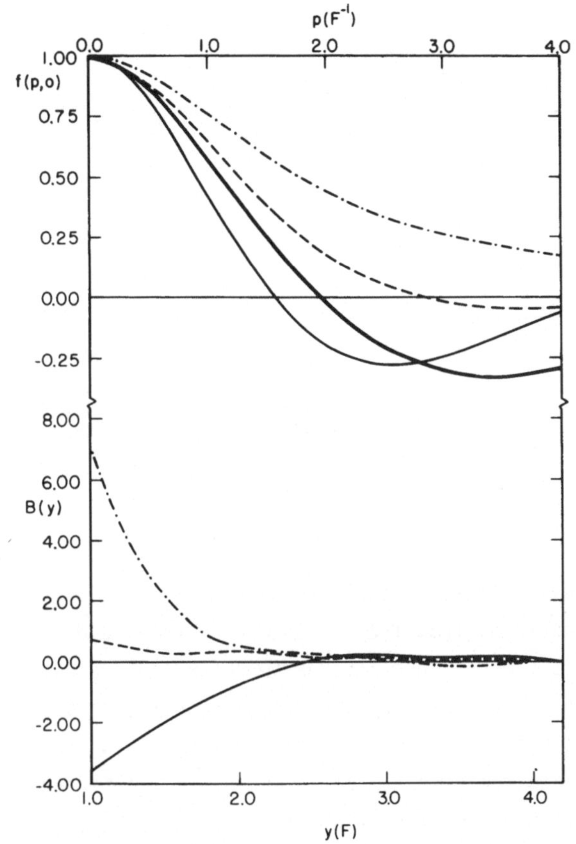

Fig.4.1. Behavior of the function B(y) defined in (4.127) (lower curves) for various choices of $f_0(p,0)$ (upper curves). The heavy solid line represents the Reid (soft core) potential

at p≃m). These choices can be characterized by plotting the function $f_0(q,0)$, as shown in the upper half of Fig.4.1 (note that the solid, dashed, and dashed-dot curves correspond to a), b) and c), respectively). For comparison, the corresponding function for the Reid soft core potential is also plotted (heavy solid line).

The lower half of Fig.4.1 illustrates the behavior of B(y) for the three models. Positivity of B(y) is associated with attraction in our construction, and hence the trend is as one would expect given our experience with separable forms. However, it is clear that the BCF representation strikingly emphasizes the differences between such models, making it particularly suitable for characterizing the off-shell structure. (Note that the minor fluctuations which appear beyond y=3 fm arise due to the approximate nature of (4.127)). The corresponding curves for C(y) are shown in Fig.4.2 (same notation) and are rather structureless, although the magnitude is distinctive of the particular model. We thus conclude that the off-shell features usually embodied in potential models have a simple and straightforward representation in terms of smooth functions $\hat{B}$ and $\hat{C}$ in the BCF.

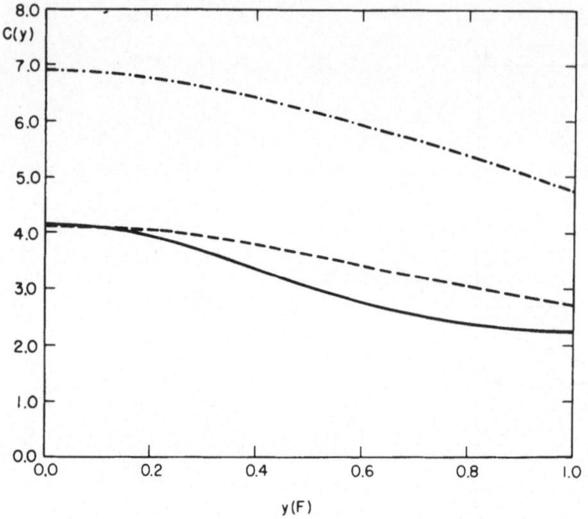

Fig.4.2. Behaviour of the function C(y) defined in (4.127) for the choices of $f_0(p,0)$ shown in Fig.4.1 (same notation)

In practice, virtually all three-body scattering problems involve two particles bound together in the initial state. Providing that the corresponding threshold conditions are satisfied, the physical processes then include *break-up* (to a state of three free particles), and *rearrangement collisions*, involving transitions to other bound state-plus-spectator configurations. In particular, if the initial and final (pairwise) bound states are the same, we have *elastic* bound state scattering, e.g., pd→pd. We thus conclude this section by relating the solution $X_k(p_k)$ of (4.112) to the corresponding physical amplitudes.

For simplicity, we shall assume that only a single two-particle bound state exists. Suppose, for definiteness, that particles 2 and 3 bind at an energy $E_1^{(b)} = -E_B<0$, corresponding to a (positive imaginary) momentum $\kappa_1^{(b)} = i(2\mu_{23}E_B)^{1/2}$. In the vicinity of $\kappa_1 = \kappa_1^{(b)}$,

$$t_1(\kappa_1) \to g_1^2/\left(E_1 - E_1^{(b)}\right) \quad , \tag{4.129}$$

where $E_1 = \kappa_1^2/(2\mu_{23})$. We note that with this convention g_1 is either a positive real or positive imaginary number, depending on the sign of the residue (g_1^2). The amplitudes for break-up and elastic scattering are defined in the usual manner by identifying appropriate residues of the three-particle t-matrix with respect to this bound state pole. Thus

$$\begin{aligned} T^k(\kappa_k,p_k) &= b_{k1}(p_k,E)g_1/\left(E_1^0 - E_1^{(b)}\right) + \cdots \quad , \\ b_{k1}(p_k,E) &= \delta_{k1}e_{11}(E)g_1/\left(E_k - E_1^{(b)}\right) + \cdots \quad , \end{aligned} \tag{4.130}$$

where the neglected terms are regular at the pole. For real values of κ_k corresponding to $p_k^2 \leq 2\mu_k E$, $b_{k1}(p_k,E)$ is the channel break-up amplitude, whereas $e_{11}(E)$ is the amplitde for elastic bound state scattering. The latter corresponds to the momentum $p_1^{(b)} = [2\mu_1(E - E_1^{(b)})]^{1/2}$.

Recalling (4.38), (4.111) and (4.112), we observe that

$$\begin{aligned}\Omega_k(p_k') &= (2/\pi) \int_{y_k^0}^{\infty} dy_k y_k^2 j_0(y_k p_k')\Omega_k(y_k) \\ &= -\delta_{k1} N_1^c(\kappa_1^0)\delta(p_k' - p_1^0)p_k'^{-2} + \cdots \quad ,\end{aligned} \tag{4.131}$$

where the remaining terms do not contribute to the pole at $E_1^0 = E_1^{(b)}$. The bound state scattering terms arise upon iterating (4.112), whereupon the driving term becomes

$$-\overline{K}_{k1}(p_k', p_1^0)D_1^c(\kappa_1^0)t_1(\kappa_1^0) + \cdots \tag{4.132}$$

using (4.113). Defining $\hat{X}_{k1}(p_k')$ to be the solution of the equation

$$\begin{aligned}\hat{X}_{k1}(p_k') &= -\overline{K}_{k1}\left(p_k', p_1^{(b)}\right)D_1^c\left(\kappa_1^{(b)}\right)g_1 \\ &+ \sum_j \int_0^{\infty} dp_j p_j^2 K_{kj}(p_k', p_j)\hat{X}_{j1}(p_j) \quad ,\end{aligned} \tag{4.133}$$

and comparing (4.38), (4.111), (4.130) and (4.132), we deduce that

$$\begin{aligned}b_{k1}(p_k, E) &= -t_k(\kappa_k)\hat{X}_{k1}(p_k)/N_k^c(\kappa_k) \quad , \\ e_{11}(E) &= -g_1\hat{X}_{11}\left(p_1^{(b)}\right)/N_1^c\left(\kappa_1^{(b)}\right) \quad .\end{aligned} \tag{4.134}$$

4.9 Discussion

In Section 4.1 we motivated our description of the boundary condition approach with a discussion of the difficulties and ambiguities inherent in extracting off-shell information from three-body experiments. We now consider this question from the standpoint of the formalism summarized in Section 4.7. As is evident from (4.109), we have achieved a clean separation between the on-shell parameters which specify the minimal kernel ($K^{(0)}$), and the off-shell parameters which specify the A operator. This means that it is now trivial to hold the two-particle data fixed while freely varying the off-shell degrees of freedom. Furthermore, although the input phase shifts may be arbitrary, the resultant integral equation is never more difficult to solve than the Faddeev equation with a rank *one* separable potential. This point is of great practical importance in including states such as the coupled $^3S_1$-$^3D_1$ partial-waves, which apparently require a separable potential of rank *four* (or greater) for an adequate description [4.16]. This corresponds to a significant increase in the dimension of the finite matrix one must ultimately compute (and store) in solving such equations numerically, and severely handicaps realistic calculations of N-d polarization by Faddeev techniques [4.17].

An advantage of even greater importance stems from the linearity of the full kernel in the off-shell term. Thus, we may write (4.112) schematically as

$$X = \Omega + KX \quad , \qquad K = K^{(1)} + AK^{(2)} \quad , \tag{4.135}$$

where

$$K^{(1)}_{kj}(p'_k,p_j) = K^{(0)}_{kj}(p'_k,p_j)\rho_j(\kappa_j) \quad , \qquad K^{(2)}_{kj}(p'_k,p_j) = \delta_{kj}\delta(p'_k - p_j)p_j^{-2}D_j^c(\bar{\kappa}_j)/\tilde{D}_j(\kappa_j) \quad , \tag{4.136}$$

using (4.109), (4.113) and (4.114). Since $A_{kj}(p'_k,p_j)$ is known to be a smooth function of p'_k and p_j with no singularities in a strip $|\mathrm{Im}\{p'_k\}|<\mu$, $|\mathrm{Im}\{p_j\}|<\mu$ (cf. Sec.4.8 and [4.13]), we may expand it in terms of some complete set of functions. Thus

$$A_{kj}(p'_k,p_j) = \sum_{n,m} A_{nm}(kj)\phi_n(p'_k)\psi_m(p_j) \quad ; \qquad A = \sum_{n,m} A_{nm}|\phi_n\rangle\langle\psi_m| \quad , \tag{4.137}$$

schematically. Specific choices of the sets $|\phi_n\rangle$, $\langle\psi_m|$ can be made at our convenience; one possible choice is $\langle p'_k|\phi_n\rangle=(p'^2_k+\mu_n^2)^{-1}$ for some sequence $\mu<\mu_1<\mu_2<\ldots$, as evidenced by the success of (4.127). Substitution of this form for A into (4.135) results in the equation

$$X = \left[1 + Z^{(1)} \sum_n |\phi_n\rangle\langle X_n|K^{(2)}\right]Z^{(1)}\Omega \quad , \tag{4.138}$$

where $Z^{(1)} = (1-K^{(1)})^{-1}$, and $\langle X_n|$ satisfies

$$\langle X_n| = \langle\tilde{\psi}_n| + \sum_m K_{nm}\langle X_m| \quad . \tag{4.139}$$

Here

$$\langle\tilde{\psi}_n| = \sum_m A_{nm}\langle\psi_m| \quad , \qquad K_{nm} = \langle\tilde{\psi}_n|K^{(2)}Z^{(1)}|\phi_m\rangle \quad . \tag{4.140}$$

One can thus tabulate $_k\langle p_k|Z^{(1)}|\phi_n\rangle$ and the set of complex numbers $\langle\psi_n|K^{(2)}Z^{(1)}|\phi_m\rangle$ by solving the minimal equation ($A\equiv 0$) in order to determine $Z^{(1)}$. With this information any choice of the real numbers A_{nm} specifies a value for X via the solution of the coupled *algebraic* equations stated in (4.139). That is, one need solve an integral equation only *once* for a given energy and a prescribed set of phase shifts; the complete set of off-shell variations can then be generated very rapidly by the procedure. The effectiveness of this technique has been demonstrated in an analysis of n-d elastic scattering and breakup at 14.4 MeV [4.18]. In particular, it was shown that all "off-shell" effects in this region are uniquely determined to the level of a few percent once the value of ^2a is specified, i.e., one must go to significantly higher energies in order to isolate independently sensitive parameters.

In concluding this chapter, we note several generalizations of the boundary condition method which are quite straightforward, but which significantly enhance the scope of the theory. The first concerns the relativistic three-body problem, and has already led to a variety of very interesting results. In this case the BCF possesses a great advantage over Faddeev-like theories in that only a description of the *asymptotic* state is required. This means that the particles may be taken on the mass-shell, thus avoiding the usual ambiguities in reducing from a four-vector to a three-vector representation. Except for the use of relativistic kinematics, the resultant equations are quite similar to the nonrelativistic equations; all of the subtleties associated with covariant dynamics are incorporated into the A operator. Despite its complexity, the latter is again a smooth function of its variables, and the techniques (and approximations) described above may be readily employed. The relativistic BCF has led to a marked improvement in the theoretical description of the three-pion system [4.15,19], and has also been applied to π-d scattering [4.20], and a covariant calculation of the N-N interaction [4.21].

Another interesting prospect involves the four-body scattering problem. In this case, for the description of the asymptotic (on-shell) state, the analoque of (4.42) requires a *two-variable* function $\overline{X}_{kj}(p_k,p_j)$ in order to describe the various ways in which the kinetic energy can be shared among the four particles. Application of the boundary conditions thus leads to a set of coupled two-variable integral equations from which the scattering amplitudes can be computed [4.22]. In contrast with four-body equations of the Faddeev type (which are *three dimensional*), the resultant equations are amenable to exact solution, and in fact are no more difficult numerically than current Faddeev calculations with local potentials. Finally, Coulomb effects in systems such as p-d (where only two of the particles are charged) can be treated exactly by building the asymptotic wave function out of Coulomb (Whittaker) functions instead of Bessel functions. While retaining the one-dimensional character of the (three-body) equation, this would permit finer comparisons of theory with the more accurate experimental data. Such extensions of the technique are currently under study.

References

4.1 L.M. Delves, A.C. Phillips: Rev. Mod. Phys. 41, 497 (1969)

4.2 S.A. Moszkowski, R.P. Haddock, W.T.H. van Oers, I. Slaus (eds): *Few Particle Problems in the Nuclear Interaction*, (North-Holland, Amsterdam 1972)

4.3 R.J. Slobodrian, B. Cujec, and K. Ramavataram, editors, *Few Body Problems in Nuclear and Particle Physics* (Les Presses de L'Université Laval, Quebec 1975)

4.4 W.M. Kloet, J.A. Tjon: Phys. Lett. 49B, 419 (1974)

4.5 D.D. Brayshaw: Phys. Rev. C7, 1731 (1973)

4.6 A.C. Phillips: Nucl. Phys. A107, 209 (1968)

4.7 D.D. Brayshaw: Phys. Rev. D8, 952 (1973)

4.8 P.M. Morse, H. Feshbach: *Methods of Theoretical Physics* (McGraw-Hill, New York 1953), pp.1573-1576

4.9 K.L. Kowalski: Phys. Rev. Lett. 15, 798 (1965)

4.10 H.P. Noyes: Phys. Rev. Lett. 15, 538 (1965)

4.11 D.D. Brayshaw: Phys. Rev. C3, 35 (1971)

4.12 D.D. Brayshaw: Phys. Rev. 167, 1505 (1968)

4.13 D.D. Brayshaw: Phys. Rev. C13, 1024 (1976)

4.14 D.D. Brayshaw: invited talk presented at the Intern. Conf. Few Body Problems in Nuclear and Particle Physics, Laval University, Quebec City, Canada (August 27-31, 1964, pp.28-43 of Ref.4.3)

4.15 D.D. Brayshaw: Phys. Rev. D11, 2583 (1975)

4.16 S.C. Pieper: Phys. Rev. C9, 883 (1974)

4.17 P. Doleschall: Nucl. Phys. A220, 491 (1974)

4.18 D.D. Brayshaw: Phys. Rev. Lett. 32, 382 (1974)

4.19 D.D. Brayshaw: Phys. Rev. Lett. 36, 73 (1976)

4.20 D.D. Brayshaw: Phys. Rev. C11, 1196 (1975)

4.21 D.D. Brayshaw, H.P. Noyes: Phys. Rev. Lett. 34, 1582 (1975)

4.22 D.D. Brayshaw: (unpublished)

5. A Relativistic Three-Body Theory

R. Aaron*

With 9 Figures

In the past decade it has become possible to solve certain three-body problems with strong interactions between pairs. In most cases, the essential trick has been to reduce the number of coordinates needed to specify intermediate states by the use of separable interactions; calculations of n-d scattering and breakup, and of d-α scattering using separable two-body potentials have been remarkably successful [5.1]. In this chapter we extend such methods to the relativistic domain, deriving and applying equations of the type first proposed by BLANKENBECLER and SUGAR [5.2], and by FREEDMAN et al. [5.3]. Because of difficulties in formulation and practical application we shall not discuss more general approaches to relativistic three-body equations such as that of ALESSANDRINI and OMNES [5.4]. In our relativistic theory we assume that the two-body interactions are dominated by a few bound states or resonances (isobars) and write down linear integral equations for the scattering amplitudes which include Lorentz invariance, two- and three-body unitarity and the cluster property. In essence, the result is an isobar model which incorporates unitarity and a significant amount of analyticity. Such a formalism may be applied, for example, to the πN and KN systems below ~1 GeV where single pion production is the dominant inelastic process and we are thus dealing mostly with three-body final states. Also, the two-body systems (πN, $\pi\pi$, and Kπ) are dominated by low-lying elastic resonances [Δ(1236), ρ, ε, K*] and hence the two-body subsystems entering the three-body calculation can be described by separable interactions. It is a fact that our equations emphasize unitarity at the expence of crossing symmetry, but at medium energies in the πN and KN systems one might expect unitarity to be the more important of the two principles since the relatively large mass of the nucleon ensures that crossing singularities are "far" away and hence (hopefully) not important. In Sections 5.1, 2, 4 and 5 we derive and discuss applications of these relativistic three-body equations.

In the nonrelativistic case the number of physical systems amenable to description by the three-body Schrödinger equation with two-body separable interactions is

* Supported in part by National Science Foundation grant MPS71-03134A04.

very limited. This is also the case in the relativistic situation. Here there is a further difficulty; besides the inherent limitations of description by separable interactions, the underlying three-body equations no longer describe nature exactly. Nevertheless, by judicious application, these three-body equations have been able to explain important features of several elementary particle systems. However, in view of their limited applicability, recent research has concentrated on extracting from the relativistic three-body equations general features based on unitarity and analyticity which must be included as part of any valid three-body phenomenology. In some sense the three-body equations developed in Section 5.1 represent the minimal formalism consistent with unitarity and analyticity, and in Section 5.4 we shall show how they may be used to develop a phenomenology which is (hopefully) better and more generally applicable than the usual isobar model.

5.1 Derivation of the Basic Equations - Spinless Case

There seem to be two avenues approaching dynamical calculations - off-shell and on-shell. The classic example of the former is the Schrödinger equation or, equivalently, the Lippmann-Schwinger equation in nonrelativistic quantum mechanics. Its most common relativistic manifestation is the Bethe-Salpeter equation. We shall present the Blankenbecler-Sugar [5.2] off-shell form, which we believe suits certain problems better. The on-shell approach is that of S-matrix theory; it has had some limited success in the two-body problem, but we know of no tractable method for including the dynamical effects of higher particle sectors in it. One feature of it, however, we do borrow. That is its emphasis on unitarity. For example, in discussing π-N elastic scattering and π production in π-N collisions in a domain where both are large, it is essential that the constraints of unitarity on these amplitudes be accurately imposed. This is, of course, precisely what the Schrödinger equation or Lippmann-Schwinger equation manages to do (in spite of being a linear equation). The Bethe-Selpeter equation in its usual truncated form does not do so well with the multi-particle states, and this is our main reason for replacing it with one more of the Schrödinger or Lippmann-Schwinger spirit. It turns out that this replacement also reduces the dimensionality of the equation, without violating Lorentz invariance, and hence makes the technical problems of obtaining numbers easier.

Our method for obtaining relativistic three-body equations follows closely the approach of BLANKENBECLER and SUGAR [5.2] and also that of FREEDMAN et al. [5.3] who construct linear integral equations describing the scattering of a particle from a two-body bound state or resonance by combining the isobar idea with two- and three-body unitarity. We begin by assuming that the two-body interactions are dominated by isobars (bound states or resonances), which assumption leads to separable two-body interactions. We then write down integral equations which describe particle-isobar reactions by analogy with procedures used in nonrelativistic problems. At this point the Born terms and propagators which enter the equations are not specified; only Lorentz invariance and the cluster property are imposed. We now make the "isobar assumption" which relates the production amplitudes back to those describing scattering from the isobars, and the imposition of two- and three-body unitarity gives the *discontinuities* of the Born terms and propagators in terms of the interaction parameters and mass-shell delta functions. Finally, the assumtion that these functions have no further discontinuities beyond those required by unitarity allows one to write dispersion relations which fully determine the three-body equations. The solutions are Lorentz invariant and satisfy two- and three-body unitarity at all energies. They do not have the same (and presumable correct) left-hand cut structure as do the Bethe-Salpeter amplitudes, but they do take better account of unitarity and particularly of multiparticle states, and this is an important feature for many problems.

It should be clear that the equations obtained by the above procedures are not unique; rather, they are the simplest which incorporate Lorentz invariance, two- and three-body unitarity, and the cluster property. As an added bonus, the integral equations are expressed in terms of a single *three*vector and after partial wave analyses become one-dimensional Fredholm equations which can be solved by standard methods on a computer.

5.1.1 Unitarity; Relativistic Two-Body Problem

Since unitarity is our strongest tool, we first review our notation for it. Using conventions of PILKUHN [5.6] we define the transition (T-) matrix in terms of the S-matrix by

$$S_{fi} = \delta_{fi} + (2\pi)^4 i\delta^4(P_f - P_i)T_{fi} \quad , \tag{5.1}$$

where the unitarity statement is

$$SS^{\dagger} = S^{\dagger}S = 1 \quad . \tag{5.2}$$

In order to follow the literature on relativistic equations we are using the normalization convention

$$\langle p|p'\rangle = (2\pi)^4\delta^4(p - p') \tag{5.3}$$

which differs from that used in the rest of this book. Thus, the n-body phase space element is given by

$$d\rho^{(n)} = (2\pi)^4\delta^4(P_f - \sum_{i=1}^{n} p_i) \prod_{i=1}^{n} (2\pi)^{-4}d^4q_i 2\pi\delta^+(q_i^2 - m_i^2) \quad , \tag{5.4}$$

where

$$\delta^+(q_i^2 - m_i^2) = \theta(q_{i0})\delta(q_i^2 - m_i^2) \quad . \tag{5.5}$$

Substituting (5.1), (5.4) and (5.5) into (5.2) we obtain our working expression of unitarity; schematically we write

$$T_{fi} - T_{fi}^{\dagger} = i\sum_n d\rho^{(n)} T_{fn}^{\dagger} T_{ni} = i\sum_n d\rho^{(n)} T_{fn} T_{ni}^{\dagger} \quad . \tag{5.6}$$

To demonstrate the Blankenbecler-Sugar techniques, we first apply them to a two-body equation of the Bethe-Salpeter type which we write in the form

$$T_{pq}(s) = V_{pq} + (2\pi)^{-4}\int d^4k V_{pk} G_k(s) T_{kq}(s) \quad , \tag{5.7}$$

where one usually writes

$$G_k(s) = [(k_1^2 - m_1^2)(k_2^2 - m_2^2)]^{-1} \quad , \tag{5.8}$$

but we take $G_k(s)$ arbitrary for the moment. The variables are chosen so that

$$k_1 + k_2 = P, \quad k_1 - k_2 = 2k, \quad P = (W,0,0,0) \quad , \tag{5.9}$$

and $s = W^2$ is the square of the total energy in the CM system. V_{pq} may be thought of as the usual ladder potential, but, in fact, any real symmetric V_{pq} will do. Using the fact that V_{pq} is real and symmetric and that $T_{pq}{}^*(s^+) = T_{qp}(s^-)$, we obtain from (5.7)

$$T_{pq}(s^+) - T_{pq}(s^-) = (2\pi)^{-4} \int d^4k T_{pk}(s^+)[G_k(s^+) - G_k(s^-)]T_{kq}(s^-) \quad . \tag{5.10}$$

A more detailed derivation of (5.10) is given in Appendix A. If we allow only two-body intermediate states in the unitarity relation, (5.6), we obtain a similar equation, i.e.,

$$T_{pq}(s^+) - T_{pq}(s^-) = (2\pi)^{-4} \int d^4k T_{pk}(s^+) \times [(2\pi)^2\delta^+(k_1^2 - m_1^2)\delta^+(k_2^2 - m_2^2)]T_{kq}(s^-) \quad . \tag{5.11}$$

From direct comparison of (5.10) and (5.11) with p on the energy shell, it is clear that the choice

$$G_k(s^+) - G_k(s^-) = i(2\pi)^2\delta^+(k_1^2 - m_1^2)\delta^+(k_2^2 - m_2^2) \tag{5.12}$$

will give an integral equation that satisfies two-body unitarity for all energies. To obtain a Green's function $G_k(s)$, we then write the dispersion relation

$$G_k(s) = \frac{1}{2\pi i} \int_{(m_1+m_2)^2}^{\infty} ds' \frac{\text{disc}\,\{G_k(s')\}}{(s' - s)} \quad , \tag{5.13}$$

where

$$\text{disc}\,\{G_k(s)\} = G_k(s^+) - G_k(s^-) \quad . \tag{5.14}$$

The integral of (5.13) is performed in the following manner: (5.5) and (5.9) yield

$$\delta^+(k_1{}^2 - m_1{}^2)\delta^+(k_2{}^2 - m_2{}^2) = (4\omega_1\omega_2)^{-1}\delta(k_0 + \frac{W}{2} - \omega_1)\delta(W - \omega_1 - \omega_2) \tag{5.15}$$

with

$$\omega_1 = (k^2 + m_1{}^2)^{1/2} \quad , \qquad \omega_2 = (k^2 + m_2{}^2)^{1/2} \quad . \tag{5.16}$$

Substituting (5.15) into (5.13) and performing the dispersion integral, we obtain the result

$$G_k(s) = \frac{\pi}{\omega_1\omega_2}\, \delta(k_0 - \tfrac{1}{2}\omega_1 + \tfrac{1}{2}\omega_2) \frac{\omega_1 + \omega_2}{(\omega_1 + \omega_2)^2 - s} \quad . \tag{5.17}$$

The remaining δ function in (5.17) allows one to evaluate one integral in (5.7) so that one ends up with a three-dimensional integral equation, but since the steps leading to it have been convariant, so is the equation. It need hardly be mentioned that the above prescription for obtaining $G_k(s)$ is not unique. In particular, further cuts can be added to $G_k(s)$. We have taken the simplest choice within the constraints of unitarity and Lorentz invariance. The final equation that we obtain by substituting (5.17) in (5.7) is

$$T_{pq}(s) = V_{pq} + (2\pi)^{-3} \int \frac{d^3k}{2\omega_1\omega_2} V_{pk} \frac{(\omega_1 + \omega_2)}{(\omega_1 + \omega_2)^2 - s} T_{kq} \quad . \tag{5.18}$$

This equation may be thought of as the relativistic analog of the Lippmann-Schwinger equation. It has many faults. Some of them, like not treating crossing correctly, it shares with the two-body Bethe-Salpeter equations; others, like improper treatment of the left-hand cuts, are special to it. Its main forte is that it does not take into account only parts of the multiparticle states, and hence does not violate unitarity. This seems an obvious advantage in the scattering region. Even below a multiparticle threshold the inconsistent treatment of virtual states by the Bethe-Salpeter equation may lead to incorrect results. Moreover, (5.18) has the technical advantage of being only a three-dimensional equation, and one with rather good convergence properties at that.

5.1.2 Relativistic Three-Body Equations: Spinless Particles

We now use techniques similar to those described in Subsection 5.1.1 to obtain a set of relativistic three-body equations. We consider, for simplicity, the case of three identical spinless particles and consider bound state scattering, that is, the elastic scattering of one particle from a bound state of the other two. Just as in the two-body case, we start by assuming a form for the equation, which we take as

$$\langle p|T(s)|q\rangle = \langle p|B(s)|q\rangle + (2\pi)^{-4} \int d^4k \langle p|B(s)|k\rangle \tau(\sigma_k) \langle k|T(s)|q\rangle \quad , \qquad (5.19)$$

with

$$\sigma_k = (P - k)^2 \quad . \qquad (5.20)$$

Eq.(5.19) is a straightforward generalization of the nonrelativistic three-body equation originally proposed by AMADO [5.7]. Its essential ingredient is the assumption that the two-body interaction proceeds via a quasi-particle, or separable interaction, or, in the more usual relativistic language, via an isobar. The function $\tau(\sigma_k)$ is the propagator of that isobar. The external bound states are also that isobar, so that the Born term B is just particle exchange between the external isobars. Eq. (5.19) may be represented diagrammatically as shown in Fig. 5.1, where

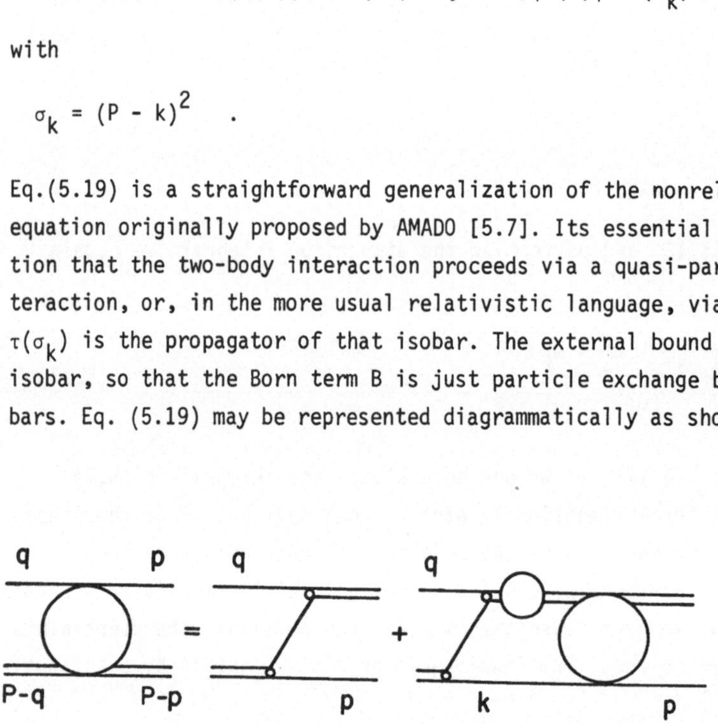

Fig.5.1. Diagrammatic representation of (5.19)

the variables are also defined. For a systematic derivation of the nonrelativistic equivalent of (5.19) in the context of formal scattering theory we refer the reader to Section 1.8. To use unitarity, we must also define the production (two-body → three-body amplitude $\langle q_1q_2q_3|T(s)|q\rangle$; we choose the form

$$\langle q_1q_2q_3|T(s)|q\rangle = \frac{1}{\sqrt{3!}} \sum_{n=1}^{3} \langle q_n|T(s)|q\rangle S(\sigma_{q_n}) v(p_n{}^2) \quad , \qquad (5.21)$$

where, for example, $p_1{}^2 = (q_2-q_3)^2$, where v is the vertex for quasi-particle dissociation, and where S is a propagator function whose relation to τ will be determined. The form of (5.21) is the same as that of (1.161) and the normalization factor $(3!)^{-1/2}$ is chosen so that in the nonrelativistic limit our conventions are consistent with the second quantized formalism used by AMADO [5.1]. Such a formalism is a useful guide for treating problems associated with identical particles, spin and statistics, etc.

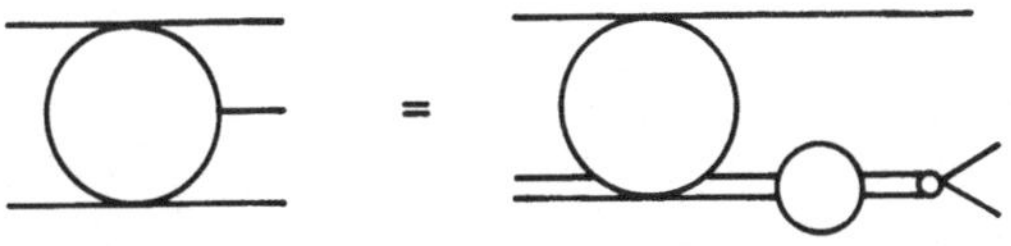

Fig.5.2. Diagrammatic representation of (5.21)

The diagrammatic interpretation of (5.21) is given in Fig.5.2. With p and q on the bound state energy shell, by methods very similar to those needed to get (5.10), we show in Appendix A that for equations of the form of (5.19), the discontinuity of T satisfies the relation

$$\begin{aligned} \langle p|T(s^+)|q\rangle - \langle p|T(s^-)|q\rangle &= (2\pi)^{-4} \int d^4k \langle p|T(s^+)|k\rangle [\tau(\sigma_k{}^+) - \tau(\sigma_k{}^-)] \\ &\times \langle k|T(s^-)|q\rangle + (2\pi)^{-8} \int d^4k \int d^4k' \langle p|T(s^+)|k\rangle \\ &\times \tau(\sigma_k{}^+)[\langle k|B(s^+)|k'\rangle - \langle k|B(s^-)|k'\rangle]\tau(\sigma_{k'}{}^-)\langle k'|T(s^-)|q\rangle \quad . \end{aligned} \tag{5.22}$$

This equation is represented diagrammatically in Fig.5.3. To obtain an equation that satisfies two- and three-body unitarity, we compare (5.22) with (5.6), using (5.21) for the break-up amplitude. This *isobar ansatz* for the production is pivotal

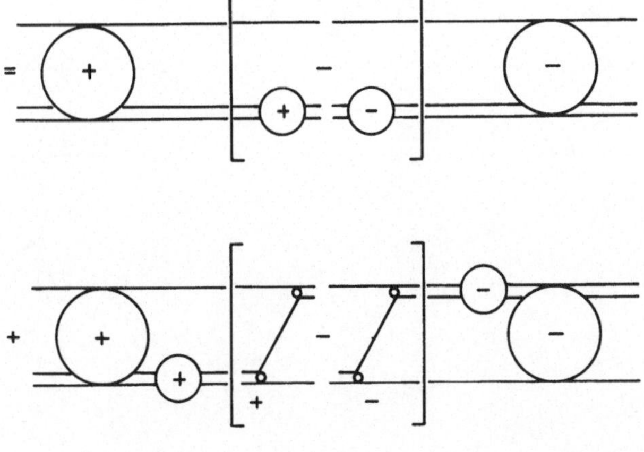

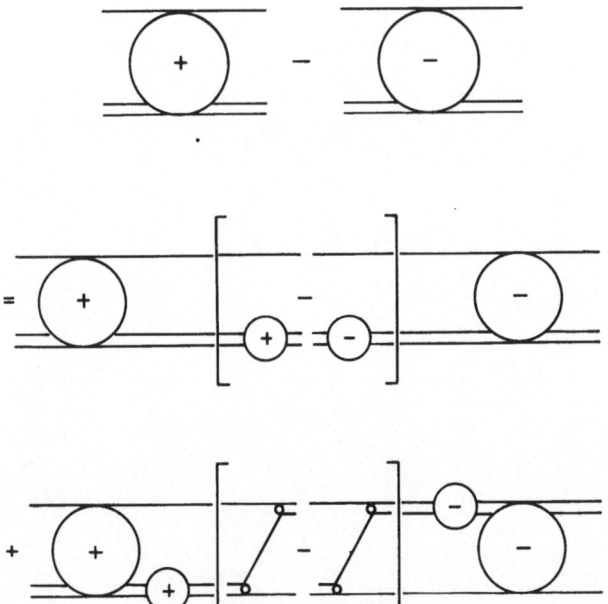

Fig.5.3. Diagrammatic representation of (5.22)

to the analysis, since it relates the 2→3 amplitude back to the 2→2 and gives a closed set of equations. The unitarity relation gives

$$\begin{aligned}
&\langle p|T(s^+)|q\rangle - \langle p|T(s^-)|q\rangle = \frac{i}{(2\pi)^4}\int d^4k\langle p|T(s^+)|k\rangle\langle k|T(s^-)|q\rangle(2\pi)^2 \\
&\quad\times \delta^+(\sigma_k - \mu^2)\delta^+(k^2 - m^2) + \frac{i}{(2\pi)^5}\int\delta^4(P - k_1 - k_2 - k_3) \\
&\quad\times \delta^+(k_1{}^2 - m^2)\delta^+(k_2{}^2 - m^2)\delta^+(k_3{}^2 - m^2)d^4k_1d^4k_2d^4k_3 \\
&\quad\times \frac{1}{3!}\sum_{n,m=1}^{3}\langle p|T(s^+)|k_n\rangle S(\sigma_{k_n}{}^+)v(p_n{}^2)v(p_m{}^2)S(\sigma_{k_m}{}^-)\langle k_m|T(s^-)|q\rangle \quad .
\end{aligned} \tag{5.23}$$

The first term is the contribution to unitarity of the elastic bound state scattering, μ is the bound state mass, and m is the mass of one of the identical particles. From (5.22) the first term will clearly contribute to the discontinuity of τ. The second term comes from the break-up. It makes two kinds of contributions - those for which $m = n$ will contribute to the discontinuity of τ; they come from cutting the propagator "bubble", and those for which $m \neq n$ involve the exchange of a particle between the bound states and contribute to the discontinuity of B. Comparison of (5.23) and (5.22) gives for the discontinuities

$$\begin{aligned}
&\tau(\sigma_k{}^+)[\langle k|B(s^+)|k'\rangle - \langle k|B(s^-)|k'\rangle]\tau(\sigma_{k'}{}^-) \\
&= iv[(P - k - 2k')^2]S(\sigma_k{}^+)(2\pi)^3\delta^+(k^2 - m^2)\delta^+(k'^2 - m^2) \\
&\quad\times \delta^+[(P - k - k')^2 - m^2]S(\sigma_{k'}{}^-)v[(P - 2k - k')^2]
\end{aligned} \tag{5.24}$$

and

$$\begin{aligned}
&\tau(\sigma_k{}^+) - \tau(\sigma_k{}^-) = i(2\pi)^2\delta^+(k^2 - m^2)\delta^+(\sigma_k - \mu^2) - i\,\frac{\delta^+(k^2 - m^2)}{2(2\pi)^3} \\
&\quad\times S(\sigma_k{}^+)S(\sigma_k{}^-)\int d^4p_{12}v^2(4p_{12}{}^2)(2\pi)^2\delta^+(p_1{}^2 - m^2)\delta^+(p_2{}^2 - m^2) \quad ,
\end{aligned} \tag{5.25}$$

where

$$P = k + p_1 + p_2 \quad , \qquad p_{12} = \tfrac{1}{2}(p_1 - p_2) \quad . \tag{5.26}$$

If we choose

$$\tau(\sigma_k) = 2\pi\delta^+(k^2 - m^2)S(\sigma_k) \quad , \tag{5.27}$$

(5.24) and (5.25) yield the following formulas for the discontinuities of B and S:

$$\begin{aligned} \langle k|B(s^+)|k'\rangle - \langle k|B(s^-)|k'\rangle &= iv[(P - k - 2k')^2] \\ &\times 2\pi\delta^+[(P - k - k')^2 - m^2]v[(P - 2k - k')^2] \end{aligned} \tag{5.28}$$

subject to the constraint

$$k^2 = m^2 \quad , \qquad k'^2 = m^2 \quad , \tag{5.29}$$

which comes from the δ functions we have factored off, and

$$\begin{aligned} S(\sigma_k^+) - S(\sigma_k^-) &= 2\pi i\delta^+(\sigma_k - \mu^2) + \frac{S(\sigma_k^+)S(\sigma_k^-)}{2(2\pi)^4}\, i\int d^4p_{12}v^2(4p_{12}^2) \\ &\times (2\pi)^2\delta^+(p_1^2 - m^2)\delta^+(p_2^2 - m^2) \quad . \end{aligned} \tag{5.30}$$

Note that (5.28) is not a unique expression for Disc $\{k|B(s)|k'\}$ since we may add any function which vanishes when $k^2 = m^2$ or $k^2 = m'^2$. In the absence of motivation to the contrary we have made the simplest choice which introduces no s-singularities in addition to those required by unitarity. (A similar problem in a different context has been discussed by AARON and AMADO [5.8,9]). There are ambiguities related to the possible ways of choosing the energies for which the discontinuities are taken [5.2]. We have already made these choices in (5.19). It is clear that all three particles contribute equally to the discontinuity of B; hence its discontinuity is in the total-energy variables s, whereas the discontinuity of τ or S comes from the pair interacting in τ. Hence S depends on $\sigma_k = (P-k)^2$. By making this choice, we assure ourselves the correct cluster-decomposition properties for the intermediate states.

We now wish to obtain B and S from their discontinuities, for the moment assuming no cuts in addition to those required by unitarity. To obtain B from (5.28) we first note that in the three-body CM system,

$$\begin{aligned} \delta^+[(P - k - k')^2 - m^2] &= (2\omega_{k+k'})^{-1}\delta(W - \omega_k - \omega_{k'} - \omega_{k+k'}) \quad , \\ \omega_k &= (\underline{k}^2 + m^2)^{1/2} \quad , \end{aligned} \tag{5.31}$$

and then write a dispersion relation in $s = W^2$ (as was done in (5.13)) assuming no cut contribution from v. We obtain (in the three-body CM system)

$$\begin{aligned} \langle k'|B(s)|k\rangle &= \frac{v(t_{k'k}^2)(\omega_k + \omega_{k'} + \omega_{k+k'})v(t_{kk'}^2)}{\omega_{k+k'}[(\omega_k + \omega_{k'} + \omega_{k+k'})^2 - s]} \\ t_{kk'} &= (\omega_k - \omega_{k+k'})^2 - (\underline{k} - 2\underline{k})^2 \quad . \end{aligned} \tag{5.32}$$

In the expression for Disc $\{B\}$ (5.28), the δ-functions in the expression for three-body phase space ensure that energy and momentum are conserved and that all three intermediate particles are on their mass shells. We then put (5.31) into a dispersion integral and integrated over s to obtain (5.32). In general, in the latter expression $W \neq \omega_k + \omega_{k'} + \omega_{k+k'}$ and thus energy is not conserved at vertices in the Born term, and, furthermore, the exchanged particle is not on its mass shell. However, the external legs k and k' are still on their mass shells, even in intermediate states. In the corresponding Feynman diagram energy and momentum are conserved at each vertex and thus, in general, *neither* the exchanged particle *nor* the external legs k and k' will be on their mass shells. We speak of our theory as being off-energy-shell and of a Feynman diagram theory as being off-mass-shell.

Let us now use the expression for Disc $\{S\}$ to obtain the propagator S. The inverse of S is obtained more easily than S itself, so we write

$$S(\sigma_k) = - D^{-1}(\sigma_k) \tag{5.33}$$

and substitute in (5.30), obtaining

$$\begin{aligned} 2 \text{ Im } \{D(\sigma)\} = 2\pi\delta^+(\sigma_k - \mu^2) + [2(2\pi)^2]^{-1} \int d^4p_{12} v^2(4p_{12}^2) \\ \times \delta^+(p_1^2 - m_1^2)\delta^+(p_2^2 - m_2^2) \quad . \end{aligned} \tag{5.34}$$

We may evaluate the above integral in the 1-2 CM system using (5.15):

$$\int d^4p_{12} v^2(4p_{12}^2)\delta^+(p_1^2 - m_1^2)\delta^+(p_2^2 - m_2^2) = \pi(\sigma_k)^{-1/2} k_{12} v^2(4k_{12}^2) \quad , \tag{5.35}$$

where k_{12} is the relative momentum of particles 1 and 2 in their own CM system with total CM energy σ_k, i.e., for equal masses, $m_1 = m_2 = m$

$$k_{12}^2 = \sigma_k/4 - m^2 \quad . \tag{5.36}$$

We finally obtain

$$2 \text{ Im } \{D(\sigma_k)\} = 2\pi\delta^+(\sigma_k - \mu^2) + [8\pi^2]^{-1} k_{12}(\sigma)^{-1/2} v^2(4k_{12}^2) \quad . \tag{5.37}$$

In the absence of a bound state (i.e., if the isobar is unstable) one can directly verify that *a* solution of (5.37) is

$$D(\sigma_k) = \sigma_k - M_0^2 - \frac{1}{2(2\pi)^3} \int \frac{d^3p v^2}{\omega_p(\sigma_k - \sigma_p)} \quad , \tag{5.38}$$

where the constant M_0 may be interpreted as a bare mass. If there is a bound state at $\sigma = \mu^2$ it is convenient to subtract twice, first to explicitly display the zero at $\sigma_k = \mu^2$, and second, to ensure unit residue for S. Thus we may write

$$D(\sigma_k) = (\sigma_k - \mu^2)\left[1 - \frac{\sigma - \mu^2}{2(2\pi)^3} \int \frac{d^3p v^2}{\omega_p(\sigma_k - \sigma_p)(\sigma_p - \mu^2)^2}\right] \quad . \tag{5.39}$$

When there is a two-body bound state at $\sigma_k = \mu^2$, $\mathrm{Im}\{[D(\sigma_k)]^{-1}\}$ has a discrete piece $\pi\delta^+(\sigma_k - \mu^2)$ in addition to the continuum contribution from the integral; therefore, (5.39) is a solution of (5.34). If one compares (5.38) or (5.39) with (5.18), because $S(\sigma)$ is essentially a two-body scattering amplitude, there seems to be an "extra" factor of 1/2 multiplying the integral. However, we obviously could renormalize by letting $v \to \sqrt{2}v$ so the factor depends on the normalization chosen in (5.21). Our results are consistent with the nonrelativistic second quantized approach (see Ref.5.7, Eq.(14)) and ensure that the Born term has residue $v^2/2\omega_{k+k'}$ in conformity with the usual Feynman rules (see (5.79)). Finally, from a physical point of view, (5.38) is a reasonable solution for a propagator (i.e., it resembles propagators normally encountered in elementary particle physics) but is not unique. Obviously we may add to (5.38) a function of σ with arbitrary left-hand cut structure (not required by unitarity). Similarly B in (5.32) may contain additional terms with left-hand cuts in s. In applications we never tamper with $D(\sigma)$, (5.38), but the freedom to add terms to B which do not affect unitarity is an extremely useful one, since it enables us to include phenomenologically particle exchanges normally neglected by the three-body formalism.

We get our working integral equation by substituting (5.33) into (5.27), and (5.27) and (5.32) into (5.19), and after making the replacement $T(s) \to -T(s)$ so that in the non-relativistic limit our T is the same as that normally defined from Schrödinger equation, we finally obtain

$$\begin{aligned} \langle p|T(s)|q\rangle = {} & \frac{v(\omega_p + \omega_q + \omega_{p+q})v}{\omega_{p+p}[s - (\omega_p + \omega_q + \omega_{p+q})^2]} \\ & + \frac{1}{(2\pi)^3} \int \frac{d^3k}{2\omega_k} \frac{v(\omega_p + \omega_k + \omega_{p+k})v}{\omega_{p+k}[s - (\omega_p + \omega_k + \omega_{p+k})^2]} \frac{\langle k|T(s)|q\rangle}{D(\sigma_k)} \quad , \end{aligned} \tag{5.40}$$

where we have suppressed the arguments of the vertex function v, but they may be determined from (5.19) and (5.32). Eq. (5.40) ends our quest. It is a linear, three-dimensional, Lorentz-invariant integral equation for the elastic scattering of one particle from the bound state of two. Its solutions are constructed to satisfy two- and three-body unitarity and to have no higher particle contributions at all. Furthermore, from a knowledge of the solution of (5.40) one can construct the production or break-up amplitude by using (5.21). No new equations need be solved. It is clear that, just as in the nonrelativistic case, if one wishes to put in more bound states, separable interactions, or isobars, one will just get a coupled set of such equations. As discussed before (5.38), unstable quasi-particles are handled on the same footing as stable ones. In subsequent sections, we shall show how to include spin and fermions, but effectively this will only change numerators. The denominators, which are the real seat of unitarity, will remain as in (5.40). It should finally be noted that the usual nonrelativistic three-body scattering equations with separable potentials can be derived using the methods of this section. The crucial assumption is the isobar ansatz which introduces the separable potential "v" via (5.21).

5.2 Spin and Isospin

Internal quantum numbers can be included in the three-body formalism derived in the previous section. In Subsection 5.2.1 we discuss isospin which can be taken into account trivially, and in Subsections 5.2.2, 3 and 4 we consider those spin complications which have arisen in recent applications of the three-body integral equations. In particular, we discuss the case of integral spin, as well as spins 1/2 and 3/2. For these cases we derive a special formalism which greatliy simplifies the angular momentum decompositions. Important angular momentum decompositions for the πN problem are given in Appendix B.

5.2.1 Isospin

The effects of isospin are taken into account in (5.16) by including in the decay vertex a Clebsch-Gordan coefficient $\langle i_1 i_2 \tau_1 \tau_2 | i'\tau' \rangle$ which describes an isobar with isospin i' and z component τ' decaying into particles of isospin and z components i_1, τ_1 and i_2, τ_2. A Born term describing the process $i_1 + i'' \to i_3 + i'$ exchanging a particle with isospin i_2, and initial and final states having total isospin I and z component I_z will now be of the form derived in (5.26), but multiplied by a recoupling coefficient

$$\begin{aligned} x_I &= \sum_{\tau_1, \tau_2} \langle i'' i_1 \tau'' \tau_1 | I I_z \rangle \langle i_1 i_2 \tau_1 \tau_2 | i'\tau' \rangle \langle i_3 i_2 \tau_3 \tau_2 | i''\tau'' \rangle \langle i' i_3 \tau' \tau_3 | I I_z \rangle \\ &= (-1)^{i_1+i_2+i_3-I} [(2i'+1)(2i''+1)]^{1/2} W(i_1 i_2 I i_3; i'i'') \quad , \end{aligned} \tag{5.41}$$

where $W(i_1 i_2 I i_3; i'i'')$ is a Racah coefficient (see, e.g., Ref.5.10 p.109, Eq.6.6a). In the two outer Clebsch-Gordan coefficients in (5.41) we have adopted the convention that the bound state isospin appears first. To be consistent we must use the same convention in the angular momentum decomposition when we add channel spin and orbital angular momentum. In our πN calculations described in App.B we adopt the convention that in positions one and two in a Clebsch-Gordan coefficient, a baryon always appears before a meson.

5.2.2 Integral Spin

In the nonrelativistic three-body problem the inclusion of spin is a considerable complication; on the other hand, the treatment is straightforward and requires no more background than a first year graduate course in quantum mechanics. In the relativistic domain the situation is seemingly more complicated since one must now consider the behavior of spin wave functions under Lorentz transformations. In this section we shall demonstrate a formalism which reduces the relativistic problem to one in which the procedures followed are identical to those in the nonrelativistic case. We begin by considering a nonrelativistic separable two-body potential which has the form

$$\begin{aligned} \langle \underline{k}|V|\underline{q}\rangle &= v(k^2)v(q^2)\underline{k}\cdot\underline{q} \\ &= \sum_m k_m v(k^2)v(q^2)q_m^* \quad , \end{aligned} \tag{5.42}$$

where k_m is a spherical component of the relative momentum $\underline{k}$ (see, e.g., Ref.5.10, pp.7 and 63)

$$k_m = \sqrt{\frac{4\pi}{3}}\,|\underline{k}|Y_{1m}(\hat{k}) \quad . \tag{5.43}$$

The quantity $k_m v(k^2)$ is interpreted as the vertex function for a particle of spin 1 and z component m decaying into two spinless particles. Since the relative momentum is a Galilean invariant, this interpretation is valid in all frames, and the nonrelativistic version of (5.21) for a P-wave isobar becomes ($\hbar = 1$)

$$\langle \underline{q}_1\underline{q}_2\underline{q}_3|T(E)|\underline{q},\mu\rangle = \frac{1}{\sqrt{3!}}\sum_{\substack{n=1\\ \mu'}}^{3} \langle \underline{q}_n,\mu'|T(E)|\underline{q},\mu\rangle S\left(E - \frac{3}{4}\frac{q_n^2}{m}\right)v(\underline{p}_n^2)(p_n)_{\mu'} \quad . \tag{5.44}$$

Turning to the relativistic situation, one must now think in terms of Lorentz invariance rather than Galilean invariance. Consider two particles with four momenta k_1 and k_2, respectively; the four vector

$$\begin{aligned} P &= \frac{(k_2\cdot K)k_1 - (k_1\cdot K)k_2}{K^2} \\ &= k - \frac{k\cdot K}{K^2}K \end{aligned} \tag{5.45}$$

with

$$K = k_1 + k_2$$
$$2k = k_1 - k_2 \quad (5.46)$$

clearly reduces to the relative three-momentum in the CM momentum in that frame. Therefore, $\underline{k}\cdot\underline{k}'$ evaluated in the two-body CM system can be replaced by the Lorentz scalar $-P\cdot P'$ in all frames. This form is not very convenient for performing a partial wave decomposition, but AARON et al. [5.5] show that $-P\cdot P'$ can be written as a three-dimensional dot product in *all* frames:

$$- P \cdot P' = \underline{M} \cdot \underline{M}' \quad , \quad (5.47)$$

where the *special vector* $\underline{M}$ is

$$\underline{M} = \underline{P} - (\underline{P} \cdot \underline{K})\underline{K}/[K_0 + W)]$$
$$K_0 = (k_1)_0 + (k_2)_0 \quad (5.48)$$
$$W^2 = K^2 - K_0^{\,2} \quad .$$

The relativistic analogue of (5.42) is a Born term corresponding to s-channel exchange of a P-wave isobar. With k_1, k_2 and k related by (5.46) we have

$$\langle k|B(s)|q\rangle = \frac{v(k^2)v(q^2)}{s - (k_1 + k_2)^2}\, \underline{M}_k \cdot \underline{M}_q \quad (5.49)$$

which reduces to the result obtained by standard Feynman rules for vector meson exchange between scalar particles when $(k_1+k_2)^2 = m_v^{\,2}$ (m_v = mass of stable vector meson) and the form factors are replaced by coupling constants. Note that unlike the Feynman term, (5.49) describes pure P-wave scattering whether or not the vector meson is on its mass shell. The quantity

$$v(k^2)(M_k)_m = v(k^2)\sqrt{\frac{4\pi}{3}}\, |\underline{M}_k| Y_{1m}(\hat{M}_k) \quad (5.50)$$

is now identified as the relativistic vertex function. As we shall show below, the *special vector* formalism allows one to write down a vertex function in any Lorentz frame in terms of the *three* vectors entering the vertex and immediately proceed to a standard nonrelativistic style angular momentum decomposition. Ref.5.5 shows how the above formalism can be extended to arbitrary integral spin; for example, to construct the vertex $\langle q|\Gamma|K,\ell m\rangle$ for two spinless particles of relative four momentum q and total momentum K forming an isobar of spin ℓ, with z component m, viz.,

$$<q|\Gamma|K,\ell m> = V_\ell(q^2)Y_{\ell m}{}^*(\hat{M}_q) \quad , \tag{5.51}$$

where V_ℓ has threshold behavior q^ℓ. If the quantization axis is taken along $\underline{K}$, then m is the helicity.

We shall now show the relation of the vertex function of (5.50) to the usual Feynman vertex function. Consider Fig.5.4 which describes two pions (the pion with

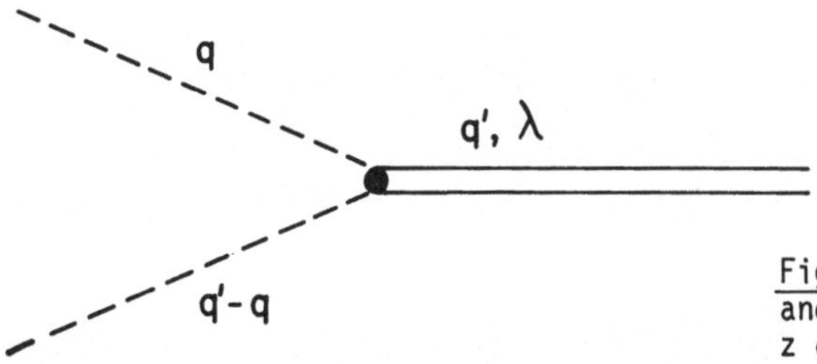

Fig.5.4. Two pions with four momenta q and q'-q combine to form stable ρ with z component of spin λ.

momentum q is on its mass shell) combining to form a *stable* ρ-meson with helicity λ in the final state. The Feynman helicity wave function for the vertex is

$$\varepsilon_\mu(q',\lambda)q^\mu \quad , \tag{5.52}$$

where ε is a polarization vector defined, for example, in Ref.5.11, p.462. According to the previous discussion our corresponding spin wave function is a component of the special vector, i.e., M_λ^* (we have complex conjugated because we are dealing with a final state). In order to compare the special vector with (5.52) we take the z axis along the direction $\underline{p}'$; furthermore, we choose our coordinate system so that $\underline{q}'$ is characterized by spherical angles (0,0) and $\underline{q}$ by spherical angles (θ,0). Using (5.48) for the special vector and remembering that we are dealing with a Feynman diagram so that energy and momentum are conserved, e.g.,

$$\begin{aligned} W &= m_\rho \\ \omega_\rho^2 - m_\rho^2 &= \underline{q}'^2 \end{aligned} \tag{5.53}$$

we obtain

$$\underline{M} = \underline{q} + \underline{q}'\{\underline{q}\cdot\underline{q}'/[m_\rho(\omega_\rho + m_\rho)] - \omega_\rho/m_\rho\} \quad . \tag{5.54}$$

We now evaluate M_1^*, M_{-1}^* and M_0^* using the relation $M_\lambda^* = (-1)^\lambda M_\lambda$ and obtain

$$M_1{}^* = \sin\theta$$

$$M_0{}^* = \omega_\rho q \cos\theta/m_\rho - \omega_q q'/m_\rho \tag{5.55}$$

$$M_{-1}^* = -\sin\theta \quad .$$

By comparing (5.55) with Ref.5.11, Eq.(27.26) we see that

$$M_\lambda{}^* = -\varepsilon^\mu(q',\lambda)q_\mu \quad . \tag{5.56}$$

The above formalism has immediate application to the 3π system where one might wish, for example, the Born term describing the quasi-two-body reaction $\pi+\rho(\lambda)\to\pi+\rho(\mu)$ where λ and μ are the z components of the ρ spins in the initial and final states respectively. Omitting isospin factors, it has a form very similar to (5.32), i.e., in the three-body CM system, following the methods of Sec.5.1.2 we would first obtain

$$<\underline{p}\mu|B(s^+)|\underline{q}\lambda> - <\underline{p}\mu|B(s^-)|\underline{q}\lambda> = \lambda M_\mu{}'^* v[2\pi\delta^+((P-p-q)^2 - m^2)]vM_\lambda \tag{5.57a}$$

and then after dispersing

$$<\underline{p}\mu|B(s)|\underline{q}\lambda> = M'_\mu{}^* \frac{v(\omega_p + \omega_q + \omega_{p+q})v}{\omega_{p+q}[(\omega_p + \omega_q + \omega_{p+q})^2 - s]} M_\lambda \quad , \tag{5.57b}$$

where $\underline{M}$ and $\underline{M}'$ are the special vectors associated with the initial and final vertices and are linear combinations of $\underline{p}$ and $\underline{q}$. The mass shell δ-function in Eq.(5.57a) permits us to put the four-momentum of the exchanged particle on its mass shell in the special vector. For reasons discussed below Eq.(5.39) and also in Sec.5.3 this choice is not unique; however, it is the simplest choice consitent with unitarity and total energy analyticity, and we thus use it in all our calculations. We now write

$$<\underline{p}\mu|B(s)|\underline{q}\lambda> \equiv a_{pp}p_\lambda p_\mu{}^* + a_{pq}p_\lambda q_\mu{}^* + a_{qp}q_\lambda p_\mu{}^* + a_{qq}q_\lambda q_\mu{}^* \quad , \tag{5.57c}$$

where a_{pp}, a_{pq}, etc., are functions of $\underline{p}^2$, $\underline{q}^2$ and $\underline{p}\cdot\underline{q}$. One now uses standard non-relativistic techniques to perform an angular momentum decomposition:

$$<pJM\ell'|B(s)|qJM\ell> = \sum_{\substack{\mu\lambda\\ mm'}} \int d\Omega_{\hat{p}} \int d\Omega_{\hat{q}} Y_{\ell'm'}(\hat{p})<\ell'1m'\mu|JM> \times <\underline{p}\mu|B(s)|\underline{q}\lambda> Y_{\ell m}{}^*(\hat{q})<\ell 1 m\lambda|JM> \tag{5.58}$$

and the final result after intergrating over solid angles and summing over Clebsch-Gordan coefficients is

$$\langle pJM\ell'|B(s)|qJM\ell\rangle = [(2\ell+1)(2\ell'+1)]^{1/2}\{\langle \ell 100|J0\rangle\langle \ell' 100|J0\rangle$$

$$\times (2J+1)^{-1}[p^2(a_{pp})_\ell + q^2(a_{qq})_{\ell'} + pq(a_{pq})_J] \qquad (5.59)$$

$$+ qp(-1)^{J-\ell-1} \sum \langle \ell 100|\rho 0\rangle\langle \ell' 100|\rho 0\rangle (a_{qp})_\rho W(1\rho J1;\ell\ell')\} \quad ,$$

where we have, for example, expanded

$$a_{qq}(\underline{p}^2,\underline{q}^2,\underline{p}\cdot\underline{q}) = \sum_{\rho\varepsilon} (a_{qq})_\rho Y_{\rho\varepsilon}{}^*(\hat{p})Y_{\rho\varepsilon}(\hat{q})$$
$$= \sum_{\rho\varepsilon} (a_{qq})_\rho Y_{\rho\varepsilon}(\hat{p})Y_{\rho\varepsilon}{}^*(\hat{q}) \qquad (5.60)$$

and thus

$$(a_{qq})_\rho = 2\pi \int_{-1}^{1} dz P_\ell(z) a_{qq} \quad . \qquad (5.61)$$

In terms of the partial wave projections (5.40) can now be written schematically as the following set of coupled equations [5.12]:

$$\langle pJM\ell'|T(s)|qJM\ell\rangle = \langle pJM\ell'|B(s)|qJM\ell\rangle$$

$$+ \sum_{\ell''} \int \frac{d^3k}{(2\pi)^3} \frac{\langle pJM\ell'|B(s)|kJM\ell''\rangle\langle kJM\ell''|B(s)|qJM\ell\rangle}{2\omega_k D(\sigma_k)} \quad . \qquad (5.62)$$

5.2.3 Spin-1/2 and the Pion-Nucleon Interaction

To consider the πN problem we must include spin-1/2 particles in the relativistic formalism previously developed. There is no essential difference from the spinless or integer-spin discussion. The prescription is to rationalize all Feynman denominators appearing and use the Blankenbecler-Sugar (BS) procedure discussed earlier on the remaining scalar propagators. The Dirac matrices appearing in the numerators can be absorbed into the form factors and the Dirac algebra can be done independently of the BS procedure. In the following material, only the pseudo-scalar coupling of pions to nucleons shall be studied. To obtain the $\pi N \rightleftarrows N$ vertices, consider the Feynman diagramm of Fig.5.5a corresponding to the amplitude

$$b_{r,r'}(kp;k'p') = \bar{u}_r(p)\gamma_5 i(\not{K} - M)^{-1}\gamma_5 u_{r'}(p') \quad , \qquad (5.63)$$

where

$$K = k + p = k' + p' \qquad (5.64)$$

and r and r' are spinor indices. In the Blankenbecler-Sugar procedure we choose to define the spin wave function from the expression for Im {B}, (5.28); B itself is

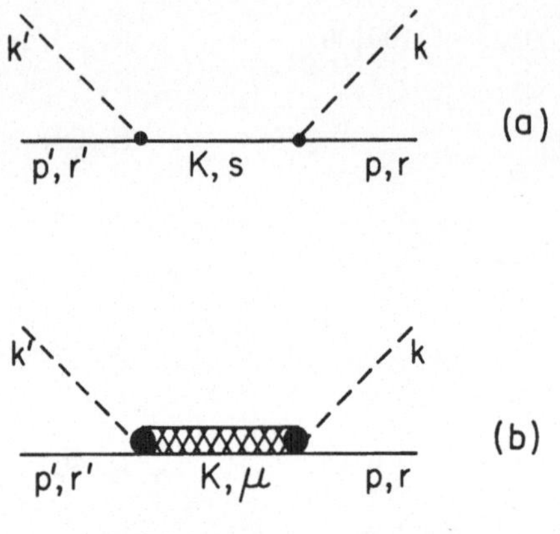

Fig.5.5. s-channel Born terms in πN scattering. Nucleon is represented by solid line, pion by dashed line and Δ by cross-hatched line. r, r', s and μ refer to z components of spin

then obtained from a dispersion integral (e.g., (5.13)). We thus have an operational procedure for obtaining spin wave functions with the proper Lorentz transformation properties, and the mass shell δ-functions which appear in Im {B} enable us to side-step the question of off-shell spinors. We now take (5.63), replace $(\not{K}-M)^{-1}$ by $2\pi\delta^{+}(K^2-M^2)$, insert a complete set of spinors of three-momentum $\underline{K}$ and mass M and obtain

$$\mathrm{Im}\{b_{r,r'}(kp;k'p')\} = 2Mi(2\pi)\delta^{+}(K^2-M^2) \sum_t \bar{u}_r(p)\gamma_5 u_t(K)\bar{u}_t(K)\gamma_5 u_{r'}(p') \quad , \qquad (5.65)$$

where only positive energy spinors contribute to the sum because $\not{K}+M$ projects onto positive energy states. In the two-body CM system the amplitude defined above has only a $P_{1/2}$ projection, and thus it is the "separable potential" for the π-N system. As for the case of integer spin in the previous subsection, we now read off from it the rules for π-N scattering in the formalism

a) For a nucleon propagator write

$$2Mi/(K^2 - M^2) \quad , \qquad (5.66)$$

where K is the nucleon four momentum and M is the nucleon mass. Since $(K^2-M^2)^{-1}$ is the usual propagator for scalar particles, the work of Subsection 5.1.2 is changed for the propagator only by the appending of 2Mi to every intermediate nucleon line.

b) At a vertex, shown schematically in Fig.5.5a, write

$$\gamma_{NN\pi}\bar{u}_r(p + k)\gamma_5 u_s(p)v[(p - k)^2] \quad , \qquad (5.67)$$

where $\gamma_{NN\pi}$ is the pion-nucleon coupling constant and v is a scalar vertex or cutoff function. ${\gamma_{NN\pi}}^2$ is related to the usual pion-nucleon pseudo-vector coupling constant f^2 by

$$f^2/4\pi = \gamma_{NN\pi}\mu^2/48\pi M^2 \approx 0.08 \quad . \tag{5.68}$$

(We have taken units in which $\hbar = c = 1$ throughout).

5.2.4 Spin-3/2 ($\Delta(1236) \to \pi + N$)

To obtain vertex functions describing, for example, $\Delta(1236)\rightleftarrows\pi+N$, by analogy with the integral-spin and spin-1/2 cases we factorize a coveriant Born term (Fig.5.5b) which describes scattering only in a $P_{3/2}$ state. MANDELSTAM et al. have shown that such an amplitude may be written (Ref.5.13, p.208)

$$b_{r,r'}(kp;k'p') \propto \bar{u}_r(p)k_\mu P_{\mu\nu}k_\nu' u_{r'}(p') \tag{5.69}$$

with

$$P_{\mu\nu} = \frac{\not{K} + W}{2W}\Pi_{\mu\nu}(K) \quad ,$$

$$\frac{1}{i}\Pi_{\mu\nu}(K) = \delta_{\mu\nu} - \frac{4}{3}\frac{K_\mu K_\nu}{K^2} - \frac{1}{3}\gamma_\mu\gamma_\nu + \frac{1}{3}\left(\frac{K_\mu\not{K}\gamma_\nu + \gamma_\mu\not{K}K_\nu}{K^2}\right) \quad , \tag{5.70}$$

$$K = p + k = p' + k' \quad ,$$

$$K^2 = W^2 \quad .$$

Furthermore, $\Pi_{\mu\nu}$ and $P_{\mu\nu}$ satisfy the relations

$$\sum_\mu K_\mu\Pi_{\mu\nu} = \sum_\nu \Pi_{\mu\nu}K_\nu = 0 \quad ,$$

$$\sum_\mu P_{\nu\mu}P_{\mu\rho} = P_{\nu\rho} \quad . \tag{5.71}$$

After considerable algebraic manipulation of the above equations [5.14] one obtains for the vertex function $\Delta_\mu(\underline{K},\underline{p}r)$ describing a $\Delta(1236)$ with *three*-momentum $\underline{K}$ and z component of spin μ decaying into a pion and a nucleon with *three*-momentum $\underline{p}$ and z component of spin r:

$$\Delta_\mu(\underline{K},\underline{p}r) = \sum_{\lambda,t} \langle\tfrac{1}{2}1t\lambda|\tfrac{3}{2}\mu\rangle\bar{u}_t(\underline{K},W)V_\lambda u_r(\underline{p},M) \quad , \tag{5.72}$$

where $W^2 = K_0{}^2 - \underline{K}^2$, $u(\underline{p},M)$ is the usual Dirac spinor of three-momentum $\underline{p}$ and mass M, and V_λ is the special vector defined in (5.48). In almost all applications of (5.72) one writes

$$\Delta_\mu(\underline{K},\underline{p}r) = \sum_\lambda N(K,W)N(p,M)\langle\tfrac{1}{2}1r\lambda|\tfrac{3}{2}\mu\rangle V_\lambda + O\left(\frac{p^2}{4M^2}\right) \quad , \tag{5.73}$$

where $N(p,M)$ is the normalization for the spinor $u(\underline{p},M)$, and neglects the terms of $O(p^2/4M^2)$. This neglect normally introduces small errors at the energies under consideration and tremendously simplifies the angular momentum decomposition [5.15]. In order to be consistent with the Feynman type rule for the πNN vertex, $\underline{V}$ of reduce to $\underline{p}_\pi$ in the πN CM system.

5.3 Physical Input: Blankenbecler-Sugar vs. Feynman, etc.

Vertex functions: Phenomenology enters first through the vertex functions whose role is to provide convergence in the three-body integral equations. At a minimum, a vertex function introduces two parameters - a coupling constant which measures the strength of the two-body force, and a cutoff parameter which is related to the range. In general, these parameters are not free, but constrained to fit the two-body data, e.g., the width and position of a resonance in the case of a resonant isobar or the binding energy and scattering length in the case of a bound isobar. When there are more parameters than necessary to fit the two-body input, the remaining ones may be juggled to get the best fit to the three-body data. An important aspect of the relativistic three-body equations is that they enable one to learn about experimentally unattainable two-body systems (e.g., unstable particles) by studying the effects of varying two-body parameters on the three-body solutions.

The vertex function is related to the two-body wave function in the nonrelativistic limit, so, for example, form factors in the vertex functions of Fig.5.5a are chosen to be functions of the variable x where

$$\begin{aligned} x &= (y^2 - M^2\mu^2)/(M^2 + \mu^2 - 2y) \quad , \\ y &= \tfrac{1}{2}[(p - k)^2 - M^2 - \mu^2] \quad . \end{aligned} \tag{5.74}$$

x reduces to $\underline{q}^2$ the relative three-momentum squared in the CM system. (x is obviously also the square of the magnitude of the *special vector* defined in (5.48)). The simplest Yamaguchi form is usually chosen for the form factor v; for example, for an S-wave vertex

$$v(x) = g(\beta^2)[x + \beta^2]^{-1} \tag{5.75}$$

where g is a coupling constant and β a cutoff parameter.

Propagators: A simple nonrelativistic separable potential style propagator, described completely in terms of an integral over the square of the vertex function,

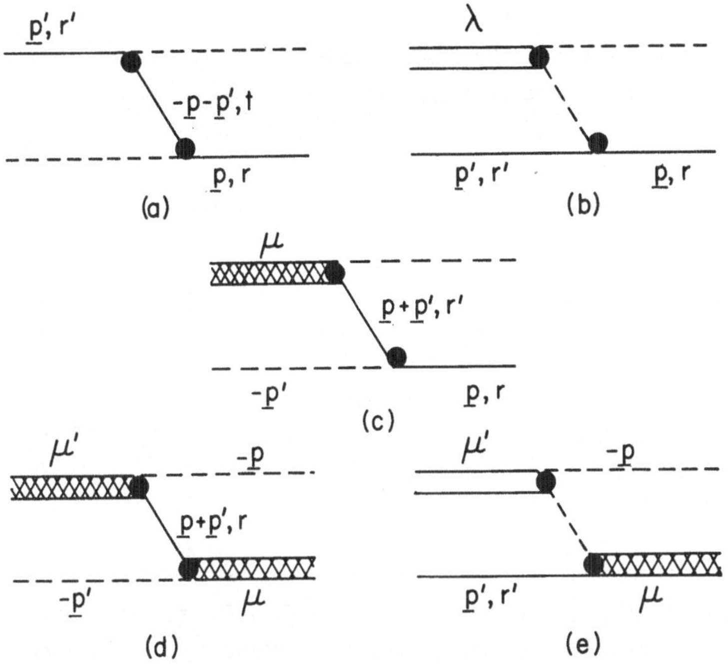

Fig.5.6. Born terms for Aaron-Amado calculation; notation is that of Figs.5.4 and 5.5

is not adequate for most relativistic cases. At least one more parameter, usually in the form of a bare mass (see (5.38)), is required to make the propagator behave like normally used field theoretic ones. In general, one may think of the isobar propagator as a phenomenological construction which is supposed to represent certain subsums of Feynman diagrams. For example, Fig.5.5b contains much of Fig.5.6a since nucleon exchange is an important ingredient of the Δ, and thus by including the Δ-propagator in the three-body problem we are phenomenologically including four-body intermediate states. An important point to realize is that one must always worry about things like double counting and never use the three-body formalism blindly.

Another problem encountered in our relativistic propagators is that they introduce a square root singularity at s = 0 as can be seen from (5.20),

$$\sigma = s - 2\sqrt{s}\omega_k + m^2 \quad . \tag{5.76}$$

This singularity is a penalty for imposing the cluster decomposition without proper treatment of anti-particles and, as pointed out by BASDEVANT and OMNES [5.16], the difficulty will appear in any n-body claculation where one reduces a Bethe-Selpeter equation to a linear integral equation involving three momenta. In πN and KN reactions, the large mass of the nucleon ensures that this singularity and other crossing singularities are far from the physical region and hence (hopefully) not important.

Born terms: It is in the Born terms where one has the most latitude in introducing phenomenological effects. For example, the three-body Blankenbecler-Sugar (BS) procedure omits from the $\pi\pi N$ system such important potential effects as ρ and Δ exchange since they contain four-body intermediate states. However, by using the freedom to add terms with only left-hand cuts in s to the BS Born terms (B_{BS}), such terms can be included phenomenologically. We may add to B_{BS} a t-channel exchange of the form

$$\frac{\lambda}{(p-q)^2 - M_{ex}^2} = \frac{\lambda}{(E_p - E_q)^2 - (\underline{p} - \underline{q})^2 - M_{ex}^2} , \tag{5.77}$$

where the denominator in (5.77) is negative definite and never vanishes as $\underline{p}$ and $\underline{q}$ run through all values in the kernel of the integral equation, and thus the addition of this term to B_{BS} cannot affect unitarity. The use of such terms in the πN problem is discussed in Subsection 5.4.1.

The above freedom is used in another situation. If should not be fargotten that the true scattering amplitude is the sum of all *Feynman* diagrams. The BS procedure which effectively puts all particles on their mass shells and ensures that the Born terms have the correct cut structure to satisfy unitarity does not necessarily give the correct potential strength. We shall elaborate on these ideas in the following manner: Consider the Feynman Born term describing a process shown in Fig.5.6 expressed in terms of old-fashioned (time-ordered) perturbation theory diagrams [5.17]. We have converted to the time-ordered formalism in order to be able to go off-shell in the same manner that is done in the Blankenbecler-Sugar (BS) Born term. The Feynman Born term B_F and the BS Born terms corresponding to energy W and evaluated in the three-body CM system become (the particles are labeled as in Subsec.5.2.1)

$$\begin{aligned} B_F &= \frac{g_{23}g_{12}}{2\omega_2}\left(\frac{1}{\omega_1 + \omega_3 + \omega_2 - W} + \frac{1}{\omega' + \omega'' + \omega_2 - W}\right) , \\ B_{BS} &= \frac{g_{23}g_{12}}{\omega_2} \; \frac{(\omega_1 + \omega_2 + \omega_3)}{[(\omega_1 + \omega_2 + \omega_3)^2 - W^2]} , \end{aligned} \tag{5.78}$$

where particles prime and double prime are composite and we have used coupling constants at the verticiles (unit form factors). First note that near the three-body pole ($W = \omega_1+\omega_2+\omega_3$)

$$B_F \approx B_{BS} \approx \frac{g_{23}g_{12}}{2\omega_2} \; \frac{1}{W - \omega_1 - \omega_3 - \omega_2} . \tag{5.79}$$

In general, however, B_F and B_{BS} will not be the same. In addition to the fact that B_{BS} is missing the pole corresponding to the "other" time ordering, it also contains additional kinematic factors arising from the requirement of Lorentz invariance. It seems that how closely B_{BS} approximates B_F is a reasonable criterion to justify the use of the BS procedure in physical problems. An example of a Born term for which $B_F \approx B_{BS}$ is the crossed pole describing $\pi N \to \pi N$ via nucleon exchange in Fig.5.6a. In the static limit, $M \to \infty$, B_F and B_{BS} are *identically* equal, and they are reasonably similar for physical values of μ/M and energies $W \lesssim 2$ GeV. However, let us now turn to the case $\pi N \to \rho$(stable)N via pion exchange (in Fig.5.6b). Unfortunately, in this case, B_{BS} and B_F differ by roughly a factor of two at the energies involved. In fact, in the static limit and on the energy shell, B_{BS} actually equals $1/2B_F$. In order to mock up the correct potential, AARON et al. [5.18] take B of (5.32) as

$$B = B_{BS} + \frac{\lambda}{(p - q)^2 - \mu^2} \tag{5.80}$$

with $\lambda \approx 1/2$; clearly they have added a t-channel exchange of the form (5.77). This procedure gives a reasonable potential strength over a wide energy range ($W \lesssim 2$ GeV), and adding λB_F, of course, does not affect unitarity. The above difficulty has arisen because of the small mass of the pion. In the sense of old-fashioned diagrams, the pion going backward in time is just as important as the pion going forward in time, and the BS procedure omits the former.

5.4 Application of Relativistic Three-Body Formalism

The relativistic three-body equations derived in the previous sections must be applied with judgment and care. Ideally, one should apply them to a true three-body system in which the two-body subsystems are dominated by a few bound states or resonances and intermediate production is very important. Unfortunately, the only system in nature that really satisfies these requirements is that of three nucleons at low energies, and not surprisingly the nonrelativistic version of the three-body equations beautifully describes this system. One must not forget that, in this case, solving the equations is equivalent to solving the Schrödinger equation with separable two-body potentials. On the other hand, Shanley successfully described d-α scattering (a six-body problem) by identical methods. The point is that, in this case, one is dealing with a pseudo-three-body problem in the sense that one can neglect the internal structure of the α-particle. In the relativistic situation one is almost always dealing with pseudo-three-body problems because multiple pion production can begin at relatively low energies. In addition, global symmetries such as SU(3) couple in other systems of particles not being treated explicitly-hopefully, these effects can be absorbed in parameters in the form factors. Finally, the underlying three-body equations do not describe exactly the relativistic world as the Schrödinger equation does the nonrelativistic world. For all these reasons, as stated earlier, the relativistic three-body equations must be used with discretion.

It is a fact that our equations emphasize unitarity at the expense of crossing symmetry, but as far as we know, no practical formalism can treat both principles exactly, and in many situation it is clear that unitarity may be the more important one to take into account. The πN and KN systems below ~1 GeV where almost all the partial waves are highly inelastic and single pion production is dominant are obvious candidates for application of the relativistic three-body formalism.

5.4.1 Inelastic Effects and π-N Resonances

A rapidly opening inelastic process can produce a resonance. In the πN system the connection of the $D_{13}(1520)$ resonance with ρ production has been known for some time [5.19]. Similarly, the importance of particle-exchange or potential terms has been known since CHEW and LOW first related the $\Delta(1236)$ and nucleon exchange [5.20]. However, there have been few attempts in the past to combine such effects in a systematic dynamical calculation in the πN system, in part because of the absence of a tractable dynamical framework in which to imbed these effects. The relativistic three-body formalism of Section 5.1 is well suited for such a calculation. AARON and AMADO [5.21] solve the relevant three-body equations by first assuming the production amplitude corresponding to (5.21) (in the three-body CM system) to be

$$\begin{aligned}\langle \underline{k}_1\underline{k}_2,\underline{p}_3 t|T_{23}|\underline{p}r\rangle = (2)^{-1/2}\Big[\sum_{\substack{i=1\\ r'}}^{2} &\langle -\underline{k}_i r'|{}_{\pi N}T_{\pi N}|\underline{p}r\rangle D_N{}^{-1}(\rho_{k_i})\bar{u}_t(\underline{p}_3,M)\gamma_5 u_{r'}(-\underline{k}_i,M)\\ &+\sum_{\substack{i=1\\ \mu}}^{2}\langle -\underline{k}_i\mu|{}_{\pi\Delta}T_{\pi N}|\underline{p}r\rangle D_\Delta{}^{-1}(\sigma_{k_i})\Delta_\mu(-\underline{k}_i,\underline{p}_3 t)\\ &+\sum_{\mu}\langle \underline{p}_3 t\mu|{}_{\rho N}T_{\pi N}|\underline{p}r\rangle D_\rho{}^{-1}(\sigma_{p_3})M_\mu(\underline{k}_1\underline{k}_2)\Big] \quad ,\end{aligned} \tag{5.81}$$

where $\underline{p}$ is the initial nucleon momentum and $\underline{k}_1\underline{k}_2$ and $\underline{p}_3$ the momenta of the final state pions and nucleon, respectively, and r', r, t and μ the appropriate z components of spin; isospin factors have been omitted. In (5.81) ${}_{\pi N}T_{\pi N}$, ${}_{\pi\Delta}T_{\pi N}$ and ${}_{\rho N}T_{\pi N}$ are the quasi-two-body amplitudes which satisfy coupled integral equations of the form (5.40). The driving (Born) terms of the integral equations are shown schematically in Fig.5.6; the detailed functional form of the Born terms and propagators that appear in the integral equations are given in Appendix B. It should be noted that Born terms for processes such as $\pi N \to \pi N$ via ρ or Δ exchange are not included automatically in our formalism since $\pi\pi\Delta$ or $\pi\rho N$ correspond to four-body intermediate states and their appearance would not be consistent with truncation of unitarity at the three-body level. However, these exchanges are important in certain cases and their effect may be included phenomenologically as suggested in (5.77).

The calculation of Aaron and Amado is most successful for the D_{13} and D_{33} partial waves where S-wave production of ρ and Δ may occur, and thus the associated inelastic process are large and open rapidly at their thresholds. As is well known, in such cases the inelastic processes, through unitarity, feed back and strongly affect the elastic amplitudes. The three-body equations excel in taking such effects properly into account. It is true that S-wave production of ρ and Δ also can occur from initial πN S-states, but here unknown short-range effects can be relatively more important, and perhaps for this reason the three-body calculations are much less successful. Moderate success is attained in the P_{33}, P_{13}, and P_{31} channels. The

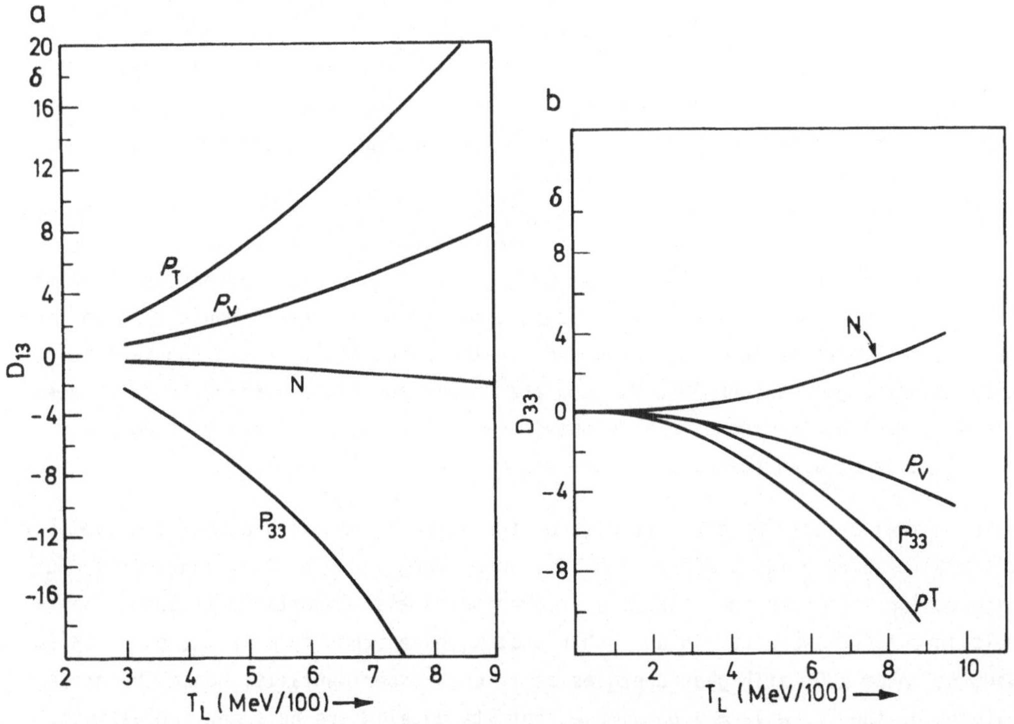

Fig.5.7. Born terms due to the exchange of various nucleonic states and the ρ-meson are shown expressed in terms of an equivalent phase shift (in degrees for the D_{13} channel (7a) and the D_{33} channel (7b). ρ_V and ρ_T label the contributions of the vector and tensor coupling of the ρ to the nucleon

behavior of the remaining partial waves seemingly cannot be explained by the three-body equations with separable interactions (corresponding to the mechanisms shown in Fig.5.6) and a few phenomenological potential terms due to relatively low mass particle exchange.

We shall now discuss in detail the treatment of the D_{13} and D_{33} partial waves. In addition to the Blankenbecler-Sugar Born terms shown in Fig.5.6, ρ and Δ exchanges must somehow be included in the calculation. For example, in Fig.5.7 are shown Born terms for various particle exchange mechanisms calculated by CARRUTHERS [5.22] for the D_{13} and D_{33} channels. The phenomenological potential used in the three-body integral equations should represent the effects of these terms. Note that for the D_{13} case the terms roughly cancel in the physical region over a wide range of energy, and one would expect that the three-body equations including just the mechanisms of Fig.5.6 would adequately describe the physics. These expectations are realized - the result of such a calculation of the πN D_{13} partial wave is shown in Fig.5.8a. In the D_{33} channel Carruthers shows that the sum of the potential terms is rather strongly repulsive. The calculation including ρ and Δ production with phenomenological repulsion of the size suggested by Carruthers by the method of (5.77) gives the results shown in Fig.5.8b. The attractive effects of the production mechanism below threshold are cancelled by the repulsive left-hand cut. Above threshold this cut cannot cancel production and the inelasticity dips. In both the D_{13} and D_{33} cases the denominator function has a zero on the second sheet and there is a πN resonance present.

In the P_{13} channel once again results are obtained which agree well with phase shift analyses (see [5.19]). As in the case of the D_{33}, ρ and Δ production combine with repulsive exchange effects to produce these results. Here Aaron and Amado find that the real part of the three-body denominator function is nearly constant near unity in the region around 1900 MeV and are unable to locate a nearby second sheet pole which would correspond to a resonance. Recent phase shift analyses do claim the existence of a πN resonance in the P_{13} channel at ~1860 MeV.

We comment briefly on other results of the Aaron-Amado calculation. The familiar $P_{33}(1236)$ resonance is obtained with nucleon exchange an important ingredient, but Δ production itself in addition to ρ production is also important. It seems impossible in our framework (as in any other that we have seen) to make the phase shift "hang up" near 180^{0} at higher energies as it does experimentally. Hence the gross features of the P_{33} wave may be simple, but its details are not, and inelasticity seems to play an important role. In the S_{11} channel there is experimental evidence of resonances at 1535 MeV and 1700 MeV. The former has been related to η production [5.23] (threshold 1480 MeV) which we do not include in our formalism. However, we find that the 1700 MeV resonance is probably driven by ρ production, and can fit

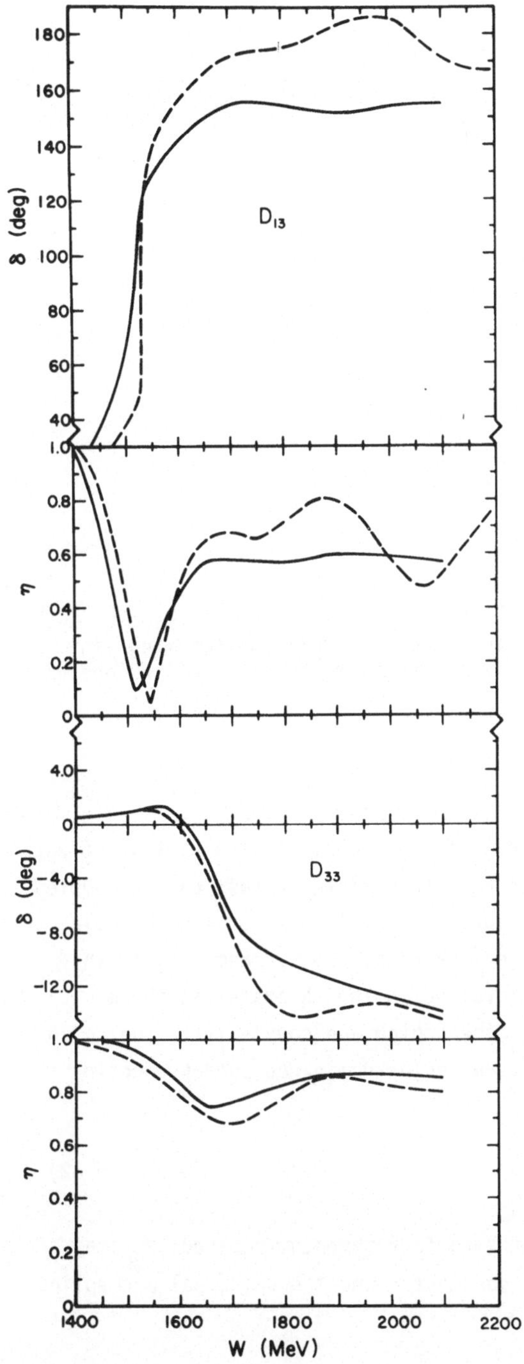

Fig.5.8. Real phase shift (δ) and inelasticity (η) for (a) D_{13} partial wave and (b) D_{33} partial wave. The theoretical results (solid line) are compared with the analysis of DONNACHIE et al. [5.39] (dashed line)

the phase shift reasonable well with a three-body calculation that includes πN and ρN intermediate states.

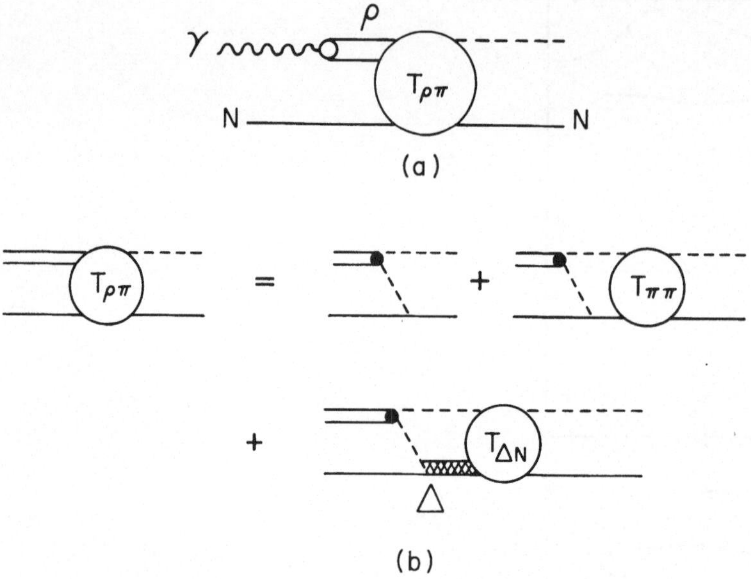

Fig.5.9. (a) Schematic representation of photoproduction amplitude assuming isovector dominance; (b) previous amplitude expressed in terms of off-shell πN and πΔ scattering

5.4.2 Further Applications

The work on πN scattering in the previous section which exploits the virtual effects of π production (particularly through ρN and πΔ intermediate states) can be coupled with the notion of ρ dominance to study photoproduction. To pass from πN elastic scattering to photoproduction we use vector dominance as shown schematically in Fig.5.9a. Since the ω is not significantly involved in the πN system at these energies, we take only the isovector part of the photon and couple it to the ρ. In terms of the off-shell ρN→πN amplitude $T_{\rho\pi}$, the vector-dominance photoproduction amplitude $T_{\gamma\pi}$ is given by

$$T_{\gamma\pi} = \Gamma_{\gamma\rho} D_\rho^{-1}(0) T_{\rho\pi} \quad , \tag{5.82}$$

where $D_\rho(0)$ is the ρ propagator evaluated at zero four momentum squared. $\Gamma_{\gamma\rho}$ may be expressed in terms of the fine-structure constant α and the universal ρ coupling $g_{\rho NN}$ [5.24]. That is

$$\Gamma_{\gamma\rho} = \alpha/[4D^{-1}(0) g_{\rho NN}{}^2/4\pi] \quad . \tag{5.83}$$

We take $T_{\rho\pi}$ from the calculation of πN scattering; its relation to the off-shell elastic πN amplitude is shown in Fig.5.9b. Note that in (5.82) $T_{\gamma\pi}$ is independent of $D_\rho(0)$. Using the above model AARON and AMADO [5.25] calculate those photoproduction amplitudes which couple strongly to the $D_{13}(1520)$ and $D_{33}(1670)$ πN resonances and are able to explain major features of photoproduction in the CM energy range 1400 MeV-1600 MeV.

Our relativistic three-body equations have also been used to study I = 0 KN system. This system is particularly interesting because it is "exotic" in the sense that resonances are forbidden by the simple quark model. However, we have already seen that the inelastic process $\pi+N\to\rho+N$ drives the D_{13} πN resonance and would expect similar effects in the KN systems from the $K^*(890)$ where the inelastic reactions $K+N\to K^*+N$ may occur strongly. Detailed calculations, in fact, predict the existence of exotic [5.26] resonances and their existence is further confirmed by a modified phase shift analysis [5.27] and recent experimental data [5.28].

Another interesting application of the formalism of Ref.[5.5] is a recent calculation of πd scattering *neglecting nucleon spin* by WOLOSHYN et al. [5.29]. In this work detailed comparisons of the exact solutions of the three-body equations are made with the results of a conventional fixed-scatter approach. The fixed nucleon calculation does not adequately reproduce the three-body results demonstrating the importance of treating the three-body kinematics (i.e., including nucleon recoil and isobar propagation). Also, they find that the multiple scattering expansion converges more rapidly in the three-body approach than in the fixed-scatterer calculation. Intermediate nucleon-nucleon interactions play an important role, giving contributions to the scattering amplitude of the same order as those given by pion multiple scattering; these effects are especially significant for back-angle scattering. Finally, the results are compared with the available experimental data for πd total and integrated elastic cross sections and good agreement is obtained.

5.5 Present Developments - Three-Body Phenomenology

The usual method of analysis of three-body final states is the isobar model where a form such as (5.21) is chosen to describe the amplitudes for two-particle to three-particle transitions. The quasi-two-body amplitude $\langle \underline{p}|T(s)|\underline{q}\rangle$ which appears in the equation is expanded in terms of partial waves

$$f^J_\alpha(W,\sigma_\alpha,\ell',\ell) \quad , \tag{5.84}$$

where J is the total angular momentum, ℓ the orbital angular momentum of the initial system, and ℓ' the orbital angular momentum of the final particle-isobar system. An important feature in almost all applications of the isobar model is that the f^J_α's of (5.84), except for centrifugal barrier factors, are taken to be independent of the two-body sub-energy variable σ_α for fixed overall center-of-mass energy $W = \sqrt{s}$.

If the isobar resonances are very narrow, only the value of f_α^J at the (well-defined) isobar mass is relevant and constant f_α^J is a reasonable choice. However, one can show that if the resonance bands are wide and overlap, f_α^J can be a rapidly varying function of σ_α; for example, unitarity forces f_α^J to have a physical (square-root) branch cut in σ_α [5.30]. For the nonrelativistic case, Amado has discussed this subject in detail in Chapter 3, and his arguments follow almost exactly in the relativistic problem. In previous material we have discussed the problem of three-particle states and derived relativistic integral equations with solutions that satisfy two- and three-body unitarity and have proper sub-energy dependence. However, only a limited number of physical systems can be described quantitatively by these equations, and recent research has concentrated on extracting from such equations general features based on unitarity and analyticity which must be included as part of any valid three-body phenomenology. The aim is to establish the domain of applicability of the usual isobar model and (hopefully) to develop a better phenomenology which is more generally applicable.

In two-body analyses the well-known phase shift parametrization permits one to satisfy unitarity automatically, while analyticity can be implemented with relative ease because of the simple analytic structure of the scattering amplitude; i.e., in terms of the energy variable, the imaginary part carries a square-root singularity and the real part is an entire function. We would like a similar parametrization for the three-body problem. Since the analyses of three-body final states are done at fixed overall center-of-mass energy, we are particularly interested in a theoretical framework which incorporates *sub-energy* unitarity and analyticity. The "minimal" relativistic theory which includes these principles has been obtained by AARON and AMADO [5.9] following the methods discussed by AMADO [5.8]. Reviewing briefly, according to (5.21) the amplitude for the reaction $a(k)+b(k')\to c(p_1)+d(p_2)+e(p_3)$ in the three-body center of mass may be written in the form

$$\langle p_1p_2p_3|T|p\rangle = \sum_{\substack{\alpha,\beta,\gamma=1\\ \text{cyclic}}}^{3} \langle p_\alpha|T|p\rangle G(p_\beta,p_\gamma) \tag{5.85}$$

with

$$2p = k - k' \tag{5.86}$$

and where the quasi-two-body amplitude $\langle p_\alpha|T|p_k\rangle$ describes the production from the initial state of particle α and the (β,γ) isobar, and $G(p_\beta,p_\gamma)$ describes the subsequent propagation of the isobar and its decay into $p_\beta+p_\gamma$. For simplicity, in (5.85) we have suppressed internal quantum numbers and assumed that each pair of the final state particles is dominated by a single isobar. The partial wave expansion coefficients of $\langle p_\alpha|T|p\rangle$ are given in (5.84). Because f_α has a physical branch cut in σ_α, it is convenient to decompose f_α into a dispersive and an absorptive part according to (3.39), (3.40) and (3.41). The constraints imposed by sub-energy unitarity appear as integral equations of the form (3.42), i.e.,

$$\mathrm{Abs}\{f_\alpha(\sigma_\alpha)\} = \sum_{\beta\neq\alpha} \int_{\substack{\text{Phase}\\ \text{Space}}} K(\sigma_\alpha,\sigma_\beta) f_\beta(\sigma_\beta) \quad , \tag{5.87}$$

where the kernel $K(\sigma_\alpha,\sigma_\beta)$ is made up of known kinematical factors and two-body t-matrices. Because we know that $\mathrm{Abs}\{f_\alpha\}$ has the singularity in σ_α required by sub-energy unitarity, it is tempting to use (5.87) in a phenomenological fitting procedure by taking a simple analytic expression for Disp $\{f_\alpha\}$ (e.g., a constant) and then solving (5.87) which becomes an inhomogeneous Fredholm integral equation for Abs $\{f_\alpha\}$. This temptation must be resisted because, in general, *both* dispersive and absorptive parts have unphysical logarithmic singularities in sub-energy which cancel one another in the full amplitude [5.31]. Choosing f_α = constant results in these singularities propagating into the physical amplitude. Unfortunately, as shown by AITCHISON and GOLDING [5.32], these singularities are present in most cases of interest when resonance bands overlap in the Dalitz plot. To avoid the above difficulties one must include analyticity as well as unitarity as a constraint on the f_α's; we do so by dispersing (5.87) in the manner discussed by AMADO for the non-relativistic problem [5.8]. The "minimal" equation satisfied by the f_α's which are consistent with sub-energy unitarity and analyticity, not surprisingly, turn out to be almost identical to those described in Section 5.1. Our working equation becomes

$$\langle \underline{p}|T(s)|\underline{p}_\alpha,\ell,m\rangle = \langle \underline{p}|R(s)|\underline{p}_\alpha,\ell,m\rangle + \frac{1}{(2\pi)^4} \sum_{\substack{\ell'm'\\ \beta\neq\alpha\\ \alpha,\beta,\gamma\\ \text{cyclic}}} \int \frac{d^3p_\beta}{2\omega_{p_\beta}} \times \frac{\langle \underline{p}|T(s)|\underline{p}_\beta,\ell',m'\rangle\langle \underline{p}_\beta,\ell',m'|B(s)|\underline{p}_\alpha,\ell,m\rangle}{D_{\ell'}(\sigma_\gamma)} \quad , \tag{5.88}$$

where the Born term B and the propagator D are exactly those quantities given by (5.32) and (5.38). In (5.88), ℓ and m refer to the angular momentum of an isobar in the α-channel. For fixed s one can choose R to be a simple function of sub-energy without fear of a gross violation of analyticity, and thus (5.88) is a useful phenomenological tool. Experience gained solving equations of this type in both relativistic and nonrelativistic problems indicates that approximate solutions will give a reasonable picture of the sub-energy dependence of the isobar amplitudes even though they do not describe very well the total energy behavior [5.33]. Eq.(5.88) is minimal in the sense that it was the simplest equation that we could derive that was soluble, incorporated sub-energy unitarity and analyticity, and did not have spurious W singularities in the physical region.

5.5.1 Data Analysis of Single Pion Production in πN Collisions in the CM Energy Range of 1300 - 2000 MeV

Single pion production data at intermediate energies is an important source of information concerning meson-baryon resonances. For example, one can obtain partial widths of known resonances and, perhaps, discover new resonances which might de difficult to identify in an elastic phase shift analysis. Recent theoretical advances have generated considerable interest in the process $\pi N \to \pi\pi N$. In particular, a proposed connection between current and constituent quarks [5.34] can be tested through the magnitudes and signs of amplitudes for pionic transitions between hadrons [5.35]. Equivalently, modified versions of $SU(6)_W$ classify baryon resonances and at the same time predict amplitudes for reactions of the type $\pi N \to \pi\Delta$, $\pi N \to \rho N$, $\pi N \to \varepsilon N$, etc. [5.36]. These same amplitudes appear in the extensive phenomenological analysis of HERNDON et al. [5.15] which, at the moment, represents the *only* analysis of the world's pion production data. Agreement of the above-mentioned theoretical predictions with the results of the Herndon et al. analysis is extremely impressive; however, we feel that comparison of theory and experiment at this time is, perhaps, premature. First, the analysis is so complex that it is important that several groups working independently confirm one another. Second, the world's pion production data in the resonance region is now ~50% more than was available to Berkeley/SLAC. Finally, there are sound theoretical grounds on which to question some of the present results. For example, D-wave production of $\pi\Delta$ is comparable to S-wave production near threshold (1450 MeV $\leq$ 1550 MeV). This seems unreasonable in view of the ranges of the forces involved, the nearness to $\pi\Delta$ threshold, and the fact that all obvious Feynman diagrams enhance rather than suppress S-wave production of $\pi\Delta$. AARON and AMADO obtain very small DD13 contributions (see Ref.5.9, Fig.4). Furthermore they find that the large DS13 $\pi\Delta$ amplitude is rapidly varying as a function of subenergy rather than constant as assumed by Berkeley/SLAC. In general imposition of unitarity and analyticity may force isobar amplitudes to be rapidly varying functions of subenergy [5.30] and this effect must be explored in all the partial waves.

5.5.2 Three-Pion System

The 3π system is studied through the four-body final state in the reaction $\pi+p \to \pi+\pi+\pi+p$. An isobar model is developed assuming that the production and decay of the 3π system can be described independently, and from a partial wave analysis one tries to isolate the contribution to the full amplitude of the three-pion final state. In the usual phenomenology, for fixed mass, the amplitude describing propagation and decay of the three-pion system is taken proportional to either the ρ or ε propagators, and the (complex) proportionality constants become fitting parameters. As in the pion-nucleon case, there is the possibility of sub-energy dependence of these "constants". However, the 3π problem is more complex because there are unstable particles in both initial and final states and we are thus dealing with a piece of a 3→3 amplitude. Using methods discussed earlier in this section, we are attempting to fit the data of ASCOLI et al.[5.37] for 3π production in the region of the A and B resonances ($1\ \mathrm{GeV} \lesssim W_{3\pi} \lesssim 2\ \mathrm{GeV}$). Of particular importance is an understanding of the $A_1(J^P=1^+)$ channel. Finally, we should like to note that the actual model we are using in the analysis is somewhat more complicated than anything discussed so far in this chapter since we include phenomenologically the possibility of a short-range 3π interaction in a manner suggested by BRONZAN [5.38].

Appendix A: Discontinuity Formulae

In order to derive (5.11), (5.25) and (5.28), we first consider the operator version of (5.19) assuming that this equation possesses a symmetry of the Lippmann-Schwinger or Bethe-Salpeter equations; namely,

$$T(s) = B(s) + B(s)\tau(\sigma)T(s) \quad , \tag{5.89}$$

or

$$T(s) = B(s) + T(s)\tau(\sigma)B(s) \quad . \tag{5.90}$$

Both of these equations can be solved for B(s) to obtain

$$B(s) = T(s)[1 + \tau(\sigma)T(s)]^{-1} \quad , \tag{5.91}$$

and

$$B(s) = [1 + T(s)\tau(\sigma)]^{-1}T(s) \quad . \tag{5.92}$$

We now let $s = s+i\varepsilon$ in (5.91) and $s = s-i\varepsilon$ in (5.92), and subtract the equations with the result

$$B(s^+) - B(s^-) = T(s^+)[1 + \tau(\sigma^+)T(s^+)]^{-1} - [1 + T(s^-)\tau(\sigma^-)]^{-1}T(s^-) \quad , \tag{5.93}$$

where σ^+ is a value of σ in (5.20) corresponding to $s = s+i\varepsilon$. Multiplying (5.93) from the left by $[1+T(s^-)\tau(\sigma^-)]$ and from the right by $[1+\tau(\sigma^+)T(s^+)]$ we finally have

$$\begin{aligned} T(s^+) - T(s^-) = {} & [B(s^+) - B(s^-)] + T(s^-)\tau(\sigma^-)[B(s^+) - B(s^-)] \\ & + [B(s^+) - B(s^-)]\tau(\sigma^+)T(s^+) + T(s^-)[\tau(\sigma^+) - \tau(\sigma^-)]T(s^+) \\ & + T(s^-)\tau(\sigma^-)[B(s^+) - B(s^-)]\tau(\sigma^+)T(s^+) \quad . \end{aligned} \tag{5.94}$$

If B is independent of s as is the case in (5.7), then

$$\text{disc}\,\{B\} = B(s^+) - B(s^-) \tag{5.95}$$

is zero and if we take matrix elements of (5.94) in momentum space we have derived (5.11) which leads to (5.12). Even if B is a function of s, there are matrix elements in $\underline{p}$, $\underline{q}$ space such that disc $\{B\} = 0$ if *either* $\underline{p}$ *or* $\underline{q}$ are external momenta, and thus taking the $\underline{p}$, $\underline{q}$ matrix element of (5.94) yields (5.22) and ultimately after comparison with the unitarity relation, (5.25) and (5.28). We may check after the fact that there are indeed values of the external momenta for which disc$\{B\} = 0$. For example, consider stable isobars with either the initial or final state on the mass and energy shell. From (5.32) we see that for this case the statement disc $\{B\} = 0$ is equivalent to

$$W - \omega_k - \omega_{k'} - \omega_{k+k'} \neq 0 \tag{5.96}$$

with

$$\begin{aligned} W &= E_k + \omega_k \quad , \\ E_k &= (\underline{k}^2 + \mu^2)^{1/2} \quad , \\ \omega_k &= (\underline{k}^2 + m^2)^{1/2} \quad , \\ \mu &< 2m \quad . \end{aligned} \tag{5.97}$$

We shall now prove that

$$(E_k - \omega_{k'})^2 < \omega_{k+k'}^{\ 2} \quad , \tag{5.98}$$

which guarantees the validity of (5.96). Inserting (5.97) into (5.98) and squaring yields

$$\mu^2 - 2E_k\omega_{k'} + 2kk'z < 0 \quad . \tag{5.99}$$

Note that $E_k\omega_{k'}-kk'z$ is the Lorentz scalar $k^\mu k'_\mu$ and may be evaluated in any frame. If we choose the frame in which $\underline{k}' = 0$ we obtain

$$\mu^2 - 2\sqrt{\underline{k}^2 + \mu^2}\, m < 0 \tag{5.100}$$

and this is a true inequality since $\mu<2m$; we have thus completed our proof.

Appendix B

For the πN calculations of Section 5.4, we now give the partial wave projections of the Born terms, and the propagators that enter the three-body integral equations such as (5.40). The relative normalization of the Born terms and propagators is determined by two- and three-body unitarity and the form of (5.81). In our Clebsch-Gordan coefficients we have adopted the conventions of Ref.5.15: 1) When adding orbital angular momentum ℓ and channel spin j to form total angular momentum J, ℓ appears in the first position. 2) In all other cases the baryon appears first.

B.1 *Born Terms*

Setting form factors equal to unity, in the three-body CM system we consider the Blankenbecler-Sugar Born terms for the following processes (using rules developed in Subsec.5.2.1):

1) $\pi+N\to\pi+N$ through nucleon exchange (Fig.5.6a)

$$\begin{aligned}\langle \underline{p},r|B_1^I(s)|\underline{p}',r'\rangle = \sum_t C_1^I\gamma_{NN\pi}^2\bar{u}_r(\underline{p},M)\gamma_5 u_t(\underline{p}+\underline{p}',M)J_1(p,p',s)\\ \times\,\bar{u}_t(\underline{p}+\underline{p}',M)\gamma_5 u_{r'}(\underline{p}',M)\quad,\end{aligned}\tag{5.101}$$

$$J_1(p,p',s) = \frac{2Mi(E_{\underline{p}+\underline{p}'}+\omega_p+\omega_{p'})}{E_{\underline{p}+\underline{p}'}[s-(E_{\underline{p}+\underline{p}'}+\omega_p+\omega_{p'})^2]}\quad,\tag{5.102}$$

where C_1^I is an isotopic spin coefficient (described in Subsec.5.2.1),

$$\begin{aligned}C_1^{1/2} &= -1/3\quad,\\ C_I^{3/2} &= +2/3\quad.\end{aligned}\tag{5.103}$$

It is convenient to decompose (5.102) according to

$$\langle \underline{p},r|B_1^I(s)|p',r'\rangle \equiv \chi_r^\dagger[E^I(\underline{p},\underline{p}',s) + i\underline{\sigma}\cdot\underline{p}\times\underline{p}'F^I(\underline{p},\underline{p}',s)]\chi_{r'}\quad.\tag{5.104}$$

We may then write the partial wave projection

$$\begin{aligned}\langle pJ\ell|B_1^I|p'J\ell\rangle = E_\ell^I(p,p',s) + \sum_\lambda 6pp'F_\lambda^I(p,p',s)\\ \times(2\lambda+1)|\langle 10\lambda 0|\ell 0\rangle|^2(-1)^{\ell+1/2-J}W(\ell\ell 11;1\lambda)W(\tfrac{1}{2}\tfrac{1}{2}\ell\ell;1J)\quad,\end{aligned}\tag{5.105}$$

where J is the total angular momentum, ℓ the orbital angular momentum, and where E_ℓ^I and F_ℓ^I are defined by the relations

$$\begin{aligned}E^I(\underline{p},\underline{p}',s) &= \sum_{\lambda\mu} E_\lambda^I(p,p',s)Y_{\lambda\mu}(\hat{p})Y_{\lambda\mu}^*(\hat{p}')\quad,\\ F^I(\underline{p},\underline{p}',s) &= \sum_{\lambda\mu} F_\lambda^I(p,p',s)Y_{\lambda\mu}(\hat{p})Y_{\lambda\mu}^*(\hat{p}')\quad,\end{aligned}\tag{5.106}$$

so (from Ref.5.10, Eq.(4.28)) one finds

$$E^I_\ell(p,p',s) = 2\pi \int_{-1}^{1} dz P_\ell(z) E^I(\underline{p},\underline{p}',s) \quad , \text{ etc.} \tag{5.107}$$

In deriving (5.105) we have used the relations (Ref.5.10, Sec.19, Eq.(5.19c))

$$<s|\underline{\sigma} \cdot \underline{k}|r> = \sum_\mu (3)^{1/2} <\tfrac{1}{2}\ell r\mu|\tfrac{1}{2}s> k^*_\mu \quad , \tag{5.108}$$

and

$$<s|\underline{\sigma} \cdot \underline{k} \times \underline{q}|r> = \sum_{M\nu\nu'} (6)^{1/2} <1\tfrac{1}{2}Ms|\tfrac{1}{2}r><11\nu\nu'|1M> k_\nu q_{\nu'} \quad . \tag{5.109}$$

2) $\rho+N\to\pi+N$ through pion exchange (Fig.5.6b)

$$\begin{aligned} <\underline{p},r|B^I_2(s)|\underline{p}',r',\lambda> &= C^I_2 \gamma_{NN\pi}\gamma_{\rho\pi\pi}\bar{u}_r(\underline{p},M)\gamma_5 u_{r'}(\underline{p}',M) \\ &\quad \times J_2(p,p',s)(cp_\lambda + dp'_\lambda) \quad , \end{aligned} \tag{5.110}$$

where we have written the special vector of (5.48) in the form $cp_\lambda+dp'_\lambda$ where c and d are functions of $\underline{p}^2$, $\underline{p}'^2$ and $\underline{p}\cdot\underline{p}'$.

$$J_2(p,p',s) = \frac{(\omega_{\underline{p}-\underline{p}'} + \omega_p + E_{p'})}{\omega_{\underline{p}-\underline{p}'}[s - (\omega_{\underline{p}-\underline{p}'} + \omega_p + E_{p'})^2]} \tag{5.111}$$

and

$$\begin{aligned} C^{1/2}_2 &= -(2/3)^{1/2} \quad , \\ C^{3/2}_2 &= +(1/6)^{1/2} \quad . \end{aligned} \tag{5.112}$$

We now write

$$\begin{aligned} <\underline{p},r|B^I_2(s)|\underline{p}',r',\lambda> \equiv\ & a_{pp}p_\lambda <r|\underline{\sigma} \cdot \underline{p}|r'> + a_{pp'}p_\lambda <r|\underline{\sigma} \cdot \underline{p}'|r'> \\ & + a_{p'p}p'_\lambda <r|\underline{\sigma} \cdot \underline{p}|r'> + a_{p'p'}p'_\lambda <r|\underline{\sigma} \cdot \underline{p}'|r'> \quad . \end{aligned} \tag{5.113}$$

The partial wave projection of B^I_2 becomes

$$\begin{aligned} <pJ\ell|B^I_2(s)|p'J\ell'j> =\ & (-1)^{1/2-j} \sum_\rho [6(2\ell+1)(2\ell'+1)(2j+1)]^{1/2} \\ & \times <\ell 100|\rho 0><\ell' 100|\rho 0> \Big\{ W(\tfrac{1}{2}1J\ell;\tfrac{1}{2}\rho)W(\tfrac{1}{2}1J\ell';j\rho) \\ & \times [p^2(a_{pp})_{\ell'} + p'^2(a_{p'p'})_\ell + p'p(a_{p'p})_\rho] \end{aligned} \tag{5.114}$$

$$+\, pp'(a_{pp'})_\rho \begin{pmatrix} \rho & 1 & \ell' \\ 1 & \frac{1}{2} & j \\ \ell & \frac{1}{2} & J \end{pmatrix} \Bigg\} \quad , \qquad (5.114)$$

where J is the total angular momentum, ℓ the orbital angular momentum of the initial πN system, ℓ' the orbital angular momentum of the final ρN system and j the ρN channel spin, and

$$(a_{pp})_\ell = 2\pi \int_{-1}^{1} dz P_\ell(z) a_{pp} \quad , \text{ etc.} \qquad (5.115)$$

The 9-j symbol is as defined in Ref.5.40, Eq.(6.4.4).

3) $\pi+\Delta\to\pi+N$ through nucleon exchange (Fig.5.6c)

$$\begin{aligned} \langle \underline{p},r|B_3^I(s)|\underline{p}',\mu\rangle = \sum_{r'\lambda} & C_3^I \gamma_{NN\pi}\gamma_{\Delta N\pi}\bar{u}_r(\underline{p},M)\gamma_5 u_{r'}(\underline{p}+\underline{p}',M)J_1(p,p's) \\ & \times N(\underline{p}+\underline{p}',M)N(p',W_{p'})(cp_\lambda + dp'_\lambda)\langle \tfrac{1}{2}1r'\lambda|\tfrac{3}{2}\mu\rangle \quad , \end{aligned} \qquad (5.116)$$

where J_1 is defined in (5.102), $cp_\lambda+dp'_\lambda$ is the special vector at the Δ-vertex, the N's are spinor normalizations, and

$$W_{p'}{}^2 = (\omega_{\underline{p}} + E_{\underline{p}+\underline{p}'})^2 - p'^2 \quad . \qquad (5.117)$$

For the Δ-vertex we have used the approximation (5.73), and

$$\begin{aligned} C_3^{1/2} &= -(8/9)^{1/2} \quad , \\ C_3^{3/2} &= +(5/9)^{1/2} \quad . \end{aligned} \qquad (5.118)$$

We now define

$$\begin{aligned} \langle \underline{p},r|B_3^I(s)|\underline{p};\mu\rangle \equiv \sum_{t\lambda} & \langle 1\tfrac{1}{2}\lambda r'|\tfrac{3}{2}\mu\rangle[a_{pp}p_\lambda\langle r|\underline{\sigma}\cdot\underline{p}|r'\rangle + a_{pp'}p_\lambda\langle r|\underline{\sigma}\cdot\underline{p}'|r'\rangle \\ & + a_{p'p}p'_\lambda\langle r|\underline{\sigma}\cdot\underline{p}|r'\rangle + a_{p'p'}p'_\lambda\langle r|\underline{\sigma}\cdot\underline{p}'|r'\rangle] \quad . \end{aligned} \qquad (5.119)$$

The partial wave projections of B_3^I are then given by

$$\begin{aligned} \langle pJ\ell|B_3^I(s)|p'J\ell'\rangle = \sum_\rho & [24(2\ell+1)(2\ell'+1)]^{1/2}\langle \ell 100|\rho 0\rangle\langle \ell' 100|\rho 0\rangle \\ \times \Big\{ & W(\tfrac{1}{2}1J\ell;\tfrac{1}{2}\rho)W(\tfrac{1}{2}1J\ell';\tfrac{3}{2}\rho)[p^2(a_{pp})_{\ell'} + p'^2(a_{p'p'})_\ell + p'p(a_{p'p})_\rho] \\ & + pp'(a_{pp'})_\rho \begin{pmatrix} \rho & 1 & \ell' \\ 1 & \frac{1}{2} & \frac{3}{2} \\ \ell & \frac{1}{2} & J \end{pmatrix} \Bigg\} \end{aligned} \qquad (5.120)$$

with J the total angular momentum, ℓ the angular momentum of the initial πN system and ℓ' that of the final $\pi\Delta$ system.

4) $\pi+\Delta\to\pi+\Delta$ through nucleon exchange (Fig.5.6d)

$$
\begin{aligned}
<\underline{p},\mu|B_4^I(s)|\underline{p}',\mu'> &= N(p,W_p)N^2(\underline{p}+\underline{p}',M)N(p',W_{p'})C_4^I\gamma_{\Delta N\pi}^{\ 2} \\
&\times \sum_{\substack{r\\ \lambda\rho}} (ap_\lambda + bp'_\lambda)<\tfrac{1}{2}1r\lambda|\tfrac{3}{2}\mu>J_1(p,p',s)(cp_\rho^* + dp'^*_\rho)<\tfrac{1}{2}1r\rho|\tfrac{3}{2}\mu'> \\
&\equiv \sum_{r\lambda\rho} <\tfrac{1}{2}1r\lambda|\tfrac{3}{2}\mu><\tfrac{1}{2}1r\rho|\tfrac{3}{2}\mu'>(a_{pp}p_\lambda p_\rho^* + a_{pp'}p_\lambda p'^*_\rho \\
&+ a_{p'p}p'_\lambda p_\rho^* + a_{p'p'}p'_\lambda p'^*_\rho) \quad ,
\end{aligned} \tag{5.121}
$$

where we have used the previous notation, including the usual approximation for the Δ-vertex. Also

$$
\begin{aligned}
C_4^{1/2} &= +\ (1/3) \quad , \\
C_4^{3/2} &= -\ (2/3) \quad .
\end{aligned} \tag{5.122}
$$

The partial wave projections become

$$
\begin{aligned}
<pJ\ell|B_4^I(s)|p'J\ell'> &= \sum_\rho 4[(2\ell+1)(2\ell'+1)]^{1/2}<\ell100|\rho0><\ell'100|\rho0> \\
&\times \Big\{W(\tfrac{1}{2}1J\ell;\tfrac{3}{2}\rho)W(\tfrac{1}{2}1J\ell';\tfrac{3}{2}\rho)[p^2(a_{pp})_{\ell'} + p'^2(a_{p'p'})_\ell + p'p(a_{p'p})_\rho] \\
&+ pp'(app')_\rho \begin{pmatrix} \rho & 1 & \ell' \\ 1 & \frac{1}{2} & \frac{3}{2} \\ \ell & \frac{3}{2} & J \end{pmatrix} \Big\} \quad .
\end{aligned} \tag{5.123}
$$

5) $\pi+\Delta\to\rho+N$ through pion exchange (Fig.5.6e)

$$
\begin{aligned}
<\underline{p},\mu|B_5^I(s)|\underline{p}'\mu'r> &= N(p,W_\rho)N(p',M)C_5^I\gamma_{\Delta N\pi}\gamma_{\rho\pi\pi} \\
&\times \sum_\lambda (a_\lambda p_\lambda + bp'_\lambda)<\tfrac{1}{2}1r\lambda|\tfrac{3}{2}\mu>J_2(p,p',s)(cp^*_{\mu'} + dp'^*_{\mu'}) \\
&\equiv \sum_\lambda <\tfrac{1}{2}1r\lambda|\tfrac{3}{2}\mu>(a_{pp}p_\lambda p^*_{\mu'} + a_{pp'}p_\lambda p'^*_{\mu'} + a_{p'p}p'_\lambda p^*_{\mu'} + a_{p'p'}p'_\lambda p'^*_{\mu'}) \quad .
\end{aligned} \tag{5.124}
$$

In (5.124)

$$
\begin{aligned}
C_5^{1/2} &= -(1/3)^{1/2} \quad , \\
C_5^{3/2} &= -(5/6)^{1/2} \quad .
\end{aligned} \tag{5.125}
$$

The partial wave projections become

$$\langle pJ\ell|B_5^I(s)|p'J\ell'j\rangle = \sum_\rho 2[(2\ell+1)(2\ell'+1)(2j+1)]^{1/2}\langle\ell 100|\rho 0\rangle\langle\ell' 100|\rho 0\rangle$$

$$\times \Big\{W(\tfrac{1}{2}1J\ell;\tfrac{3}{2}\rho)W(\tfrac{1}{2}1J\ell';j\rho)[p^2(a_{pp})_{\ell'} + p'^2(a_{p'p'})_\ell + p'p(a_{p'p})_\rho]$$

$$+ pp'(a_{p'p})_\rho \begin{pmatrix} \rho & 1 & \ell' \\ 1 & \frac{1}{2} & j \\ \ell & \frac{3}{2} & J \end{pmatrix}\Big\} \quad . \tag{5.126}$$

B.2 *Propagators*

1) Nucleon propagator

$$D_N(\sigma) = \frac{\sigma - M^2}{2Mi}\left[1 - (\sigma - M^2)\frac{M\gamma_{NN\pi}^2}{2\pi^2}\int_0^\infty \frac{dk}{E_k\omega_k}\frac{Xk^4 v_N^2(k^2)}{(E_k+M)(\sigma-X^2)(M^2-X^2)^2}\right] \quad , \tag{5.127}$$

$$X = E_k + \omega_k \quad ,$$

where v_N is a form factor, and $D_N(\sigma)$ of (5.127) is a twice-subtracted expression corresponding to (5.39). The condition that (5.127) have no ghost zeros is

$$Z = 1 - \frac{M\gamma_{NN\pi}^2}{2\pi^2}\int_0^\infty \frac{dk}{E_k\omega_k(E_k+M)}\frac{Xk^4 v_N^2(k^2)}{(M^2 - X^2)^2} \geq 0 \quad . \tag{5.128}$$

2) Δ-propagator

$$iD_\Delta(\sigma) = \sigma - M_\Delta^2 - \frac{M\gamma_{\Delta N\pi}^2}{6\pi^2}\int_0^\infty \frac{dk}{\omega_k E_k}\left(\frac{E_k + M}{2M}\right)\frac{Xk^4 v_\Delta^2(k^2)}{\sigma - X^2} \quad , \tag{5.129}$$

$$X = E_k + \omega_k \quad .$$

The i in Eq.(5.129) reflects the similar factor that appears in Eq.(5.70). If we choose a smooth form factor, e.g.,

$$v_\Delta(k^2) = \exp(-k^2/2\beta^2) \tag{5.130}$$

The following set of parameters will yield a πN resonance at $\sqrt{\sigma}\approx 1236$ MeV and width $\Gamma\approx 120$ MeV: $M\gamma_{\Delta N\pi}^2 = 4020$, $\beta_\Delta^2 = 60\mu^2$, $M_\Delta = 33.2\mu$. A Yamaguchi form factor (Eq.5.75) which gives a Δ (1236) is discussed in some detail in Ref.5.29.

3) ρ-propagator

$$D_\rho(\sigma) = \sigma - M_\rho^2 - \frac{\gamma_{\rho\pi\pi}^2}{4\pi^2} \int_0^\infty \frac{dk}{\omega_k} \frac{k^4 v_\rho^2(k^2)}{(\sigma - 4\omega_k^2)} \quad . \tag{5.131}$$

For v_ρ of the form

$$v_\rho(k^2) = (\beta_\rho^2)(\beta_\rho^2 + k^2)^{-1} \tag{5.132}$$

the choice $\gamma_{\rho\pi\pi}^2=356$, $\beta_\rho^2=50\mu^2$, $M_\Delta=7.05\mu$ will yield a P-wave $\pi\pi$ resonance at $\sqrt{\sigma}\sim760$ MeV, width $\Gamma\sim140$ MeV.

References

5.1 R. Aaron, R.D. Amado: Phys. Rev. 150, 857 (1966); R. Aaron, R.D. Amado, Y.Y.Yam: Phys. Rev. 140, B1291 (1965); P.E. Shanley: Phys. Rev. 187, 1328 (1969)

5.2 R. Blankenbecler, R. Sugar: Phys. Rev. 142, 1051 (1966)

5.3 D. Freedman, C. Lovelace, J. Namyslowski: Nuovo Cimento 43, 248 (1966)

5.4 V.A. Alessandrini, R.L. Omnes: Phys. Rev. 139, B167 (1965). With the insertion of two-body separable amplitudes which satisfy the cluster property, their equations reduce to the ones we shall derive

5.5 R. Aaron, R.D. Amado, J.E. Young: Phys. Rev. 174, 2022 (1968)

5.6 H. Pilkuhn: The *Interactions of Hadrons* (John Wiley and Sons, Inc., New York 1967). Throughout this chapter we use the metric, Dirac matrices, phase space convention, etc., of Pilkuhn

5.7 R.D. Amado: Phys. Rev. 132, 485 (1963)

5.8 R.D. Amado: Phys. Rev. C12, 1134 (1975)

5.9 R. Aaron, R.D. Amado: Phys. Rev. D13, 2581 (1976)

5.10 M.E. Rose: *Elementary Theory of Angular Momentum* (John Wiley and Sons, Inc., New York 1957), pp.108 - 109

5.11 S. Gasiorowicz: *Elementary Particle Physics* (John Wiley and Sons, Inc., New York 1967)

5.12 Compare the simplicity of our approach to that of J.L. Basdevant, R.E. Kreps: Phys. Rev. 141, 1398 (1966)

5.13 S. Mandelstam, J.E. Paton, R.F. Peierls, A.Q. Sarker: Ann. Phys. (New York) 18, 198 (1962)

5.14 R. Aaron, R.D. Amado: (to be published)

5.15 D.J. Herndon, R. Longacre, L.R. Miller, A.H. Rosenfeld, G. Smadja, P. Söding: Phys. Rev. D11, 3183 (1975)

5.16 J.L. Basdevant, R.L. Omnes: Phys. Rev. Letters 17, 775 (1966)

5.17 S. Weinberg: Phys. Rev. 150, 1313 (1966)

5.18 R. Aaron, R.D. Amado, D.C. Teplitz: Phys. Rev. 187, 2047 (1969)

5.19 J.S. Ball, W.R. Frazer: Phys. Rev. Letters 7, 204 (1961); L.F. Cook, B.W. Lee: Phys. Rev. 127, 297 (1962)

5.20 G.F. Chew, F.E. Low: Phys. Rev. 101, 1570 (1956)

5.21 R. Aaron, R.D. Amado: Phys. Rev. Letters 27, 1316 (1971)

5.22 P. Carruthers: in *Lectures in Theoretical Physics*, ed. by W.E. Brittin, et al. (Univ. of Colorado Press, Boulder, Colo. 1965), Vol. VII, pp. 83-135

5.23 Y. Lemoigne, et al.: "$\pi^- p \to \eta n$ up to $P_\eta^* = 400$ MeV/c," *Baryon Resonances-73* (Purdue University Press, West Lafayette, Indiana 1973), p.93

5.24 Through universality, $g_{\rho NN} = \gamma_{\rho\pi\pi}$, where $\gamma_{\rho\pi\pi}$ is defined in M. Gell-Mann, D. Sharp, W.G. Wagner: Phys. Rev. Letters 8, 261 (1962)

5.25 R. Aaron, R.D. Amado: Phys. Rev. D7, 1544 (1973)

5.26 R. Aaron, R.D. Amado, R.R. Silbar: Phys. Rev. Letters 26, 407 (1971)

5.27 R. Aaron, M. Rich, W.L. Hogan, Y.N. Srivastava: Phys. Rev. D7, 1401 (1973)

5.28 A.S. Carroll, T.F. Kycia, K.K. Li, D.N. Michael, P.M. Mockett, D.C. Rahm, R. Rubinstein: Phys. Letters 45B, 531 (1973)

5.29 R.M. Woloshyn, E.J. Moniz, R. Aaron: Phys. Rev. C13, 286 (1976)

5.30 R. Aaron, R.D. Amado: Phys. Rev. Letters 31, 1157 (1973)

5.31 C. Schmid: Phys. Rev. 154, 1363 (1967)

5.32 J.J.R. Aitchison, R.J.A. Golding: Physics Letters 59B, 288 (1975)

5.33 S.K. Adhikari, R.D. Amado: Phys. Rev. D9, 1467 (1974)

5.34 H.J. Melosh: Phys. Rev. D9, 1095 (1974)

5.35 F.J. Gilman, M. Kugler, S. Meshkov: Phys. Rev. D9, 715 (1974)

5.36 D. Faiman, J. Rosner: Phys. Letters 45B, 357 (1973)

5.37 G. Ascoli, et al.: Phys. Rev. D7, 669 (1973)

5.38 J.B. Bronzan: Phys. Rev. 139, B751 (1965)

5.39 A. Donnachie, R.G. Kirsopp, C. Lovelace: Phys. Letters 26B, 161 (1968)

5.40 A.R. Edmonds: *Angular Momentum in Quantum Mechanics* (Princeton University Press, Princeton, N.J., 1957)

6. Applications of Three-Body Methods to Many-Body Hadronic Systems

E. F. Redish*

With 8 Figures

The three-body methods discussed in previous chapters have been applied to hadronic systems with considerable success. The applications are of two types: calculations of the properties of three-hadron systems, and studies of three-body reaction mechanisms in many-hadron systems (N > 3). The former has been extensively reviewed at recent conferences [6.1-4] and in current books and monographs [6.5-7], while reviews of the latter have been few and far between [6.8,9]. In this chapter, we restrict our scope to the application of three-body methods to scattering and reactions in nonrelativistic hadronic systems with more than three particles.

Three-body methods have two roles to play in the many-body problem. First, the three-body system is the simplest nontrivial many-body problem for which exact solutions are possible. As such, it can serve as a theoretical laboratory for testing the approximate methods which have been developed and applied to many-body reactions. In this "ideal" laboratory, in contrast to our real ones, the full reaction mechanism is under our control and can be disassembled to reveal its inner workings, and to provide new insights as to what makes our approximations work - or not work.

The second role three-body methods play is in describing reaction mechanisms for real processes. Nature is a faster and more accurate calculator than our fastest computers, so she has no need to restrict her workings to two-body reaction mechanisms, much as we might prefer it. We therefore expect there may be some many-body reactions in which a three-body approximation is required to provide an accurate description of reality.

This chapter is divided into two parts. In the first (Sec.6.1) we consider three-body tests of standard nuclear approximations. In the second (Sec.6.2) we consider the evidence for the presence of three-body reaction mechanisms in the world of real hadrons. This section does not attempt to provide a complete review of all this evidence. To do so would require a much longer and more detailed presentation. Instead we select a number of examples which provide compelling evidence for the

* Supported by the U. S. Energy Research and Development Administration.

presence of three-body effects. Throughout the chapter we try to identify problems which seem to us to be fruitful lines for further research using the three-body methods described in this book.

6.1 Tests of Reaction Mechanisms

Since the full many-body problem with more than three particles cannot be solved exactly at the present time, approximation methods must be introduced. The most common approximations in use in nuclear reaction theory either consider the system as a problem with many degrees of freedom and treat it statistically, or consider the system as having few degrees of freedom and treat it perturbatively. A reaction which requires few degrees of freedom in its description is called "direct". In important cases these few degrees of freedom correspond to those of a three-body system. Direct reactions may be expected under two conditions. The projectile may fail to penetrate the nucleus significantly (the reaction is *peripheral*) or it may penetrate the nucleus but be moving very rapidly compared to the bound nucleons.

When a reaction is peripheral, very few degrees of freedom of the nucleus may be excited because of the low density of nucleons present. If the energy of the projectile is sufficiently low, the Coulomb and angular momentum barriers keep it outside of the nucleus. If the energy is high enough for the projectile to get over the Coulomb barrier and into the nucleus, the reaction may still be peripheral because of the suppression of the interior wave function.

If many channels are open, a deep penetration will almost certainly lead to a reaction, but rarely to the one being considered. This produces an effective absorption of the incoming beam in the nuclear interior and can be obtained in the optical model by the imaginary (absorptive) part of the optical potential.

The real part of the optical potential also plays a role in making reactions peripheral. If it is attractive, the projectile has a higher kinetic energy and a shorter wavelength inside the nucleus. The increased speed of the oscillation tends to suppress the contribution of the nuclear interior and to localize the reaction at the nuclear surface.

Finally, real nuclei are small enough that most of their volume is close to the surface. Even for a heavy nucleus like lead the matter radius is about 6 Fermis. Forty percent of the volume is within one Fermi of the surface. For a light nucleus like oxygen, the figure is closer to 80 percent.

A direct reaction may also be obtained when the projectile is moving very fast. When a very fast projectile strikes a nucleon in the target, the other nucleons do not have a chance to respond to the change in the motion of the struck nucleon. If only a small number of nucleons is involved, the reaction can be treated as direct.

A high energy direct reaction can be considered as the first term of an expansion in the number of nucleons the particle encounters at the same time. Such a cluster or correlation expansion can be expected to make sense if the nucleus is a dilute system. A dimensionless measure of the "diluteness" of a nucleus is the ratio of the range of the force d to the average spacing between nucleons in a nucleus D. The maximum density of nuclei is about 0.17 nucleons/fm^3. We may therefore estimate D as on the order of or greater than $(1/0.17)^{1/3} = 1.8$ fm. Only that part of the projectile-nucleon force which is of the same order as the average nuclear well depth (about 50 MeV) will substantially affect the motion of the target nucleons. For nucleons, this criterion gives $d \sim 1.0$ fm, while for pions and kaons, $d \sim 0.2$ fm. In either case, $d/D < 1$.

Qualitatively, we can say that direct reactions will dominate for charged particles near or below the Coulomb barrier and at high enough energies. "High enough" has been estimated both by calculating the speed of sound from collective models of nuclear motion [6.10], and by seeing where calculated compound contributions become negligible [6.11]. Both of these estimates indicate that direct processes should be important at velocities greater than about 1/4 of the speed of light. For projectiles made up of nucleons this occurs at about 25 MeV per nucleon. For pions and kaons the short range of the force and the large amount of energy available in the particle's mass should guarantee that the reaction is direct whenever absorption or emission takes place.

In this section we consider the use of three-body models to test some of the standard direct reaction approximation methods. By far the greatest number of applications of three-body theory to the many-body problem have been of this type.

6.1.1 Direct Reactions

Direct reactions and the various approximate methods of handling them are discussed in detail in numerous texts and reviews [6.11-15]. We sketch the basic arguments briefly.

The reactions usually considered to have important direct components include elastic and inelastic scattering, few-particle transfer reactions, charge exchange, and knockout.

In the case of transfer reactions such as stripping and pickup the basic mechanism assumed is three-body in character. Consider the transition from an initial state with a bound cluster i (=j+k) to a final state with a bound cluster j (=i+k). The exact matrix element is given by (cf.(1.73))

$$M_{ji} = \langle\chi_j|V^j|\Psi_i^{(+)}\rangle = \langle\chi_j|V_i + V_k|\Psi_i^{(+)}\rangle \quad . \tag{6.1}$$

This is usually treated in Distorted Wave Born Approximation (DWBA). Using the two-potential formula, the potential V_k may be used to change the plane wave for particle k in the final state into a distorted wave, giving

$$M_{ji} = \langle\chi_j^{(-)}|V_i|\Psi_i^{(+)}\rangle \quad . \tag{6.2}$$

This is exact.[1] The initial state is then approximated by a two-cluster state

$$M_{ji} \approx \langle\chi_j^{(-)}|V_i|\chi_i^{(+)}\rangle \tag{6.3}$$

where $\chi_i^{(+)}$ is a distorted wave for the initial cluster scattering in the field of particle i. This approximation treats the two-cluster part of the wave function carefully, while that part which is necessarily three-body in nature is treated as a perturbation.

For elastic, inelastic, and knockout scattering one also uses a three-body-like model. The model is not, however, truly three-body in nature so for convenience let us renumber our particles. We label the projectile by 0, and the nuclear particles by 1, ... A. We will refer to this as "multiple scattering labelling" as opposed to the "three-body labelling" used previously. If we consider the scattering from a state of the target c to an excited state c' the exact transition matrix element is given by

$$M_{oc',oc} = \langle\chi_{oc'}^{(-)}|\sum_{i=1}^{A} V_{oi} - U_o|\Psi_{oc}^{(+)}\rangle \quad . \tag{6.4}$$

The effect of an optical potential U_0 has been included in the final state and subtracted off the residual interaction. It is usually assumed not to produce any transitions so its contribution to the off-diagonal matrix element is zero. If the initial state is replaced by a distorted wave one obtains the DWBA for inelastic scattering

$$M_{oc',oc} \approx \langle\chi_{oc'}^{(-)}|\sum_{i=1}^{A} V_{oi}|\chi_{oc}^{(+)}\rangle \quad . \tag{6.5}$$

---

[1] In actual practice one is not describing a true three-body problem but a many-body system approximated by a three-body one. In this case the true effective interaction V_k is not known. The distorted wave is generated by some approximate potential U_k and a residual interaction V_k-U_k remains. We will ignore this additional level of sophistication throughout, as we will not discuss here the construction of effective three-body Hamiltonians.

This representation differs in a fundamental way from that obtained for transfer reactions. In the transfer problem, the potential V_i occurs together with the initial bound state wave function ψ_i. This includes the effect of the potential V_i to all orders. This is not the case in the inelastic matrix element above. The use of singular potentials would lead to unrealistically large matrix elements. It is necessary to go to a Distorted Wave Impulse Approximation (DWIA) which includes the correlations to all orders in a manner similar to that of transfer reactions. The matrix element is modified by (cf. Subsect.1.1.4)

$$V_{oi}|\underline{p}_i,\underline{q}_i\rangle \rightarrow V_{oi}|\underline{p}_i,\underline{q}_i^{(+)}\rangle = t_{oi}|\underline{p}_i,\underline{q}_i\rangle \quad , \tag{6.6}$$

giving

$$M_{oc',oc} = \langle \chi_{oc'}^{(-)}| \sum_{i=1}^{A} t_{oi} |\chi_{oc}^{(+)}\rangle \quad . \tag{6.7}$$

This results in reasonable matrix elements even for singular core potentials. The particular way in which the transition is carried out requires some care. It is in this step that three-body models have been found particularly useful.

Note that the final state of the nucleus c' need not be bound. If one particle (or cluster x) is free, then the reaction is described as an (a, ax) or knockout reaction (where the projectile is labelled a). Use of a distorted wave approximation for the exact state c' leads to the DWIA approximation for knockout

$$M_{oj,oc} = \langle \chi_o^{(-)} \chi_j^{(-)} | t_{oj} | \chi_{oc}^{(+)}\rangle \quad . \tag{6.8}$$

We will refer to the DWIA and the DWBA jointly as the DWA.

Elastic scattering is too strong to be treated perturbatively. Through shadowing, any reaction produces associated elastic scattering. One therefore constructs the optical potential in a perturbation (impulse) approximation and sums its effect to all orders by solving an L-S equation. Since the distorted waves will be generated in this manner, a distorted wave matrix element should not be used to generate the optical potential so as to avoid double counting. The standard approximation takes the form

$$\langle \underline{p}|U(E)|\underline{p}'\rangle = \langle \underline{p}\psi_o^A| \sum_{i=1}^{A} t_{oi} |\underline{p}'\psi_o^A\rangle \quad . \tag{6.9}$$

The derivation of this will be discussed below. We refer to (6.9) as the Impulse Approximation (IA) for the optical potential.

Since the above approximations arise from simplified treatments of three-body problems, some aspects of their accuracy can be investigated in a three-body model.

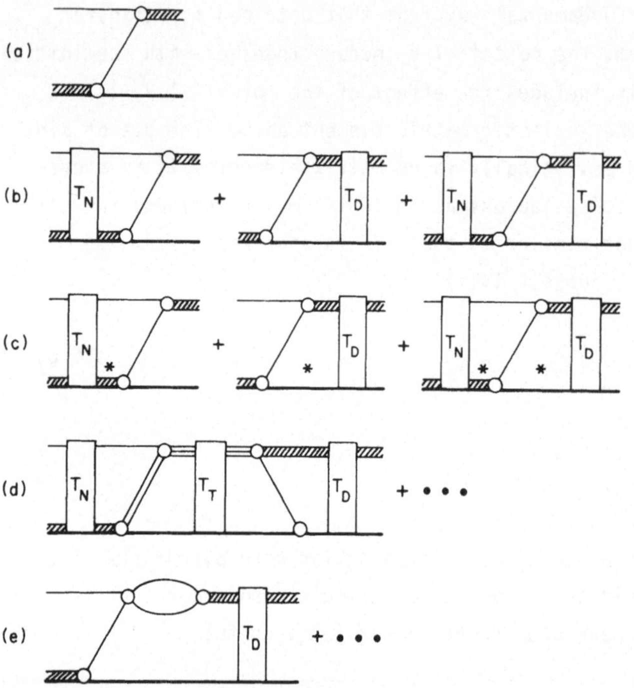

Fig.6.1a-e. Graphical structure of direct reaction amplitudes: (a) PWBA; (b) DWBA; (c) CCBA; (d) CRC; (e) three-body breakup contribution. The heavy line indicates a nucleus (nucleon number unspecified). An asterisk (*) indicates a bound excited state of the nucleus. T_j is a scattering operator for the particle j (neutron = N, deuteron = D, triton = T, etc.). Each level includes the graphs above it as well

There is another possible role for three-body methods. Instead of using a piece of a three-body model to describe the reaction mechanism, why not a full three-body model? This fits in as a part of a hierarchy of approximations in nuclear reaction theory of which the DWAs are only a small (albeit focal) part.

The hierarchy of nuclear reaction approximations is sketched for the specific example of the (p,d) reaction in Amado graphs in Fig.6.1. The successive approximations are: a) the Plane Wave Born Approximation (PWBA), b) the DWBA, c) the Coupled Channel Born Approximation (CCBA), d) Coupled Reaction Channels (CRC), and e) true three-body effects.

The trend in this hierarchy is towards increasing the number of explicitly unitarized channels. As this happens, the amount of phenomenology needed can be decreased. In each of the equations, L-S, CC, or CRC effective interactions must be utilized. In none of them can these interactions be calculated accurately from first principles at the present time. The most difficult part to calculate is the absorptive part corresponding to reactions into suppressed channels. As the number of channels included explicitly increases, the amount of phenomenological absorption decreases.

All the approximations except the last involve the inclusion of two-cluster states only. To generalize this to include the three-cluster states shown in e)

would require some version of the Faddeev equations. There is considerable evidence to indicate that this step may be more important than the previous ones. This evidence will be reviewed in Section 6.2.

6.1.2 Three-Body Formulation of the DWBA

Since direct reactions are usually treated as approximations to three-body models we begin by investigating the derivation of the DWBA and other approximations in the context of the three-body problem. This has been done by many authors [6.16-21]. The implications of these methods for the N-body problem have been discussed by BOULDIN and LEVIN [6.21] and REDISH [6.22].

Our goal in this section is to find intermediate states in the AGS equations corresponding to two-cluster states. Once this is done, the terms associated with the graphs of Fig.6.1 are easily identifiable, and we can extract a DWBA. The two-cluster intermediate states can be inserted in a number of different ways. This ambiguity can be used to advantage to improve the standard approximations.

Derivation of CRC Equations

Beginning with the transition operator $U_{ji}^{(+)}$ (see (1.117)) one transforms to the AGS operator given by (1.118a). These satisfy the AGS equations (1.121). A more convenient starting point for an investigation of the DWBA is the BENCZE-REDISH-SLOAN (BRS) [6.23-25] form of these equations. This is obtained by making the off-shell transformation of the AGS operator

$$U_{ji}^{(BRS)} = U_{ji}^{(AGS)} G_0 V_i \quad . \tag{6.10}$$

Since $G_0 V_i |\chi_i\rangle = |\chi_i\rangle$ on the energy-shell, it is only the fully-off-shell matrix elements of U that are modified by this transformation. The (right) half-on-shell matrix elements are the same; therefore, the wave functions associated with them are the same as those associated with the AGS operator. Multiplying (1.121b) on the right by $G_0 V_i$ gives

$$U_{ji}^{(BRS)} = \bar{\delta}_{ji} V_i + \sum_{\ell} \bar{\delta}_{j\ell} t_\ell G_0 U_{\ell i}^{(BRS)} \quad . \tag{6.11}$$

These equations are more convenient for a discussion of the DWBA since the PWBA for the transition operator, $U_{ji}^{(BRS)} \approx \bar{\delta}_{ji} V_i$, corresponds to that usually taken in nuclear physics. Henceforth we will write both the AGS and BRS forms as the single equation

$$U_{ji} = \bar{\delta}_{ji} B_i + \sum_{\ell} \bar{\delta}_{j\ell} t_\ell G_0 U_{\ell i} \tag{6.12}$$

where $B_i = G_0^{-1}$ for the AGS form, and V_i for the BRS.

It was observed by GRASSBERGER and SANDHAS [6.26] that the two-cluster intermediate states can be revealed by observing that the two-body T-matrix has poles with residue proportional to the propagator of the two-cluster states. Explicitly, using (1.53), if we introduce the two-body propagator

$$\Gamma_k = \int d\underline{p}_k |\underline{p}_k\psi_k\rangle (E - \varepsilon_k - p_k^2/2\mu_k)^{-1} \langle \underline{p}_k\psi_k| \quad , \tag{6.13}$$

then the two-body T-matrix has the form

$$t_k = V_k\Gamma_k V_k + t_k' \quad , \tag{6.14}$$

where t_k' is regular, and $V_k\Gamma_k V_k$ has a pole at $E = \varepsilon_k + p_k^2/2\mu_k$. Following [6.26]

$$G_0 t_k G_0 = (G_0 V_k)\Gamma_k(V_k G_0) + G_0 t_k' G_0 \quad . \tag{6.15}$$

Note that

$$G_0(E)V_k|\underline{p}_k\psi_k\rangle = \left(\frac{\varepsilon_k - T_k}{E - \dfrac{p_k^2}{2\mu_k} - T_k}\right)|\underline{p}_k\psi_k\rangle \quad , \tag{6.16}$$

where T_k is the kinetic energy operator for the relative motion of the pair k. Therefore, G_0V_k = identity operator at the pole. Any operator which is 1 at the pole may be used to extract the singularity. In general, we may write:

$$G_0 t_k G_0 = R_k'(E)\Gamma_k L_k(E) + K_t' \quad . \tag{6.17}$$

If the right and left transformation operators satisfy

$$R_k'\left(\frac{p_k^2}{2\mu_k} + \varepsilon_k\right)|\underline{p}_k\psi_k\rangle = |\underline{p}_k\psi_k\rangle \tag{6.18}$$

$$\langle \underline{p}_k\psi_k|L\left(\frac{p_k^2}{2\mu_k} + \varepsilon_k\right) = \langle \underline{p}_k\psi_k| \quad , \tag{6.19}$$

then the residual part of t will be regular. The kernel of (6.12) then takes the form

$$\bar{\delta}_{ik} t_k G_0 = \bar{\delta}_{ik} G_0^{-1} R_k' \Gamma_k L_k + \bar{\delta}_{ik} G_0^{-1} K_k' \quad . \tag{6.20}$$

The first term on the right must begin with the Born term in order to permit us to extract an L-S equation. We therefore define

$$R_k' = G_0 B_k R_k \quad . \tag{6.21}$$

Since for either the AGS or the BRS form of the Born term, $G_oB_k = 1$ at the pole, R must satisfy condition (6.18). With this (6.20) takes the form

$$t_kG_o = B_kR_k\Gamma_kL_k + G_o^{-1}K_k' \quad . \tag{6.22}$$

Putting this into (6.12) and solving the K_k' part of the kernel explicitly gives two sets of equations (written here in matrix form):

$$U = U' + U'R\Gamma LU \tag{6.23}$$

$$U' = \bar{\delta}B + \bar{\delta}G_o^{-1}K'U' \quad . \tag{6.24}$$

If we now introduce the specific off-shell transformation

$$\mathcal{U} = LUR \tag{6.25}$$

$$\mathcal{U}' = LU'R \quad , \tag{6.26}$$

the equations become

$$\mathcal{U} = \mathcal{U}' + \mathcal{U}'\Gamma\mathcal{U} \tag{6.27}$$

$$\mathcal{U}' = (L\bar{\delta}BR) + (L\bar{\delta}G_o^{-1}K'L^{-1})\mathcal{U}' \quad . \tag{6.28}$$

The first equation is the integral equation form of the CRC equations. When matrix elements between bound states are taken, they become a set of coupled L-S equations for the elastic, inelastic, and rearrangement amplitudes. Only two-cluster intermediate states are present. The effects of three-body intermediate states are obtained from solving (6.28) and are contained in the effective interactions $\mathcal{U}_{ij}'$.

Derivation of the DWBA

The DWBA can be extracted from these equations by expanding the transition amplitude in the number of rearrangements taking place. Elastic processes which do not involve rearrangement are summed to all orders. This reflects the assumption that rearrangement processes are small and may be treated perturbatively, while elastic processes are strong and must be handled by means of distorted waves.

This is done by dividing the effective interaction operator into diagonal (elastic) and off-diagonal (rearrangement) parts:

$$\mathcal{U}_{ij}' = \nu_i\delta_{ij} + v_{ij}(1 - \delta_{ij}) \quad . \tag{6.29}$$

In matrix form the CRC equations (6.27) become

$$\mathcal{U} = (\nu + v)(1 + \Gamma\mathcal{U}) \quad . \tag{6.30}$$

Bringing the $\nu\Gamma\mathcal{U}$ term to the left and solving the equation formally gives

$$\mathcal{U} = \frac{1}{1 - \nu\Gamma} [\nu + v(1 + \Gamma\mathcal{U})] \quad . \tag{6.31}$$

The operator $(1 - \nu\Gamma)^{-1} \equiv \Omega^{+}$ is an operator converting a plane wave bra into a distorted wave, since

$$\langle\chi|(1 - \nu\Gamma)^{-1} \equiv \langle\chi^{(-)}|$$

satisfies

$$\langle\chi^{(-)}| = \langle\chi| + \langle\chi^{(-)}|\nu\Gamma \quad ,$$

the standard L-S equation for a distorted wave with optical potential ν.

We want to extract distorted waves on both the left and the right. This can be done by using the form of (6.30):

$$\mathcal{U} = (\nu + v) + \mathcal{U}\Gamma(\nu + v) \quad . \tag{6.30a}$$

This leads to

$$\mathcal{U} = [\nu + (1 + \mathcal{U}\Gamma)v] \frac{1}{1 - \Gamma\nu} \quad . \tag{6.31a}$$

The operator $(1 - \Gamma\nu)^{-1} \equiv \Omega$ changes the plane wave ket which will appear on the right into a distorted wave. Distorted wave operators may be produced on both the left and the right by inserting (6.31a) into (6.31) giving (after some manipulation)

$$\mathcal{U} = \mathcal{U}_{o} + \Omega^{+}\tau\Omega \tag{6.32}$$

where

$$\tau = v + v\Gamma v + v\Gamma\mathcal{U}\Gamma v \quad . \tag{6.33a}$$

We have written

$$\mathcal{U}_{o} = \frac{1}{1 - \nu\Gamma} \nu \tag{6.33b}$$

for the transition operator produced by the elastic effective interaction alone. This has no off-diagonal matrix elements so it only contributes to elastic scattering. The operator given by (6.33a) is that operator whose matrix elements between distorted waves yield the exact transition amplitude. The expression (6.33a) for τ still does not permit us to identify powers of v since $\mathcal{U}$ has a term independent of v. Inserting (6.31) into (6.33a) and rearranging yields the equation

$$\tau = v + vg\tau \quad , \tag{6.34a}$$

where

$$g = \Gamma + \Gamma \nu g \qquad (6.34b)$$

is the distorted wave propagator. Equations (6.32) and (6.34a) are the desired result. If (6.34a) is iterated it leads to a perturbation series in which each successive term involves additional rearrangement. The lowest order approximation,

$$\tau \approx V$$

yields a DWBA of the form of (6.3) if the BRS form is used for the Born term. Now, however, the physical meaning of the initial and final optical potentials is clear and higher order (multi-step) corrections can be obtained by iterating (6.34a).

The potentials ν include the effects of three-body channels, the two-body channel effects being taken into account exactly by the use of the CRC equations. The exact optical potentials can be obtained by the formal elimination of the rearrangement channels from the CRC equations. If this is not done, the operator τ will have diagonal matrix elements giving the contribution of successive transfers to elastic scattering.

Iteration of (6.34a) for τ gives the contributions of multi-step transfers with two-cluster intermediate states. The effect of multi-step processes proceeding through breakup states is contained in (6.28) for v. Note that the division into two- and three-cluster contributions is not unambiguous due to the freedom in the off-shell transformation.

6.1.3 Stripping

Three-body models have been used as a testing ground for various stripping approximations. Virtually all of the calculations follow either the separable potential model introduced by MITRA [6.27], or the field theoretic method of AMADO [6.28]. These are equivalent if the field theoretic wave function renormalization constant Z is taken to be zero [6.29]. All the calculations limit the interactions to pairwise S-waves, and almost all to one-term separable potentials with Yamaguchi form factors. We will label the proton by 1, the neutron 2, and the core 3. We use the three-body labelling.

The calculations reviewed here include those of SHANLEY and AARON [6.30], [6.31], REINER and JAFFE [6.32], BAZ et al. [6.33], GIGNOUX [6.34], BOULDIN and LEVIN [6.35], and KING and McKELLAR [6.36-38]. The models used, and the approximations tested, are summarized in Table 6.1.

Table 6.1 Calculations testing DWBA for stripping

	V_{np}			V_{nC}			V_{pC}			E_d	Numerical methods	DWBA tested
	type	form	params.	type	form	params.	type	form	params.			
SHANLEY and AARON [6.31]	Sep	Y	ε=2.225 a=5.38	Sep	Y	ε=8.9 R=1.69 / ε=4.45 R=2.14	None			1.78, 4.45, 11.1	CD	Realistic
BAZ, DEMIN, and KUZ'MIN [6.33]	ZR	--	ε=2.2	Local	SW	ε=4.3 R=3.05	Local	SW	ε=4.3 R=3.05	2.2	Real axis	Not stated
REINER and JAFFE [6.32]	Sep	Y	ε=2.22 a=5.38	Sep	Y	ε=8.9 R=2.3	Sep	Y(+C)	ε=0.35 R=5.2	6.7, 11.2	CD	Exact
GIGNOUX [6.34]	Sep	Y		Sep	Y	ε=2 R=3.52 / ε=1,2,4 R=3.52	Sep	Y	ε=2 R=3.52 / ε=1,2,4 R=3.52	6 / 30	CD	Dodd--Greider
BOULDIN and LEVIN [6.35]	Sep	Y	ε=2.22 a=5.38	Sep	Y	ε=3.3 R=2.0	Sep	Y	ε=3.3 R=2.0	1.78, 6.7, 11.2, 15.12	CD	Exact
KING and McKELLAR [6.37]	Sep	Y	ε=2.225 a=5.38	Two Channel Sep	Y	ε=7.0 ε*=2.5 σ_{TOT}(E=0)=1b C=0.1	Two Channel Sep	Y	ε=7.0 ε*=2.5 σ_{TOT}(E=0)=1b C=0.1	15,50(5)	Real axis	Exact

Notes:

Types: SEP = separable, ZR = zero range; Forms: Y = Yamaguchi, SW = square well, C = Coulomb (partial effect); Parameters: ε = binding energy in MeV, ε* = binding energy of excited state in MeV, R = RMS radius of bound state in Fermis, C = channel coupling parameter (dimensionless); E_d = deuteron energy in MeV; Numerical Methods: CD = contour deformation.

Shanley and Aaron

In [6.30] and [6.31] SHANLEY and AARON perform a calculation of two three-body models of the type n+p+C. They consider a simplified model and a realistic model. In the simplified model the particles are taken to be spinless and distinguishable, and V_2 is taken to be zero. For V_3 the standard triplet Yamaguchi is chosen. The neutron-core potential is also chosen to be separable with a Yamaguchi form factor. Parameters are adjusted to give reasonable binding energies and RMS radii. In the more realistic model, spin and isospin are introduced, and a proton-core interaction turned on. The potentials are all taken to be charge independent and the core is taken to have spin and isospin zero.

The simplified model is used to study the validity of the DWBA and is considered in this section. The realistic model is discussed in Subsection 6.2.2. Studies concerning the three-body bound states and the elastic scattering of protons and deuterons have also been carried out. We will have more to say about these studies in our discussion of elastic scattering in Subsection 6.1.4.

Shanley and Aaron study a realistic DWBA in contrast to all subsequent authors. The exact elastic scattering output is fit with a standard Woods-Saxon search routine, and the resulting optical potentials used to generate distorted waves. These are then used in a standard stripping code with zero range approximation (i.e., $V_{np}(r)\phi_d(r) \approx \delta(r)D_0$).

The comparison of this DWBA with the exact results for the (d,p) reaction shows the DWBA to be a mediocre approximation in the model. The stripping peak is much too small in the DWBA at low energies, and improves somewhat as the energy increases. The angular structure is not well represented at any of the calculated energies.

The authors suggest that the DWBA fails here because the stripping channel is not weakly coupled to the elastic channel. One fact that argues against this explanation is that the DWBA improves with increasing energy even though the ratio of the total stripping cross section to the total elastic cross section increases.

BOULDIN and LEVIN [6.35] point out that another of the DWBA assumptions fails in this model. Usually one requires that the matrix element of V_2-U_p vanishes (see the footnote to (6.2)), where V_2 is the proton-core interaction, and U_p is the optical potential used to produce the proton final-state distortion. In the simplified Shanley-Aaron model $V_2 \equiv 0$ but U_p is not taken to be 0, nor is its matrix element calculated. This is a special property of this model and could have been remedied formally by taking no distortion in the final state. A better modification would be to include V_2-U_p explicitly. This is a serious objection and limits the conclusions we can draw from these calculations.

Reiner and Jaffe

The calculation of REINER and JAFFE [6.32] uses a more sophisticated model and tests a different DWBA.

They investigate the effect of turning on the proton-core potential and of attempting to include the effect of the Coulomb force. Their proton-nucleus form factor is modified by calculating the correction of the p-C wave function by the Coulomb potential in first-order perturbation theory. The modified wave function (in momentum space) is fit with a sum of Yamaguchi form factors and the resulting one-term separable potential used in the calculation.

They solve the Amado-Lovelace equations by contour deformation and study the effect of turning on V_2 (with and without the Coulomb modification) and the accuracy of the DWBA. They find that the effect of including the proton-nucleus interaction on the (d,p) cross section depends sensitively on the binding energy of the proton.

In testing the DWBA, they do not follow the lead of Shanley and Aaron, but rather construct an "exact" DWBA. The DWBA amplitude requires elastic proton and deuteron wave funtions. Reiner and Jaffe, and most subsequent authors, utilize the DODD and GREIDER [6.17] derivation of the DWBA. This requires that the proton distortion potential be exactly V_2 in order to cancel the V_2 in the transition operator (see the footnote after (6.2)). The final distorted wave may therefore be constructed exactly. The deuteron's distorted wave should be a projection of the full three-body wave function onto the deuteron bound state. Since the exact three-body wave function is calculated, the exact distorted wave is known. These elements are combined and an "exact" DWBA constructed by quadrature.

This represents a second distinct approach to the testing of the DWBA. Shanley and Aaron use the approximation of the practitioners. Reiner and Jaffe, on the other hand, construct an approximation which practitioners would use if they could.

They find that their exact DWBA fits the (d,p) and (d,n) cross sections well over the whole angular range. The normalization factor required varies from 1.2 for the (d,p) and 0.75 for the (d,n) at the lower energy, to almost 1 at the higher.

Bouldin and Levin

BOULDIN and LEVIN [6.35] and BOULDIN [6.39] calculate an exact DWBA in a manner similar to Reiner and Jaffe. As can be seen from Table 6.1, they use a charge independent nucleon-core force, so both the neutron-core and proton-core bound states have the same binding energy in contrast to the Reiner-Jaffe work.

They claim that the DWBA provides an extremely good fit for the angular distributions at all the energies. The normalization constants required to bring the forward

DWBA cross section into agreement with the exact decreases monotonically with energy from 1.4 to 0.8. The accuracy of the DWBA normalization therefore does not necessarily improve monotonically with increasing energy, as might have been concluded from the Reiner-Jaffe work.

The conclusion which can be drawn tentatively from the results of Reiner-Jaffe and Boulding-Levin is that, in a three-body model, the accuracy of an exact DWBA depends sensitively on the binding energy of the transferred particle, and that the energy dependence of the accuracy is affected by the binding energy somewhat. At the present time, no systematic investigation of these dependences exists.

The observed energy dependences of the normalization constant $S^2=(\text{Exact/DWBA})_{\theta=0^0}$ is of particular interest, since it is this normalization which is measured in practice and interpreted as containing the spectroscopic information. A failure of the DWBA model would be reflected in an energy dependence of the spectroscopic factor S^2, of the type observed in the model calculations. Such energy dependences have in fact been observed in real analyses of experimental data (see MacFARLANE [6.40]) and are of comparable magnitude to those observed in the model calculations.

The models can also be used in an attempt to understand what features are necessary to the accuracy of the DWBA. Bouldin and Levin investigate the partial wave decomposition of the DWBA vs. the exact amplitudes. They find that the lower partial waves disagree significantly, but the important surface partial waves are fit well. A critical factor here is the $(2\ell+1)$ in the partial wave decomposition. This cor-

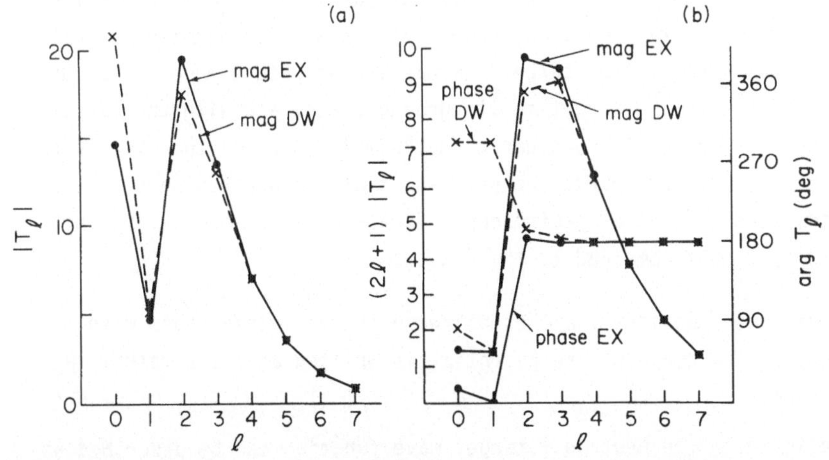

Fig.6.2a-b. Accuracy of the DWBA as a function of the deuteron-nucleus orbital angular momentum in a three-body model [N+P+C]. Calculation of [6.39]. (a) The magnitude of the exact (EX) and distorted wave (DW) amplitudes are plotted. (b) The magnitude (left scale) and phase (right scale) of the exact (EX) and distorted wave (DW) amplitudes times the geometrical factor $(2\ell+1)$

responds to the geometric enhancement of the surface discussed at the beginning of this section. The DWBA and exact partial wave amplitudes of Bouldin-Levin at 6.7 MeV are shown in Fig.6.2a. In Fig.6.2b we show the amplitudes times their geometric factor of $2\ell+1$. This indicates that even in a three-body model where the absorption of the low partial waves is not as great as in a realistic case and where the stripping channel is strongly coupled to the elastic, the dynamic and geometrical enhancement of the surface region can lead to an accurate DWBA in the higher partial waves.

King and McKellar

The primary goal of using the DWBA in (d,p) reactions is to extract spectroscopic information about the bound state of the neutron with the nucleus. This is interesting because the neutron-nucleus interaction is not expected to be well represented by a two-body problem. The one-step stripping is supposed to interact primarily with the two-body part of the true many-body wave function, thereby giving a measure of how the simple two-body state is broken up among the large number of real many-body states.[2]

In all the three-body models discussed above, the target was considered to be an inert core, so the exact neutron-target wave function was a pure two-body state. The spectroscopic factor was therefore always equal to unity. This limitation prevents the models from investigating an important facet of the DWBA; whether in fact the cross section responds only to the simple part of the wave function and thereby measures the spectroscopic factor.

An important increase in the sophistication of the model calculations was made by KING and McKELLAR [6.36-38]. They allow the target particle to have an excited state. The nucleon-target interaction then becomes a matrix which permits inelastic excitation of the target. The neutron-target now has a bound state whose wave function has components both with the target in its ground state, and with the target in its excited state. The former is considered the "simple" part of the wave function and the latter the many-body part. The parameters of the neutron-core interaction can then be adjusted to give a spectroscopic factor of less than unity, and the value extracted by a DWBA compared to the exact.

The specific model they consider includes pairwise S-wave interactions with Yamaguchi form factors. The nucleon-core potential is written as a 2x2 matrix in the states of the core.

[2] The simple state in which the neutron + target wave function can be described as a product of a neutron-target relative wave function times an intrinsic target wave function is described in the nuclear literature as a *single particle state*. This is because the presence of an infinitely heavy, nonrecoiling nucleus is usually assumed. Since, when three-body methods are used, there is no need to restrict to this case, we use the term *two-body state* to emphasize the relative nature of the wave function.

To construct a DWBA one needs optical potentials for the initial and final state. For the one-state model, the Dodd-Greider statement of the DWBA requires that the final state distortion be given by V_2. Since we now have two states, inelastic scattering is possible. The effect of the excited state must be eliminated to yield a two-body optical potential. This can be done in a straightforward manner. The exact DWBA is then calculated.

Their model parameters are summarized in Table 6.1. The parameters are adjusted to produce a spectroscopic factor of 0.65. Investigations of the parameter dependence of the nucleon-core subsystem indicate that the spectroscopic factor depends primarily on the binding energy of the nucleon-core ground state and the core's excitation energy.

They investigate the energy region from 15 to 50 MeV. This energy region begins where the previous studies leave off without any overlap. This is unfortunate as it does not permit us to make comparisons with the one-state-core work. It would be extremely interesting to use the King-McKellar model beginning with, say, the Bouldin-Levin parameters at 6.7 MeV where the DWBA gives excellent results, and then turn on the coupling slowly, reducing the spectroscopic factor from 1. The resulting extraction of spectroscopic information via the DWBA would test one of its fundamental assumptions, viz., that the stripping takes place only through the simple two-body part of the state. (It should be noted that the primary goal of the work of King and McKellar was to understand the strong energy dependence of spectroscopic factors observed in the non-DWBA method of BUTLER et al. [6.41], and that this goal was admirably accomplished. We do not discuss this application for lack of space.)

The authors construct and solve the separable three-body equations resulting by matrix inversion of the integral along the real axis, using special treatment of the regions in the neighborhood of the singularities.

King and McKellar find that the DWBA only fits the angular distribution in the stripping peak, and in the range from 25 to 45 MeV the DWBA produced a substantially broader peak than the exact calculation. The spectroscopic factors extracted from the DWBA average about 0.8 compared to the exact 0.65 and show a scatter of about ±7%. The authors assign this variation to numerical inaccuracies in the calculations. The spectroscopic factor obtained by the DWBA in this model is therefore too large by almost 25%.

Other Works

Two other calculations which should be mentioned are those of BAZ et al. [6.33] and GIGNOUX [6.34].

The former authors investigate the DWBA in a model with V_3 being a delta function, and the nucleon-core forces given by square wells. The equations derived by the same authors in an earlier paper [6.42] are used. These are somewhat like Faddeev equations, but only the part with the last interaction being a V_3 is separated. The equations are solved by a K-matrix-like method exactly at the threshold for a three-body breakup. No discussion of the effect of this singularity is given, and details of the calculation are not presented. They find that the DWBA disagrees badly with their exact angular distribution.

The calculation of GIGNOUX [6.34], is a standard Faddeev separable potential calculation performed using the method of contour deformation at 6 and 30 MeV incident deuteron energies.

The DWBA he tests is not the standard factorized DWBA described in the text above, but a more complicated DWBA suggested by DODD and GREIDER [6.17]. This approximation uses instead of the initial wave function given in (6.3), the wave function

$$|\chi_{DG}^{(+)}\rangle = \left(1 + \frac{1}{E^+ - H_0 - V^3} V^3\right)|\chi_3\rangle \quad . \tag{6.35}$$

While this cannot be easily calculated for a standard nuclear problem, it can be easily obtained in a Faddeev calculation by turning off V_3.

The Dodd-Greider DWBA is found to produce fairly good angular distributions in these calculations, but the normalizations are severely off. The discrepancy is a factor of 4.5 at the low energy, and about a factor of 2 at the higher energy. This calculation strongly suggests that the Dodd-Greider DWBA is not nearly as good as the standard DWBA for extracting spectroscopic factors.

The result is particularly interesting since Dodd and Greider show that the standard DWBA can be obtained as the inhomogeneous term of an integral equation with disconnected kernel and therefore presumably (though not necessarily) the first term of a divergent series. Their modified DWBA is the inhomogeneous term of an integral equation with connected kernel. This should be a clear warning to the three-body physicist attempting to apply his results to the many-body problem. The formal mathematical considerations are not always decisive. The standard DWBA can be formulated in many ways, some of which may be as a good approximation to the inhomogeneous term of a well-behaved (though perhaps not yet discovered) equation. In this context see the paper of REDISH [6.22] which constructs a standard DWBA using many-body connected kernel equations.

Summary

The model calculations considered here test the DWBA with a wide range of three-body models. Each of the calculations, however, considers only one or two aspects of the accuracy of the DWBA. The lack of overlap of the testing parameters limits the conclusions which can be drawn. This is particularly true with reference to the nucleon-core potential where comparable parameters are rarely used.

In spite of this limitation, we may make a few observations about the sensitivity of the model cross sections and the mechanism which produces an accurate DWBA. Specifically we may summarize the most important results as follows:

1) In these models the exact DWBA at low energies ($E<15$ MeV) provides a good representation of the contributions of the peripheral partial waves, but not of the deeply penetrating ones. The geometrical enhancement of these surface partial waves makes the overall exact DWBA a reasonably good approximation.

2) The accuracy of the exact DWBA at low energies is energy dependent to within about a factor of 1 ± 0.5. The best accuracy is not necessarily provided by the highest energy but depends on the parameters of the model.

3) At higher energies ($15<E<50$ MeV), the spectroscopic factor extracted by the DWBA appears reasonably stable, but not accurate to better than 25%.

These results, while being suggestive, cannot be taken as conclusive. Factors which should be included simultaneously, and varied to check the accuracy of the DWBA include:

1) more partial waves in the nucleon-core potential,
2) more realistic nuclear form factors,
3) the charge dependence of the nuclear potential ($V_1 \neq V_2$),
4) effect of the two-body Coulomb potential,
5) effect of inelastic core channels (non-unit spectroscopic factors),
6) effect of absorption in V_1 and V_2.

Some of these effects have been included in the above calculations, but rarely has more than one been put in at a time. Because of the parameter sensitivity, future studies of the accuracy of the DWBA should map the dependence of the accuracy on the model parameters.

Another point to note is that except for the pioneer work of Shanley and Aaron, none of the calculations tested a realistic DWBA (i.e., one involving phenomenological distorted waves). The comparison between an exact model, an exact DWBA, and a realistic DWBA could be very useful in understanding whether the phenomenological freedom in realistic DWBAs masks the true reaction mechanism.

Although considerable work on the accuracy of the DWBA has been carried out with three-body models, and much has been learned, there are still numerous questions to be answered as more sophisticated calculations become possible.

6.1.4 Elastic Scattering

Elastic scattering of a single projectile from a nucleus is less clearly a three-body problem than is stripping. Although the reaction mechanism leading to the optical potential is often treated as being three-body in character (the projectile interacts with a nucleon in the presence of a spectator core), one then sums the contribution of all the many nucleons in the target and solves an L-S equation. This appears to destroy the simple three-body character of the process. (This is not actually the case. See the discussion in Sec. 6.2.1.) Nevertheless, the three-body problem can still serve as the simplest theoretical laboratory in which the approximations involved have nontrivial content. We will discuss the results in the model of Shanley and Aaron and the extensive studies of resonances made by the Budapest group.

Shanley and Aaron

In their exhaustive study of their model, SHANLEY and AARON [6.31] investigate elastic scattering as well as stripping. Both elastic proton and elastic deuteron are investigated in the two-potential ($V_2 = 0$) and three-potential model.

In the two-potential model, they observe that elastic proton scattering shows a cusp-like anomaly at the threshold for the deuteron channel in the s-wave phase shift. When the third potential is turned on, the cusps are no longer visible. They attribute this to the increased absorption in the three-potential model.

They also observe that in their three-potential model the Levinson theorem is not satisfied. This theorem states that for potential scattering, the change in the partial wave phase shift between $E=0$ and $E=\infty$ is equal to π times the number of bound states [6.43]. In the many-body problem it is well known that the exclusion principle can produce violations of the Levinson theorem by forbidding a bound state which would otherwise exist [6.44]. Shanley and Aaron attribute the failure of the theorem to a zero of the S-matrix on the physical sheet. The results of their model calculation demonstrate that the Levinson theorem does not generalize in a simple way to the many-body problem, even when the particles are distinguishable.

A third interesting result of the Shanley-Aaron study concerns the use of phenomenological optical potentials. In conjunction with their construction of a realistic DWBA, they fit their elastic proton and deuteron cross sections with phenomenological Woods-Saxon potentials. The angular distributions are well represented (better than

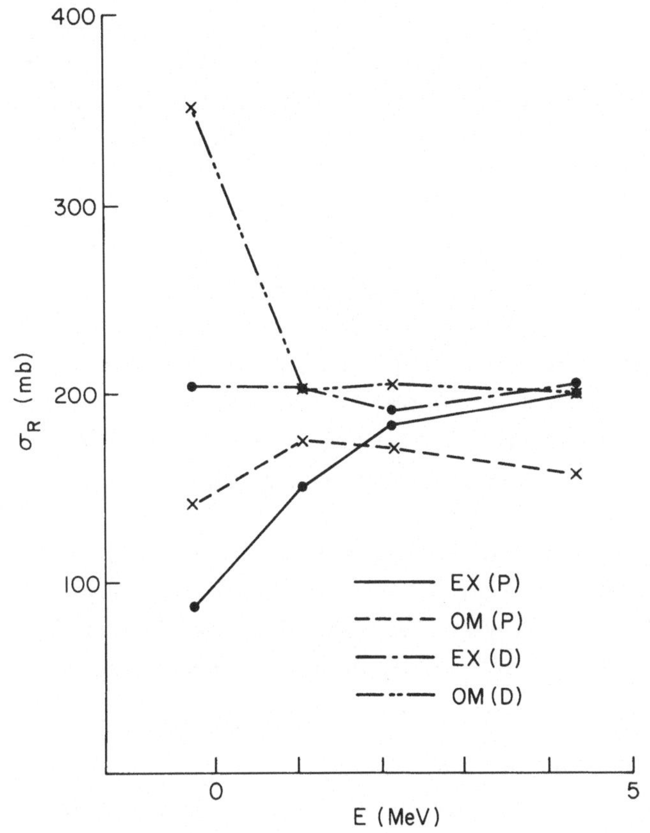

Fig.6.3. Optical model extraction cross sections in a three-body model [N+P+C]. Calculation of [6.31]. Exact reaction cross sections (EX), and those resulting from fitting the elastic cross section with a phenomenological optical potential (OP) are plotted for incident protons (P) and deuterons (D)

5% at all angles for protons, slightly worse for deuterons). Interestingly enough, the accuracy of the fit does not correspond to an accurate representation of the phase shift. The phenomenological deuteron phase parameters correspond to the exact ones better than for proton scattering, even though the angular distribution is not as good.

Optical potential fits are often used as a means for extracting total cross sections which are difficult to measure experimentally. The elastic angular distribution is used to determine the parameters of the optical potential. The two-body L-S equation then yields a forward scattering amplitude from which a total reaction cross section can be extracted. Since the exact reaction cross section can be calculated in the model, its inference via an optical potential can be tested. The results from the work of Shanley and Aaron are shown in Fig.6.3. The reaction cross sections extracted from the optical model can be off by as much as 40%.

Although these results are very interesting and quite suggestive, they cannot be taken as conclusive. The phenomenological parameters found by Shanley and Aaron are not especially smooth functions of energy; and because of the simple structure of the model, are not comparable to realistic nuclear parameters. This work points out, however, that a more sophisticated three-body model coupled with more extensive searches could provide a valuable testing ground for phenomenological models. Besides investigating phase shifts and reaction cross sections, the wave functions should be compared, as the extraction of these wave functions for use in DW matrix elements is the primary motivation for doing optical model fits.

The Budapest Model

In a series of papers, I. LOVAS and his collaborators at Budapest have extensively studied resonance phenomena in elastic scattering [6.45-52]. The model they study consists of two nucleons and a heavy core interacting in S-waves with spin-independent potentials. The unique feature of the model is that the nucleon-core interaction is taken to be a two-term separable potential, each term having the form of the difference of two Yamaguchis. The nucleon-core subsystem can therefore have either two bound states or a bound state and a resonance. This simulates the situation in the nuclear shell model where the nucleon is considered to be moving in a potential well which ordinarily supports a number of bound and resonance levels. The model is only solved below the threshold for three-body breakup. Even with this limitation, a wide variety of interesting phenomena occurs due to the presence of the second state in the two-body subsystems.

Three types of resonances can occur in the model. These are:[3]

1) *A single particle resonance.* The single incident nucleon is at an energy at which it resonates with the core.

2) *A quasi-doorway resonance.* One of the two nucleons is in a bound excited state and the other is in a single particle resonance.

3) *A doorway resonance.* This resonance is coupled to the configuration in which both particles are in bound states.

Depending on the parameters of the model, three-body spectra of differing complexity are obtained. The most interesting case is the one in which each nucleon has two bound states with the core, since the types of available resonances are the richest [6.46]. (They have also studied the case with two nucleons having one and two bound levels [6.51], and one and one bound levels [6.52]. We refer the reader to the original papers for consideration of these examples.)

With the parameters used, the three-body system has a doorway resonance when both particles are in their excited level. This resonance is examined in three

[3] The authors use the term "compound" instead of "doorway". We feel this is somewhat misleading. A compound nuclear state usually means a state where a large number of degrees of freedom is excited - sufficiently many so they can be treated statistically. In all of the resonance states of this model the number of degrees of freedom is the same - that of a three-body problem. For this reason we prefer the terms *quasi-doorway* (QDR), and *doorway* (DR).

approximations: the coupled channel (CC), coupled channel with exchange (CCE), and the isolated resonance approximation (IR). The exact three-body wave function $\Psi(p_1,q_1)$ can be expanded in terms of the eigenstates of the 2-3 pair by

$$\Psi^{EX} = \sum_{\nu=1}^{2} u_\nu(p_1)\phi_{1\nu}(q_1) + \int d^3k\, u_k(p_1)\phi_{1\underline{k}}^{(+)}(q_1) \tag{6.36}$$

where $\phi_{1\nu}(q_1)$ are the bound states of the 2-3 pair, and $\phi_{1\underline{k}}^{(+)}(q_1)$ are the scattering states with incoming relative momentum $\underline{k}$.

The CCA neglects the continuum part of this and writes

$$\Psi^{CC} = \sum_{\nu=1}^{2} u_\nu(p_1)\phi_{1\nu}(q_1) \quad . \tag{6.37}$$

It obtains coupled Schrödinger equations for the u_ν by writing

$$\langle\phi_{1\mu}|E - H|\Psi^{CC}\rangle = 0 \quad . \tag{6.38}$$

Exchange is included by writing

$$\Psi^{CCE} = \sum_{\nu=1}^{2} \left[u_\nu^d(p_1)\phi_{1\nu}(q_1) + u_\nu^e(p_2)\phi_{2\nu}(q_2)\right] \quad , \tag{6.39}$$

and using

$$\langle\phi_{2\mu}|E - H|\Psi^{CCE}\rangle = 0 \tag{6.40}$$

as well as (6.38).[4]

The IR assumes that the wave function is approximately given by the bound shell model part plus one particle in the continuum; viz.,

$$\Psi^{IR} = u_1(p_1)\phi_{11}(q_1) + \gamma\phi_{12}(p_1)\phi_{22}(p_2) \quad . \tag{6.41}$$

(Because the target is taken to be infinitely heavy, $p_1=q_2$ and $p_2=q_1$.) Eqs.(6.37) and (6.38) are applied to obtain two equations for u and γ. The constant γ can be eliminated to obtain an effective Schrödinger equation for u.

The three approximate methods are calculated and compared with the exact in the neighborhood of the resonance. The positions and widths are fit by using a single Breit-Wigner plus background term. The results are shown in Fig. 6.4. The energies and widths are plotted as a function of the n-p coupling constant λ. The range

---

[4] There are some formal difficulties in doing this because of the lack of orthogonality of the rearranged channels. The authors use the method of FRIEDMAN [6.53] to overcome them.

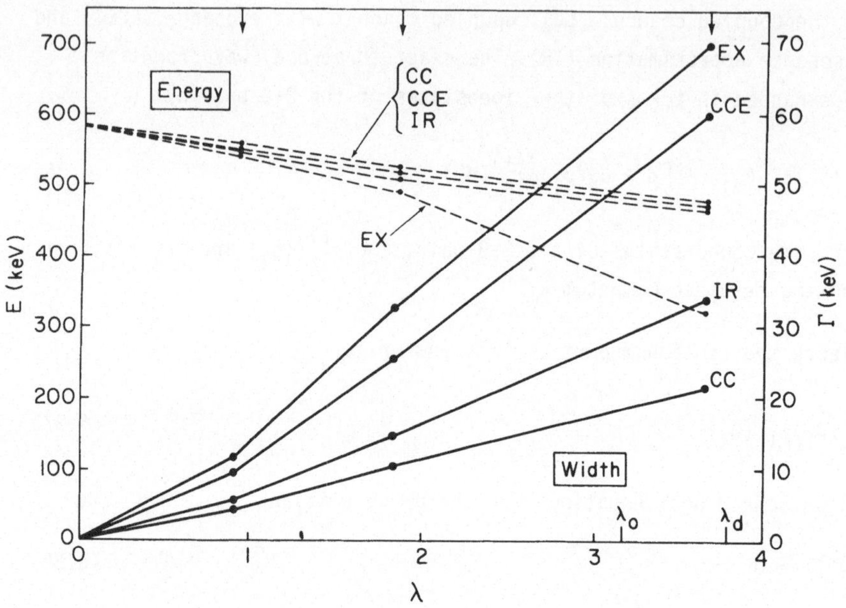

Fig.6.4. Calculation of resonance parameters for nucleon-nucleus scattering in a three-body model [N+N+C] by various approximation methods. Calculation of [6.46]. Energy (left scale) and width (right scale) of the resonance are plotted as a function of the nucleon-nucleon coupling strength. EX = exact, IR = isolated resonance, CC = coupled channel, CCE = coupled channel with exchange. λ_0 = coupling strength to produce a zero energy nucleon-nucleon bound state; λ_d = coupling strength to bind the two nucleons with 2.225 MeV

choosen is $\beta = 2.5\ F^{-1}$. This gives a shorter range force than the usual Yamaguchi triplet potential (β=1.4 F^{-1}). For this range, a zero energy bound state is obtained at λ = 3.17, while a strength of 3.78 is required to produce the correct deuteron binding energy.

We observe that all three approximate methods give essentially the same resonance energy, and one which underestimates the energy shift of the resonance position significantly. They all converge at the noninteracting value at λ=0 as they must, but realistic values of the interaction strength are much too great to yield a good value for the shift.

The width is underestimated by all the approximate methods, only the most sophisticated doing a reasonably accurate job. Note that including more "shell-model" configurations does not guarantee improving the width. The possible shell-model configurations for each particle include two bound states and the continuum. Labelling the bound states 1 and 2 and the continuum C, the configurations included in the various approximate calculations are as follows: (the first term in parentheses indicates the state of the first particle, the second term, the state of the second particle)

IR = (11) + (21) + (C1) + (22) ,

CC = IR + (12) + (C2) ,

CCE = CC + (1C) + (2C) ,

EX = CCE + (CC) .

Even though the IR includes fewer configurations than CC, it systematically gives a better width.

Some of the complexity in the full nuclear problem is present in these simple three-body models. The conclusions one can draw are similar to those we found in the study of stripping. The standard nuclear models do not do terribly well in giving quantitative fits to the resonance cross sections in the model case. However, the parameters of the model are not particularly realistic, and the validity of the models was investigated only as a function of one of the parameters (the N-N coupling strength). As a result, we can only conclude that the standard approximations may not work well, but cannot determine a region in the parameter space where they should work.

6.1.5 Other Reactions

In addition to the extensive studies that have been made of rearrangement reactions and elastic scattering in three-body models, a few other direct reactions have been studied. These include the knockout reaction (such as (p,2p)) and exchange reactions (such as (p,n)).

Knockout Reactions: Young and Redish

The knockout of a target particle by a fast projectile was suggested as a possible probe of nuclear wave functions in the early 1960's [6.54]. The energy spectrum of the residual nucleus should reflect the separation energies of the single particle orbitals in the target, and the angular distributions of the projectile and ejectile in coincidence should reflect the wave function of the orbital in momentum space. The experimental results from this reaction are generally interpreted as being the first direct proof of the validity of the nuclear shell model [6.55].

A major difficulty with this reaction is that the DW calculations for knockout are considerably more difficult than those which involve only two final state particles. Another problem is that the reaction penetrates more deeply into the nucleus than do reactions involving deuterons. (This is, of course, one of the factors which makes the reaction valuable.) One is therefore unsure whether a straightforward one-step DWA will be sufficient to describe the reaction mechanism. The accuracy of the DWIA approximation for knockout (6.8) is considered by YOUNG and

REDISH [6.56] and YOUNG [6.57] in a three-body model. The model used consists of two identical nucleons and a core. The parameters of the system are chosen to simulate the reaction $^4He(p,2p)^3H$. The standard DWIA for knockout is written

$$M_{fi} = \langle \chi^{(-)}_{p_1} \chi^{(-)}_{p_2} | t_{pp} | \chi^{(+)}_{k_o} \phi_i \rangle \quad . \tag{6.42}$$

The distorted waves $\chi^{(+)}_{k_0}$, $\chi^{(-)}_{p_1}$ and $\chi^{(-)}_{p_2}$ represent the incoming particle and the two outgoing particles, respectively. The wave function ϕ_i is the relative wave function of the ejected particle and the nuclear core.

There are a number of difficulties associated with the three-body nature of the matrix element which make it difficult to evaluate. 1) The effective interaction t_{p-p} is a two-body T matrix imbedded in a three-body space. It is a nonlocal, energy-dependent interaction. A large number of fully-off-shell matrix elements is needed. The most realistic current calculations suppress the nonlocality, energy dependent and off-shell effects for simplicity of evaluation.[5] 2) The final state should really be an eigenstate of $H_0+V_1+V_2$. Because the kinetic energy term does not separate in the coordinates of particles 1 and 2, the final state wave function should not be simply the product of distorted waves. This effect is expected to be small if the core is heavy enough [6.59].

Young and Redish evaluate an exact DWIA for knockout, and compare the results to the full calculation. The authors solve the AGS equations along a deformed contour by iterating, and summing the resulting multiple scattering series by Padê approximants.

In this method an exact DWIA can be obtained by simply "turning off" the proton-proton interaction after it has occurred once. The breakup amplitude is given on-shell by

$$U_{oi} = \Sigma_j t_j G_o U_{ji} \quad . \tag{6.43}$$

To obtain the DWIA the equations in Subsection 6.1.2 ((6.27) and (6.28)) are used to separate t_3. The specific equations obtained are

$$U'_{ji} = (1 - \delta_{ji})(\varepsilon - H_o) + \sum_{k\neq 3} (1 - \delta_{jk}) t_k G_o U'_{kj} \tag{6.44}$$

$$U_{ji} = U'_{ji} + U'_{j3} G_o t_3 G_o U_{3i} \quad . \tag{6.45}$$

The DWIA then requires the following approximations to obtain exactly one t_3 in each term contributing to U_{oi}:

[5] A code has recently been developed by KOSHEL [6.58] which should overcome some of these difficulties.

$$U_{ji} \approx U'_{j3}G_o t_3 G_o U'_{3i} \quad , \qquad j \neq 3 \tag{6.46}$$

$$U_{3i} \approx U'_{3i} \quad . \tag{6.47}$$

Substituting this in (6.43) gives

$$U_{oi}^{DWIA} = t_3 G_o U'_{3i} + \sum_{k=1,2} t_k G_o U'_{k3} G_o t_3 G_o U'_{31} \quad . \tag{6.48}$$

The full amplitude can be expressed as

$$U_{oi} = [U_{oi}^{DWIA}] + [U'_{33}G_o t_3 G_o U'_{33} G_o t_3 U_{3i} + t_3 G_o U'_{33} G_o t_3 G_o U_{3i}] + [\sum_{k=1,2} t_k G_o U'_{ki}] \quad . \tag{6.49}$$

This decomposition is similar to that of KAZAKS and KOSHEL [6.60]. These authors refer to the terms in brackets as the DWIA term (one t_3), the "resonance" term (more than one t_3), and the "recoil term" (no t_3). We retain the notation "recoil term" for the third part of the breakup amplitude since it can be shown that this term vanishes in the limit that the mass of the core goes to infinity. However, as there are no resonances in this problem, the nomenclature "resonance term" seems inappropriate. We refer to this term rather as the "multi-step term", since it describes multi-step processes going through both rearrangement and breakup states.

The calculation is carried out in the S-wave separable model at 65 and 100 MeV incident proton energies. The breakup cross sections are found for the energy-sharing geometry, that is, the angles of the outgoing protons are fixed, and the spectrum is measured as a function of the energy of one of the protons. (This fixes the kinematics of the final state completely.) Symmetric angle pairs in the neighborhood of the quasi-free peak were studied to facilitate comparison with the experimental data of PUGH et al. [6.61].

The authors find that the DWIA is reasonably good at 100 MeV (at the quasi-free peak Exact/DW = 0.94), but is considerably poorer at 65 MeV (Exact/DW = 0.66). The plane wave normalization factors (Exact/PW) are 0.69 and 0.46 at 100 and 65 MeV, respectively. The shapes of the energy-sharing spectra are very good in both the PWIA and the DWIA (and actually agree well with the experimental shape). This shows that the agreement with the shape of the quasi-free breakup cross section does not guarantee the accuracy of the impulse approximation, and certainly does not imply that spectroscopic factors can be extracted using the approximation. (A similar observation has been made by AMADO and TAKAHASHI [6.62] in the FSI region.)

The recoil term is calculated explicitly and found to give a contribution of only a few percent at the quasi-free peak. (It does not fall off as quickly as the quasi-free part does, so it is somewhat more important in the tail region.) Since a very light nucleus was used, one can expect that this term will be entirely negligible even for nuclei as light as carbon or oxygen.

The multi-step term, on the other hand, appears to become more important as the energy decreases. At these energies one can expect that a substantial part of the multi-step term is coming from successive scatterings in breakup states.

The authors chose parameters to simulate a realistic situation. The accuracy of the DWIA was not investigated as a function of the potential parameters. The effect of including P and D-waves in the p-t interaction was investigated in the exact calculation at 65 MeV and was found to reduce the cross section by about 10%. The effect on the accuracy of the DWIA was not considered.

Charge Exchange: Gignoux

The charge exchange reaction (p,n) was investigated by GIGNOUX [6.34] in his model (described in detail in Subsect.6.1.3 above). Charge exchange can be thought of in three-body terms simply as heavy particle stripping: 1(23)→2(13). As in ordinary deuteron stripping, the DWBA may be written

$$M_{fi} = \langle \chi_n^{(-)} \phi_p | V_{n-p} | \chi_p^{(+)} \phi_n \rangle \quad , \tag{6.50}$$

where ϕ_n and ϕ_p are neutron-core and proton-core bound states, respectively.

DODD and GREIDER [6.17] suggested a modification of the DWBA in order to obtain a multi-step series generated by a connected operator. This modification replaces V_{np} by t_{np} in the DW matrix element (DWTA). Gignoux compares this DWTA, the DWBA, and the PWBA with the exact results.

He finds that for all the sets of parameters chosen (see Table 6.1) none of the approximations works well. The DWTA is systematically too large by about the same amount that the DWBA is too small. The plane wave Born approximation is better than either for all the cases considered, but is not very good in an absolute sense. No explanation of this interesting result is attempted. This subject certainly deserves further study. The charge exchange reaction is one which has recently been found to have important two-step contributions from rearrangement processes (such as (p,d,n), pickup followed by stripping) [6.63]. Since rearrangement two-step processes can be treated in a three-body model, an investigation of them could be of conserable importance for understanding the kinds of reactions in which the DWBA can be expected to work.

6.1.6 Summary and Conclusions

Three-body models have been used in a number of different reactions including stripping, elastic scattering, knockout, and charge exchange. In these calculations, standard nuclear approximations have been applied to the models and the approximate results compared with the exact. In essentially all the studies the result is the same. The approximate methods are not nearly as good as one had hoped or had been led to expect from their agreement with data. The conclusion that must be drawn is that the approximate methods cannot be expected to work in all situations, or under a wide variety of conditions.

It is therefore incumbent upon students of nuclear reactions to consider carefully the conditions for the validity of their approximations. Agreement of a model with experiment, while certainly desirable, provides a tenuous grasp on a deep understanding of the process. Much of importance can be masked by the phenomenological uncertainties present in the standard DW methods. Few-body models should be able to provide testing grounds and usable guides for the further development of realistic theories. There are very few examples of this to date. Most of the models discussed in the section above have provided important information on the existence of limitations of the standard approaches to direct reactions, but have not yet led to the development of new insights for improving these methods.

6.2 Three-Body Effects in Nuclear Processes

In the previous section we have learned that perturbation theory does not always suffice for the treatment of three-body effects in a three-body model. In this section we consider the question: are three-body effects present in real nuclear processes? To this end we review some attempts to describe nuclear reactions by realistic three-body models, some relevant nuclear data, and modifications of standard semi-phenomenological methods arising from three-body arguments. The use of three-body models in nuclear reactions has been extensive, but rarely related to Faddeev methods. Due to space limitations, we select a small number of areas for careful consideration. They are specifically: elastic scattering of simple projectiles, deuteron scattering, and reactions with three-body final states. In these cases, the three-body effects are clearly present. There appears to be considerable opportunity for the application of three-body methods.

6.2.1 Elastic Scattering of Simple Projectiles: Optical Potentials

In this subsection we consider the elastic scattering of a noncomposite hadronic projectile (nucleon, pion, or kaon) from a nucleus. The standard treatments of elastic scattering rely on a multiple scattering series (MSS) for the optical potential.

We begin by reviewing the Watson MSS for the optical potential. We observe that the effective projectile-nucleon interaction appearing in this series is a many-body operator. In order to determine the type of approximation needed to handle this operator, some experimental data concerning the important inelastic states are considered. We conclude that, for nucleon projectiles at energies above 100 MeV, the states corresponding to the knockout of a single target nucleon are of primary importance. We then show how to build approximate three-body models of the process. The unitarity relation is cited to show the implicit assumptions concerning the reaction amplitudes in the model.

Although no calculations of the optical potential using a full three-body model yet exist, there are a number of calculations which include some three-body effects. These provide evidence that the three-body corrections yield important effects which are needed if the data are to be accurately reproduced.

Since the elastic process in a nuclear scattering is usually the most important one, it is useful to separate out those intermediate states in which the target remains in its ground state. The rest of the intermediate states are summed (formally) into an effective energy-dependent nonlocal interaction called the optical potential. When a two-body L-S equation is solved using this potential, the exact elastic scattering amplitude results. This method was introduced by FRANCIS and WATSON [6.64] and refined and extended in important ways by FESHBACH [6.65], KERMAN et al. [6.66], BELL and SQUIRES [6.67], and GRASSBERGER and SANDHAS [6.18]. (This list is by no means exhaustive.)

Most nuclear calculations of the optical potential for those energies where direct reactions are expected to dominate (and for which few-body methods should be relevant) are performed with some version of Watson's MSS (see GOLDBERGER and WATSON [6.68], Chap.XI). We therefore review it briefly here.

The Watson Series

We use the multiple scattering labelling described in Subsection 6.1.1. The residual interaction in the elastic channel is $V^0 = \sum_{j=1}^{A} V_{0j}$, and the elastic scattering operator satisfies the L-S equation

$$U_{00} = V^0 + V^0 G_A U_{00} \quad , \tag{6.51}$$

where G_A is the channel Green function

$$G_A^{-1} = E^+ - T_0 - H_A \quad . \tag{6.52}$$

The kinetic energy of the projectile in the CM frame is written as T_0, and the full Hamiltonian of the target is H_A. The intermediate states in which the target is not excited are extracted by the introduction of the operator

$$P_A = |\phi_0^A><\phi_0^A| \quad , \tag{6.53}$$

where ϕ_0^A is the ground state wave function of the target. Defining $Q_A=1-P_A$, the L-S equation is

$$U_{00} = V^0 + V^0 G_A (P_A + Q_A) U_{00} \quad . \tag{6.54}$$

Solving the Q_A part explicitly yields the pair of equations

$$U_{00} = U'_{00} + U'_{00} G_A P_A U_{00} \tag{6.55}$$

$$U'_{00} = V^0 + V^0 G_A Q_A U'_{00} \quad . \tag{6.56}$$

An expansion for the optical potential U'_{00} is obtained by summing the V_{0j}'s into a two-body T-matrix (imbedded in the many-body space). Defining

$$U^{j'}_{00} = V_{0j}(1 + G_A Q_A U'_{00}) \quad , \tag{6.57}$$

(6.56) implies

$$U'_{00} = \sum_{j=1}^{A} U^{j'}_{00} \quad . \tag{6.58}$$

By introducing an effective two-body scattering operator

$$\begin{aligned} \tau_{0j} &= V_{0j} + V_{0j} G_A Q_A \tau_{0j} \\ &= (1 + \tau_{0j} G_A Q_A) V_{0j} \quad , \end{aligned} \tag{6.59}$$

the V_{0j} in (6.57) may be converted to a τ_{0j} by left multiplying by the factor $(1+\tau_{0j} G_A Q_A)$. This gives

$$U^{j'}_{00} = \tau_{0j} + \tau_{0j} G_A Q_A \sum_{k \neq j} U^{k'}_{00} \quad . \tag{6.60}$$

Iterating this equation and inserting it in (6.58) yields the Watson MSS for the optical potential:

$$U'_{00} = \sum_j \tau_{0j} + \sum_{j \neq k} \tau_{0j} G_A Q_A \tau_{0k} + \sum_{i \neq j \neq k} \tau_{0i} G_A Q_A \tau_{0j} G_A Q_A \tau_{0k} + \cdots \quad . \tag{6.61}$$

The series derived above has been obtained by formal manipulations from the exact L-S equation, so if (6.60) were solved, we would expect to get the exact answer. This equation contains all the many-body aspects of the problem and cannot be solved exactly in practice.

We note that the L-S equation (6.51) has a disconnected kernel. In the separation of (6.51) into (6.55) and (6.56) all the disconnectedness is in the equation for the optical potential. This is because the ground state of the target only appears in the sum of all those graphs which are A-connected, i.e., bring in all the particles of the target. The operator V^0P_A is therefore completely connected. The kernel of (6.56), on the other hand, clearly contains disconnected graphs. We will encounter the problems associated with these in any approximation to (6.56) which attempts to include three-body degrees of freedom. These degrees of freedom are usually suppressed in the standard treatments, so the disconnectedness problem is not encountered. (The disconnectedness problem can be avoided entirely by beginning with a generalization of the Faddeev equations to the N-body problem. See, for exemple, [6.69].)

The Important Intermediate States

We wish to determine methods of approximating (6.60) so as to include the most important processes. One way of choosing a starting point is to investigate which reactions are the most important experimentally. While this is not conclusive due to the presence of off-shell amplitudes in the intermediate states, it certainly should provide a good starting point. If one investigates the inclusive (singles) spectra of the type A(p,x) it is found that above 50 MeV, the most important particles emitted are single nucleons [6.70-71].

WALL and his collaborators [6.72-74] have argued that the dominant process above about 100 MeV is knockout. This conclusion is drawn primarily from the shapes and magnitude of proton singles spectra compared with those obtained in DWIA calculations. The qualitative shapes of these spectra are displayed in Fig.6.5. They show clearly the importance of the continuum of the target excitation spectra, and show the characteristic large quasi-free bump.

These data suggest that a large part of what happens in nucleon-nucleus scattering is three-body in nature. The range of validity of this assumption appears to be $E>150$ MeV, $A<40$, with the three-body mechanism gaining importance as energy increases and A decreases. It seems likely that as A decreases, three-body methods could be extended to lower energies.[6]

---

[6] Note that the conclusions on the importance of three-body mechanisms in these processes are being drawn from a mixture of theoretical expectations and limited experimental information. More extensive data, such as broad spectrum coincidence experiments, would be extremely valuable in understanding the nature of the many-body reaction mechanisms which actually occur.

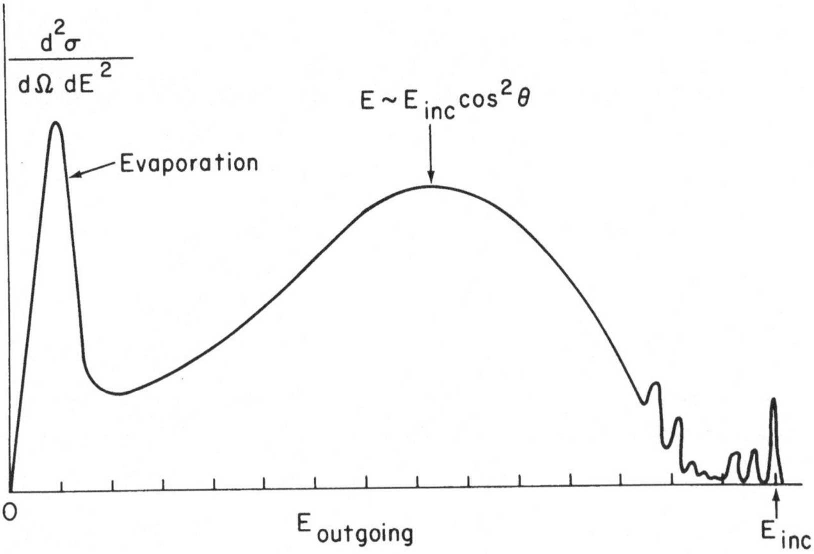

Fig.6.5. Qualitative structure of the singles spectrum at a fixed angle for the reaction (p,p') on a light nucleus at energies greater than 100 MeV. The peak energy of the large broad bump is approximately that given by the kinematics of free nucleon-nucleon scattering. From [6.136]

How can the three-body reaction mechanism be applied to the many-body problem? It is rarely possible to treat the nucleon-nucleus system simply as two nucleons plus an inert core. The problem must be considered as a superposition of A three-body problems, which then have to be woven together in a coherent manner. One such weaving can be performed using the Watson MSS.

The Three-Body Model of Single Scattering

Equation (6.59) for the effective interaction operator shows that it is actually a many-body operator. It is more complicated than the simple imbedding of a two particle operator in a many-body space (see (1.72)), since the shift of the Green function from the two-body form includes the nondiagonal operators V_{ij}. The operator τ_{0j} therefore cannot be interpreted as a two-body T matrix with a shifted energy. Many-body effects are present even if only the single scattering approximation for the optical potential, $U_{00}^{j'} \approx \tau_{0j}$, is retained in (6.60). The discussion of the previous subsection suggests that a three-body treatment of τ_{0j} might be valuable. This was done for a single scattering treatment of the optical potential by TANDY, REDISH, and BOLLÉ (TRB) [6.75]. (See also [6.76] and [6.137].)

We focus attention on the struck nucleon in (6.59) by decomposing the target Hamiltonian via

$$H_A = T_j + T_J + H_J + V_0^j \quad . \tag{6.62}$$

We label the struck nucleon by j and the rest of the nucleus by J. T_j is the kinetic energy of the jth particle, T_J the CM kinetic energy of the remaining A-1 target nucleons, H_J their internal Hamiltonian, and V_0^j the interaction of the jth particle with the rest of the nucleons in the target:

$$V_0^j = \sum_{\substack{i=1 \\ (i\neq j)}}^{A} V_{ij} \quad . \tag{6.63}$$

The decomposition (6.62) emphasizes the three-body nature of the process. It is an exact equation. If one wants to retain only the three-body processes while suppressing higher order effects, (6.62) may be written:

$$H_A = T_j + T_J + H_J + \overline{U}_0^j + (V_0^j - \overline{U}_0^j) \tag{6.64}$$

where $\overline{U}_0^j$ is a single particle potential felt by the particle j.[7] TRB then suggest expanding in the residual interaction $V_0^j - \overline{U}_0^j$. In lowest order they obtain for the single scattering operator τ_{0j} (which we write in three-body notation as τ_J)

$$\tau_J \approx V_{0j} + V_{0j} \frac{Q_A}{E - T_0 - T_j - T_J - H_J - \overline{U}_0^j} \tau_j \quad . \tag{6.65}$$

Except for the presence of the operator Q_A, this is precisely a three-body transition operator for the three-body problem with particles 0, j, and J, and interactions $V_0 = \overline{U}_0^j$, $V_j = 0$, and $V_J = V_{0j}$. After some algebraic manipulation, one obtains the Faddeev-like equations[8]

$$\begin{aligned} \tau_J &= t_J G_0^{(j)} \tau_0^{(j)} \\ \tau_0^{(j)} &= (G_0^{(j)})^{-1} + (t_0^{(j)} G_0^{(j)} - G_0^{(j)-1} \Gamma_e) \tau_J \quad , \end{aligned} \tag{6.66}$$

where

$$t_J = V_{0j} + V_{0j} G_0^{(j)} t_J \quad , \tag{6.67}$$

---

7 Note that although we use the three-body "odd-man-out" notation, the interaction between particles j and J requires an additional labelling beyond "$\overline{U}_0$", to indicate which of the A three-body problems we are considering. This is also necessary on the "free" Green function.

8 Alternatively, they can be written down as the AGS equations (1.121b) for this problem ($U_{1,1} \to \tau_J$, $U_{2,1} \to 0$, $U_{3,1} \to \tau_0^{(j)}$, etc.). One need only realize that the effect of Q_A is just to remove the bound-state pole in the nucleon-core T-matrix.

$$t_0^{(j)} = \overline{U}_0^i + \overline{U}_0^j G_0^{(j)} t_0^{(j)} \quad , \tag{6.68}$$

and

$$G_0^{(j)} = (E^+ - T_0 - T_j - T_J - H_J)^{-1} \quad . \tag{6.69}$$

This differs from the three-body equations with two interactions in the presence of the term $G_0^{(j)-1}\Gamma_e$ in the kernel. This term is necessary to correct for the over-counting which is implicit in the use of the free T matrices. Since they permit any intermediate states for the particles 0 and j, there will be some component in which the j-J wave function overlaps the ground state of the full target. When the optical potential is inserted in (6.55) and this L-S equation is solved, the part of the wave function in which the target is in its ground state is overcounted. The correction, which arises naturally from the use of the Watson's Q_A in conjunction with a three-body model, is similar to the Grassberger-Sandhas separation of the optical potential [6.18] in the three-body problem since

$$tG_0 - G_0^{-1}\Gamma = (t - G_0^{-1}\Gamma G_0^{-1})G_0 \quad . \tag{6.70}$$

Splitting t into a regular part t', and a pole part $V\Gamma V$, gives

$$tG_0 - G_0^{-1}\Gamma = (t' + V\Gamma V - G_0^{-1}\Gamma G_0^{-1})G_0 \quad . \tag{6.71}$$

But $V\Gamma V$ and $G_0^{-1}\Gamma G_0^{-1}$ agree at the pole, so the kernel is of the form arrived at in Section 6.1.2 above.

The authors show that the KMT trick [6.66] can be used to eliminate the correction in the kernel, thus yielding a true three-body equation for the optical potential. If the target nucleons are considered identical, the three-body operator τ_J may be replaced by the operator $\hat{\tau}_J$ which satisfies the standard three-body equations:

$$\begin{aligned} \hat{\tau}_J &= t_J G_0^{(j)} \hat{\tau}_0^{(j)} \quad , \\ \hat{\tau}_0^{(j)} &= (G_0^{(j)})^{-1} + t_0^{(j)} G_0^{(j)} \hat{\tau}_J \quad . \end{aligned} \tag{6.72}$$

If the L-S equation (6.55) is replaced by

$$\mathcal{U}_{00} = \mathcal{U}'_{00} + \mathcal{U}'_{00}\Gamma_e \mathcal{U}_{00} \quad , \tag{6.73}$$

where

$$\mathcal{U}_{00} = \left(\frac{A-1}{A}\right) U_{00} \tag{6.74}$$

then, in the single-scattering approximation to (6.60), one finds

$$\mathcal{U}'_{00} \cong (A-1)\,\hat{\tau}_J \cong \left(\frac{A-1}{A}\right) U'_{00} \quad . \tag{6.75}$$

(When matrix elements of $\mathcal{U}'_{00}$ are taken, all the target nucleons will contribute by identity of particles.)

This therefore shows that a simple superposition of three-body problems (6.72) yields a description of the three-body reaction mechanism for elastic scattering without overcounting, if the three-body equations (6.72) are used for the KMT optical potential, instead of the usual L-S optical potential, (6.55). (Within the framework of multiple scattering theory, three body equations similar to (6.66) have also been obtained by REVAI [6.137].)

Unitarity

The imaginary part of the optical potential arises from reactions which lead to a depletion of the elastic beam. It is important to understand the relation between the imaginary part of the optical potential, and the reaction mechanism assumed, as this provides an important consistency check on the optical potential.

This was done by TRB [6.75] in the above single-scattering model of the optical potential using the standard methods of three-body unitarity. Because of the complexity of the argument, we refer the reader to the appendix of [6.76], and quote only the results.

Extracting the imaginary part of U_{00} from (6.55) yields the equation

$$\text{disc}\,\{U_{00}\} = U^{\dagger}_{00}\,\text{disc}\,\{\Gamma_e\}U_{00} + \Omega^{\dagger}_0\,\text{disc}\,\{U'_{00}\}\Omega_0 \quad , \tag{6.76}$$

where for any operator $A(E)$

$$\text{disc}\,\{A(E)\} = A(E^+)^{\dagger} - A(E^+) \quad , \tag{6.77}$$

and

$$\Omega_0 = 1 + \Gamma_e U_{00} \tag{6.78}$$

converts the initial plane wave into the exact distorted wave. The unitarity defect is therefore given by a distorted wave matrix element of the anti-Hermitian part of the optical potential.

Since in the TRB model, U'_{00} is a sum of terms satisfying three-body equations, three-body unitarity may be used to explicate the reactive content of U'_{00}. They obtain[9]

$$\mathrm{Im}\, U_{00} = U^{\dagger}_{00}\Lambda_0 U_{00} + U^{\dagger}_{p0}\Lambda_p U_{p0} + U^{\dagger}_0 \Lambda U_0 \tag{6.79}$$

where Λ_0, Λ_p, and Λ project on the on-shell elastic, pickup and knockout states, respectively. The amplitudes U_{p0} and U_0 are the amplitudes for pickup and knockout, and are given by

$$U_{p0} = \tau_0^{(j)}\Omega_0 \sqrt{A} \quad , \tag{6.80}$$

$$U_0 = (1 + t_e G_0)\tau_J \Omega_0 \sqrt{A} \quad , \tag{6.81}$$

where t_e is one of the $t_0^{(j)}$, and G_0 is the corresponding $G_0^{(j)}$.

The amplitudes have a clear physical interpretation which can be used to suggest refinements of the model. The operators τ_J and $\tau_0^{(+)}$ are three-body operators. They include all contributions to the pickup and breakup processes which are three-body in character, including multi-step rearrangement and breakup processes. The operator Ω_0 is the full many-body elastic scattering projection. It converts the incoming plane wave into an exact distorted wave. The reactions which are implicit in the model are therefore treated in a manner similar to a DW approximation but including only the distortion in the initial state and all final state interactions with the picked-up or knocked-out particle. The distortion of the projectile on the residual nucleus in the final state is not included. One may increase the sophistication of the model by including a third interaction (projectile-residual nucleus) in the three-body equations, adjusting the equations to produce a reliable approximation for the reaction amplitudes via the three-body unitarity relation. Once this is done, it seems likely that it will be possible to include exchange effects for nucleon projectiles in a reasonably straightforward way.

TRB also point out in [6.76] that if the reaction amplitudes are also desired, it is not necessary to solve the three-body equations for the optical potential, solve the L-S equation for the elastic amplitude, and finally perform the DW quadratures implied by (6.80) and (6.81). Rather, one can combine these equations into a single three-body equation for the reaction and elastic scattering amplitudes, viz.,

---

[9] This may be derived by recognizing that (6.72,73) are a special case of the AGS equations (cf. earlier footnote), and using the unitarity relations (1.138).

$$U_{00} = \sqrt{A} t_p G_0 U_{p0} \quad ,$$

$$U_{p0} = \sqrt{A} G_0^{-1} + \frac{1}{\sqrt{A}} (t_e G_0 - (A-1) G_0^{-1} \Gamma_e) U_{00} \quad , \qquad (6.82)$$

$$U_0 = \frac{1}{\sqrt{A}} (1 + t_e G_0) U_{00} \quad .$$

These equations provide a unified three-body-like model for the elastic and reaction amplitudes in the many-body problem, which should be amenable to standard three-body numerical methods. Since the amplitudes reduce directly to well-known approximations when the three-body effects are assumed to be small (IA for the optical potential, DWIA for knockout), one is guaranteed a starting point which gives reasonably reliable answers.

The Three-Body Energy Shift:

Nucleons. Multiple scattering calculations for single nucleon optical potentials usually ignore binding and three-body effects. Equations (6.72) are approximated by the first iterate

$$\hat{\tau}_J \approx t_J \quad , \qquad (6.83)$$

and the Green function in (6.67) is approximated by the two-body Green function

$$G_0^{(j)} \approx (\varepsilon^+ - T_{0j})^{-1} \quad , \qquad (6.84)$$

where T_{0j} is the relative kinetic energy of particles 0 and j, and ε is some energy related to E (usually 1/2 the lab K.E. of the projectile). One obtains this from (6.69) by 1) ignoring the binding potential $\overline{U}_0^j$, 2) ignoring the binding energy of the struck particle, H_J (the "hole state" energy), and 3) approximating the CM energy of the pair by their on-shell value. (This last reduces E to E/2.) Most calculations also 4) ignore all off-shell effects, replacing t_J by an on-shell two-body amplitude.

These effects are due to the three-body character of the problem. Ignoring them all replaces the effective three-body problem by an effective two-body one.

The last three approximations have been investigated for nucleon projectiles by LERNER and REDISH [6.77] and LERNER [6.78]. These authors calculated the optical potential using (6.83) in (6.75)[10], with the full Green function (6.69).

---

[10] These authors do not make the KMT correction as later suggested be TRB. Since $A/(A-1) \approx 1.06$, the effect of including it should not distort their conclusions very much.

Because of the presence of T_0+T_j and H_J in the Green function, the two-body T-matrix is not on-shell. The calculations of the optical potential matrix elements

$$<\underline{k}|U'(E)|\underline{k}'> = \sum_{\lambda \text{ occupied}} <\underline{k}\psi_\lambda|t(E)|\underline{k}'\psi_\lambda> \qquad (6.85)$$

require off-shell input. The authors use the two-body T matrix generated by the realistic soft core potential of REID [6.79], including fully off-shell effects in the nucleon-nucleon S-wave states.

The conventional wisdom holds that the single scattering approximation for the optical potential is reasonably good at energies above 100 MeV. Lerner and Redish perform their calculation at 65 MeV for ^{16}O, to see whether including some three-body effects can extend the range of validity to lower energies.

Only the real part of the well is calculated and an effective local potential found. This agrees very well with the best phenomenological Woods-Saxon potential in the nuclear surface, but is significantly deeper (65 MeV vs. 37 MeV) in the interior.

The effect of this difference on the scattering is tested by adding phenomenological imaginary and spin-orbit potentials to the calculated real well, and adjusting their parameters to optimize the fit to the differential cross section. They find, using the calculated well with realistic single particle wave functions, that the fit to the data could be improved - the total χ^2 per point dropping from 4.1 for the purely phenomenological fit (nine parameters), to 2.5 for the calculated well (six parameters).

The on-shell approximation usually employed at energies greater than 100 MeV was tried and found to give a poor representation of the real part especially in the surface region. Adding phenomenological spin-orbit and imaginary parts did not permit the real well computed from the on-shell approximation to obtain a fit to the data. The best χ^2/N value obtained with this real well was 23.

These results strongly suggest that the single scattering approximation for the optical potential may have a greater range of validity than usually expected, but that at energies below 100 MeV three-body effects become important.

Another interesting result concerning the importance of three-body effects in elastic nucleon-nucleus scattering has been obtained by MACKINTOSH and KOBOS [6.80, 81]. These authors consider the scattering of 30 MeV protons from ^{40}Ca. There are excellent data for this process, both differential cross sections and polarization. They have been difficult to fit using standard phenomenological Woods-Saxon parameters, especially at back angles.

The authors use a CRC equation in which the elastic proton channel is coupled directly to (three) deuteron channels. The equations are similar to (6.27). The proton and deuteron optical potentials are adjusted to fit the proton elastic and the pick-up data.

An improvement in χ^2 of almost a factor of three was obtained by going from an L-S equation with a phenomenological optical potential to the CRC equations. This seems to indicate that the removal of a nucleon from the nucleus, followed by its restoration, is important in nucleon elastic scattering. This three-body effect would be difficult to reproduce by adjusting parameters in a simple local potential, but can be easily included in a three-body formalism. (As seen from the discussion in Section 6.1.2, the CRC equations may be considered as three-body equations with pole approximations.)

The two calculations described above indicate that both from a theoretical and a phenomenological point of view, our current best guesses for the two-nucleon force, and for phenomenological optical potentials and transfer strengths, imply that three-body effects play a significant role in elastic nucleon-nucleus scattering below 100 MeV.

Pions: Because of its small mass, relativistic effects become important in pion scattering at quite low energies. (At 50 MeV, the pion velocity is $\sim$2/3 the speed of light.) However, the simplicity of the Watson-KMT calculations and the short range of the π-N force have impelled many authors to attempt calculations of the pion-nucleus optical potential using this formalism.

There have been a number of calculations which have considered the three-body effects on the single scattering form of the optical potential. We here consider two of them: the calculations of KUJAWSKI and AITKEN [6.82] and LANDAU and THOMAS [6,83], the latter including some relativistic effects.

Kujawski and Aitken follow closely the method of Lerner and Redish discussed immediately above. They choose a potential model for the π-N interaction, and fold the off-shell T matrix over bound state single-particle wave functions (6.85). The three-body Green function of (6.69) is used. They calculate π-^{12}C scattering at 120 and 180 MeV.

The potentials they choose are the one- and two-term separable potentials fit to π-N data using nonrelativistic kinematics [6.84]. They take harmonic oscillator wave functions for the single particle states. (Note that these wave functions do not have realistic long-range behavior. This could affect the accuracy of the optical potential in the surface region severely.) The optical potential is calculated in momentum space with the full nonlocality retained. The nonrelativistic L-S equation is then solved to obtain the π-nucleus amplitudes.

They find that at 180 MeV the elastic cross section is fit quite well except in the forward direction (15^0), where the cross section obtained is too large. The fit at 120 MeV is better for the forward points, but does badly away from the forward peak.

They also compare their calculations with factorized approximations of the type

$$\langle \underline{k}|U(E)|\underline{k}\rangle \approx A\rho(\underline{q})\langle \underline{k}',\underline{p}_0-\underline{q}|t(\omega)|\underline{k},\underline{p}_0\rangle \quad , \tag{6.86}$$

where $\underline{p}_0$ is some "average" momentum (or zero), $\underline{q}=\underline{k}'-\underline{k}$, and ω is approximated by the on-shell energy with the Fermi motion terms dropped. These approximations do not work well away from the forward peak. It seems likely that a factorized approximation is precluded by the rapid variation of the off-shell T matrix in this energy region arising from the resonance. The coupling of the Fermi motion of the struck nucleon with this rapid off-shell variation is not simply representable in factorized form. The good qualitative nature of the fits is encouraging, especially at 180 MeV, since neither the best potentials nor wave functions available were used.

LANDAU and THOMAS [6.83] investigate the factorized approximation (6.86) at energies below 100 MeV. Since this is well below the resonance, the factorized approximation should work better than it did at higher energies, the off-shell π-N T matrix being more slowly varying. These authors examine $\pi^{\pm}$ scattering from ^4He and ^{12}C at 50 MeV by using (6.85) in a relativistic L-S equation. They consider a number of different choices for the energy ω including ω equal to the invariant π-N mass

$$\omega_0 = [E_\pi(k) + E_N(P_0)]^2 - (\underline{k} + \underline{p}_0)^2 \quad , \tag{6.87}$$

and the three-body choice

$$\omega^{(3)} = E_\pi(k) + \frac{k^2}{2Am_N} + m_N - |E_b| - \frac{\underline{P}^2}{2(A-1)m_N} - \frac{\underline{P}^2}{2(m_N+m_\pi)} \quad , \tag{6.88}$$

where $\underline{P}$ is the total pion nucleon momentum (to be Fermi averaged). These two choices differ by typically 30-40 MeV (the three-body prescription being lower) for ^4He and ^{12}C, if E_b is the full single-particle binding energy.

They find a sensitivity to the choice of energy, with the three-body choice being preferable. The best fits were obtained if the value $|E_b|$ were taken to be 5 MeV in all three sets of data considered, π^+-^{12}C, and $\pi^{\pm}$-^4He. These results cannot be taken as being conclusive due to the presence of numerous possible corrections (not the least of which is the reliability of the factorized approximation at these energies).

Both these calculations seem to indicate that, among the many factors that must be included in pion-nucleus scattering below 200 MeV, the effect of the implicit three-body kinematics is one which probably needs to be considered if detailed fits are desired.

Kaons: The kaon optical potential provides an extremely interesting example of the presence of three-body effects in multi-particle hadronic systems. In this case, a sub-threshold resonance requires the three-body energy shift be included with some care.

This subject has recently been reviewed by SEKI and WIEGAND [6.85] so we will keep our discussion of the background information brief. A good survey of the subject is also given in the early sections of ALBERG et al. [6.86]. The nuclear information is contained in the kaon capture rates which may be inferred from the shifts and widths of the last X-ray emitted by the kaon in its cascade to the nucleus. Early studies of this problem used perturbation theory to calculate the capture rates, but more recently it has been found that perturbation theory does not suffice [6.87,88]. The method now used is to solve the Klein-Gordon or Schrödinger equation containing both the electric potential arising from the nuclear charge distribution and a kaon-nucleus (complex) optical potential. (See, for example, KRELL and ERICSON [6.89].)

In order to construct a kaon-nucleus optical potential, we may employ the Watson or KMT single scattering formulas. Using the form (6.85) we may make use of the fact that the kaon-nucleon force is very short range. The longest range part of this interaction arises from exchange of a ρ-meson which has a range $1/m_\rho \approx 1/4$ fm. The momentum dependence of the $\bar{K}$-N fully-off-shell T-matrix should therefore be small, suggesting the low-energy approximation

$$\langle \underline{k}|t_{\bar{K}-N}(E)|\underline{k}'\rangle \approx \langle 0|t(E)|0\rangle \equiv t(E) \quad . \tag{6.89}$$

Putting this into (6.85) gives

$$\langle \underline{k}'|U'(E)|\underline{k}\rangle \approx \sum_{\lambda\ \text{occupied}} \int d\underline{q}\,\psi_\lambda^*(\underline{q}+\underline{\kappa})\langle 0|t(W)|0\rangle\psi_\lambda(\underline{q}) \quad , \tag{6.90}$$

with

$$W = E - \varepsilon_\lambda - \frac{(\underline{k}+\underline{q})^2}{2M}$$
$$\underline{\kappa} = \underline{k}' - \underline{k} \quad . \tag{6.91}$$

The integral does *not* factorize even with an approximation which is effectively zero range because of the three-body energy shift of the Green function.

If the T matrix is slowly varying in energy, then the optical potential factorizes in the form

$$\langle \underline{k}'|U'(E)|\underline{k}\rangle \approx t(E)\rho(\underline{\kappa}) \quad , \tag{6.92}$$

where ρ is the Fourier transform of the nuclear density. This form was tried using for E the $\bar{K}N$ threshold energy and was found to give an inadequate result. Attempts to fit the data by adjusting the complex constant $t(E_0)$ require the opposite sign for the real part of t from the free $\bar{K}N$ amplitude [6.90].

It was suggested [6.91] that the presence of the sub-threshold resonance could produce this effect. The energy dependence of the free $\bar{K}N$ scattering amplitude [6.92] shows a sign change of the real part at the resonance, so a downward shift in the effective energy at which the amplitude is evaluated could produce the phenomenological results.

BARDEEN and TORIGOE [6.93] studied this suggestion using a factorization of (6.90), viz.

$$\int d\underline{q}\psi_\lambda^*(\underline{q}+\underline{\kappa})\tau[W(\underline{q})]\psi_\lambda(\underline{q}) \approx \frac{\int d\underline{p}\tau[W(\underline{p})]|\psi_\lambda(\underline{p})|^2}{\int d\underline{p}'|\psi_\lambda(\underline{p}')|^2}\int d\underline{q}\psi_\lambda^*(\underline{q}+\underline{\kappa})\psi_\lambda(\underline{q}) \quad ,$$

$$= \{\int d\underline{p}\tau[W(\underline{p})]P[W(\underline{p})]\}\rho(\underline{\kappa}) \quad , \tag{6.93}$$

where P(W) is the probability of finding a nucleon on the nucleus with the momentum appropriate to give the three-body energy W. The probability distribution is plotted in Fig.6.6 (after [6.93]). This distribution is relevant for all our elastic scattering discussions, since it represents the shift of the energy appearing in the T-matrix arising from the Fermi motion of the struck nucleon. (Note, however, that

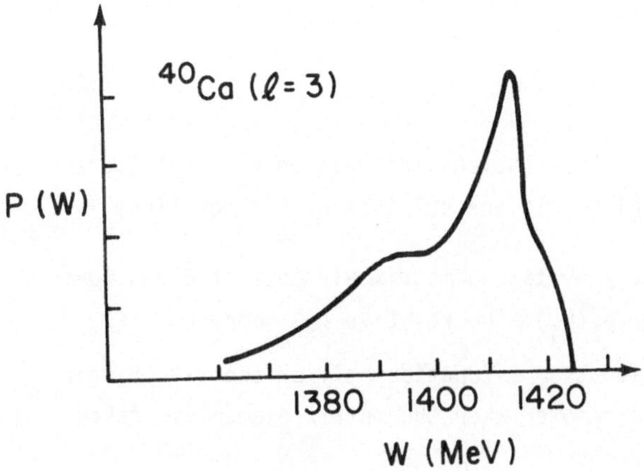

Fig.6.6. Probability of finding a given three-body energy W in scattering of a kaon from an f-shell nucleon in ^{40}Ca. From [6.93]. The kaon mass is included in W

the projectile mass is included in W in this figure.) It gives a good qualitative idea of the energy shifts involved. In particular, the peak of the probability gives a good indication of the region of W relevant for the full integral. Since the Y_0^* energy is 1405 MeV [6.94], near the peak of the distribution, the energy dependence of the resonance should be very important.

Using the energy averaged amplitude (6.93), Bardeen and Torigoe find a significant improvement over the use of the free amplitude in explaining the shifts and widths.

This approximation was investigated in detail by ALBERG, HENLEY and WILETS (AHW) [6.86]. These authors use a three-body model to describe the kaon-nucleus optical model (though the methods of solution employed are not based on the Faddeev equations).

Since the kaon and the nucleon both have I-spin=1/2, two isospin channels are possible, I=0 and I=1, with the Y_0^* resonance occurring in the I=0 channel. The I=0 state is treated as two coupled channels (KN and $\Sigma\pi$) with a separable potential matrix. The I=1 state is treated as a single channel, with a one-term separable potential. All the form factors are taken to be S-wave Yamaguchis. In the I=1 state the scattering length is complex and the effective range real. This necessitates using a complex strength and range because the $\Sigma\pi$ channel is suppressed in handling the I=1 state. The complex part of the range produces a slow oscillation in the coordinate space form factor.

AHW derive the optical model from a combination of Feshbach's method [6.65] with shell model assumptions, yielding a set of equations very similar to the ones we obtained above in the context of the Watson theory. Labelling the kaon as 1, the nucleon as 2, and the core as 3, and using the three-body notation for the potentials, their equations become

$$\tau(E) = V_3 + V_3 G^3 Q \tau(E) \tag{6.94}$$

$$G^3 = (E^+ - H_0 - V^3)^{-1} \tag{6.95}$$

$$\langle p_2 | V_2 | p_2' \rangle = \sum_{\lambda \text{ occupied}} \langle \lambda p_2 | \tau(E_K) | \lambda p_2' \rangle \quad . \tag{6.96}$$

The optical potential is taken to be the kaon-nucleus interaction V_2. This therefore introduces a self-consistency problem into the solution of the equations.

The equations are solved approximately by taking matrix elements of Green functions of the type $(E - T_{p_i} - T_{q_i} - V^i)^{-1}$ over $g_1(p_i)$. The relative K-N coordinate is replaced by zero in the potentials making them a function only of the K-N CM coordinate. This is considered as a fixed parameter when the matrix element is taken on the relative coordinate. The kinetic energy operator is also approximated by a constant $K^2/2M$, and later a Fourier transform is taken with respect to K.

(These somewhat heavy-handed approximation methods are quite commonly used by nuclear physicists when dealing with three-body problems. They arise from generalizing the two-body problem in nuclear matter. In that case, a pair of nucleons interact in the presence of an infinite, translationally invariant nucleus. Two of the three potentials in the problem then become simply constants (possibly momentum dependent). This eliminates the three-body nature of the problem and allows the use of two-body methods. These tricks are generalized to finite nuclei by only permitting the pair to interact when they are near or at the same point, and then taking the potentials arising from the third particle (the nucleus) to be constant in space, having the value the nonuniform potential has at some place between the two interacting particles. This class of approaches which transfer an infinite nuclear matter method to finite nuclei goes under the rubric of "Local Density Approximations." It is difficult to estimate their accuracy[11] and comparisons with experiment usually replace such estimations (a dangerous procedure in the author's opinion). It would be extremely interesting to study the validity of these approximations in a three-body model.)

AHW consider a number of effects on the optical potential: nonlocality, the effect of the projector, and the sensitivity of the widths and shifts to the inputs - the two-body potentials and the nuclear wave functions. Finally, a comparison with the calculation of Bardeen and Torigoe discussed above allows one to evaluate the accuracy of the factorization (6.93).

To examine the nonlocality, they compare three "effective local" potentials constructed from their nonlocal optical potential. These three forms give qualitatively similar results for the bulk of the potential (better than 10% in the interior). They have important differences from a simple factorized form like (6.92). The real part changes sign, becoming repulsive outside of about 3 fm. This implies that the "effective local potential" does not track with the nucleon density which is, of course, monotonic. The imaginary part is significantly more long range than the density. This clearly demonstrates the need for including finite range effects. The increase in range is not a reflection of the nonzero range of the K-N force, but an explicit three-body effect. The disagreement of the local potentials, however, appears to be restricted to the nuclear interior, so locality may be a good approximation in the surface.

---

[11] This is particularly true when the decomposition of the energy denominator into relative and cm coordinates of the light particles is made (Jacobi coordinates p_3, q_3). The use of the single particle energies in G-matrix calculations [6.95], treats the denominator more carefully (Jacobi coordinates p_1, p_2).

The projector Q appearing in (6.94) enforces both the Pauli principle and the optical model condition that the nuclear ground state never appear. It prevents the nucleon from returning either to the state it left or to a state occupied by another nucleon. AHW use an approximate analytic form for Q which permits them to perform all the necessary intergrals analytically (given the local density approximations discussed above). The effect of including Q on the transitions in the P-shell is on the order of 5%. The effect on the optical potential is not discussed.

The variation of the shift and width on the parameters of the problem is considered in a factorized approximation. They find that changes of a tenth of a Fermi in either r_0 (= nuclear radius/$A^{1/3}$) or the surface thickness lead to changes of 10-20% in the widths and shifts.

Since the parameters of the nuclear density are what is desired, let us compare this sensitivity to the theoretical uncertainties, for example, to the three-body energy shifts. In the factorized approximation, the effective energy involves the three-body kinematic shift. Evaluating this sensitivity they find that a 20 MeV shift in the energy denominator produce changes of 10-15% in the shifts and widths.

The sensitivity to the class of nuclear model was extremely important, the difference between harmonic oscillator and Woods-Saxon wave functions being almost 50%. This is expected since HO wave functions fall off much too rapidly in the surface region.

AHW compare the results of their finite-range factorized approximation with the zero-range factorization of Bardeen and Torigoe discussed above. They find that the finite range changes the widths and shifts from the values obtained by Bardeen and Torigoe by about 20% (occasionally by a factor of 2). In some cases, the shift is towards the data, in others away.

They consider a number of other effects and find them generally small. A number of other theoretical uncertainties were not treated. The recoil of the nucleus is suppressed, and the true many-body nature of the model not considered (see the discussion in Section 6.2.1). Both of these should be 1/A effects, but since for the P-shell nuclei considered 1/A~10%, these effects may be important. The application of an approach based on three-body methods would permit their inclusion.

The study of the kaon-nucleus optical potential for use in obtaining widths and shifts of atomic levels in kaonic atoms clearly indicates that the presence of three-body effects is extremely important. The three-body shift of the energy is required to get the sign of the optical potential, and if nuclear information is to be extracted, further three-body effects such as the increase of the range and the nuclear recoil will almost certainly need to be included. Only preliminary approaches to this problem using highly simplified three-body models have been made till now.

6.2.2 Deuteron Reactions

The scattering and reaction processes induced by incoming deuterons are perhaps the most natural three-body problems in all of many-body nuclear physics. The structure of the deuteron is so loose that its RMS radius is significantly greater than the range of the nuclear force and its binding energy per nucleon is smaller than that of any other bound nucleus. This means that much of the time the neutron and proton in the deuteron are outside the range of the interaction. When a deuteron interacts with a compact, tightly bound nucleus, one therefore expects that it will behave very much like a three-body system consisting of a proton (labelled 1), a neutron (labelled 2), and a nucleus (labelled 3).

Although deuteron reactions in nuclear physics are commonly treated by the DWBA, corrections of a three-body nature have been investigated since the earliest days of direct reactions. The most obvious three-body effects are multi-step processes proceeding through rearrangement (stripping followed by pickup), and through deuteron breakup states. Though originally there was considerable controversy about the importance of these effects, in recent years it has become clear that they are quite important, especially if good quality fits to the data are desired.

In this section we will discuss three-body effects on deuteron elastic scattering and stripping. The next subsection considers the attempts to fit these two processes with true three-body models. Because of space limitations we have omitted a discussion of the deuteron breakup reaction itself. There has been considerable work on this subject particularly by BAUR and collaborators (see [6.96] and references therein).

Elastic Scattering

The effect of rearrangement and breakup processes on elastic deuteron scattering has been considered by numerous authors [6.97-107]. The most extensive and detailed of these investigations have been the studies of JOHNSON and SOPER [6.101] and RAWITSCHER [6.104-107]. We will consider only these here.

Essentially all of the treatments of deuteron-nucleus scattering as a three-body problem begin with the assumption that the system can be described by a three-body Hamiltonian.[12] The authors considered here simply assume that the many-body Hamil-

---

[12] The extraction of an effective three-body Hamiltonian from the many-body problem is a nontrivial one. This question is beyond the scope of the present article.

tonian can be approximated by a three-body one, with V_1 and V_2 equal to nucleon-nucleus optical potentials. Since these potentials are in general energy dependent there is some uncertainty as to which to choose. A common procedure is to take the nucleon optical potentials at half the energy of the incident deuteron [6.101]. This question deserves more theoretical attention than it has received so far.

RAWITSCHER [6.104-107] considers the effect of rearrangement on the deuteron-nucleus optical potential. He begins with the presumption that the deuteron should be strongly absorbed in the nuclear interior due to the ease of breakup. Phenomenological fits to deuteron elastic scattering fail to yield this strong absorption. Rawitscher constructs a set of coupled reaction channel equations equivalent to (6.27), using phenomenological potentials which couple the elastic deuteron channel to the stripping channel. The specific nucleus considered is ^{40}Ca, with deuteron energies between 7 and 22 MeV. All the stripping strength is put into a single effective state of the 2p shell-model type. The proton optical potential is taken as the phenomenological Woods-Saxon well, fit to elastic data. The deuteron potential is chosen to be a Woods-Saxon with a strong volume absorption. The n-p force is taken as zero range.

He finds [6.104] that using coupled equations has considerable effect on the elastic cross section. The primary effect is to introduce a strong even-odd ℓ dependence into the S-matrix. The wave function produced by the coupled channel equations is much more strongly suppressed in the interior than that arising from the phenomenological optical potential. The fits to the elastic data are fair, while the fits to the stripping cross sections are good.

This result shows that stripping amplitudes of realistic magnitudes (10% of geometric in this case) can produce important effects on the optical potential, which are difficult to reproduce with a simple local optical well. Furthermore, they can be responsible for the suppression of the internal wave functions, which has been found necessary to fit stripping cross sections.

In a later paper [6.105] RAWITSCHER fits his coupled channel elastic cross sections with phenomenological optical potentials, and observes that the only parameter to require significant variation with energy is the radius of the imaginary well. This falls as energy increases with a slope of about $dr_0/dE \sim -0.05$ fm/MeV. This is comparable to what is found experimentally.

The effects of deuteron breakup on elastic scattering are considered by JOHNSON and SOPER [6.101]. These authors begin by expanding the effective three-body wave function ψ in eigenstates of the n-p channel Hamiltonian, H_3, viz.

$$\psi(\underline{p}_3,\underline{q}_3) = \phi_3(\underline{q}_3)\chi_0(\underline{p}_3) + \int d\underline{k}\phi_k^{(+)}(\underline{q}_3)\chi(\underline{k},\underline{p}_3) \quad . \tag{6.97}$$

They then construct coupled channel equations from the three-body Schrödinger equation.

Using projection operators

$$P = |\phi_3\rangle\langle\phi_3| \quad , \tag{6.98}$$

$$Q = 1 - P \quad , \tag{6.99}$$

the coupled channel equations

$$(E - H_3 - PV^3P)P\psi = PV^3Q\psi \tag{6.100a}$$

$$(E - H_3)Q\psi = QV^3P\psi + QV^3Q\psi \tag{6.100b}$$

are easily obtained. When written as an integral equation (6.100b) has a disconnected kernel. They argue that for low deuteron energies, the n-p breakup contribution is dominated by their relative S-states. If the relevant energies are low, the relative wave functions will be dominated by the deuteron pole and the shape of the scattering wave function $\phi_k^{(+)}$ will not change much. (This is essentially the pole-dominance argument familiar in three-body physics.)

This implies that the dependence on k of the coupling matrix element PV^3Q ($=\langle\phi_0|V^3|\phi_k^{(+)}\rangle$), factors out. Second, they assume that the k dependence of $Q\psi(=\chi_k)$ can be obtained from (6.100b) by ignoring the coupling term, QV^3Q. This introduces the connected operator V^3P into the expression for χ_k eliminating the disconnectedness problem. They obtain a pair of coupled equations for χ_0 and a single effective breakup wave function

$$\chi_1(p_3) = \langle p_3 q_3 = 0|V_3Q|\psi\rangle / \langle q_3 = 0|V_3|\phi_3\rangle \quad . \tag{6.101}$$

They then solve the effective coupled channel equations numerically. The nucleon-nucleus potentials are taken to be the phenomenological optical potentials at half the deuteron energy. The calculation is carried out for 21.6 MeV deuterons incident on Ni.

The authors compare the results obtained by using the Watanabe potential with those obtained using the coupled channel equations (evaluated approximately), and with phenomenological fits to the data. They find that the inclusion of breakup modifies the Watanabe cross section considerably. Neither agree particularly well with the pehnomenological cross section (which fits the data will). A 10% reduction of the Watanabe fit produces a reasonable fit to the data which is considerably improved by including the breakup effect.

Johnson and Soper suggest that a possible source for the 10% correction to the real part might be from exchange effects with the core. This is an effective three-body force which would be obtained from formally reducing the many-body problem to an effective three-body one. Their estimate of this effect shows it to be the right order of magnitude.

The main effect of the breakup on the optical potential is to redistribute the absorption away from the surface, producing a resulting *decrease* in the reaction cross section compared to the Watanabe potential. This occurs even though the Johnson-Soper potential has absorption corresponding to deuteron breakup and Watanabe's does not. This reduction improves the agreement with the reaction cross section extracted from the phenomenological optical potential.

RAWITSCHER [6.109] extends these studies by solving a coarse discretization of the continuum in the coupled channel equations (6.100). The results are qualitatively similar to those of [6.101] but he finds that rather high relative n-p energies are important, in contrast to the assumptions of [6.101]. Relative energies up to at least 10 and possibly 40 MeV seem relevant.

Deuteron breakup and rearrangement both can therefore be expected to influence the elastic scattering of deuterons significantly in scattering from real nuclei. To include both effects simultaneously a true three-body approach would be required.

Stripping

The effects of breakup in stripping and pickup processes have also been studied by a number of authors [6.101-103,109-112]. The most important and influential is the paper of JOHNSON and SOPER (J-S) [6.101], considered above in the context of elastic scattering. We will discuss only this paper here, and refer the reader to the literature cited above for more extensive information. (See also [6.138.])

The central physical point of the J-S treatment of stripping is to observe that the stripping matrix element in the three-body model may be written as in (6.2) for i=3, j=1. It is therefore only the combination $V_3\psi_3$ which enters. Since the potential V_3 is much shorter range than the size of the deuteron, the only part of the coordinate space wave function $\psi_3(r_3,R_3)$ which is important for the matrix element (6.2) is $\psi_3(0,R_3)$. If one now uses the coupled channel equation in coordinate space, and replaces the relative n-p energy ε_k by the deuteron binding energy $-\varepsilon_3$, the equation becomes an equation for $\overline{\chi}(R) = \psi_3(0,R_3)$. The stripping amplitude then becomes simply a DWBA matrix element using $\overline{\chi}^{(+)}$ instead of $\chi_3^{(+)}$ as the incoming distorted wave. This has the considerable advantage that the breakup effect can be approximately included, without any added complexity compared to existing calculations.

The distorted wave $\bar{\chi}$ is to be generated with a new effective potential (the "adiabatic" potential) not necessarily related to elastic scattering. After various approximations Johnson and Soper arrive at the form for this potential

$$\bar{U} = \frac{\langle \underline{q}_3 = 0|V^3 V_3|\phi_3\rangle}{\langle \underline{q}_3 = 0|V_3|\phi_3\rangle} \quad . \tag{6.102}$$

This should be compared with the lowest order (Watanabe) expression for the elastic optical potential, $U'_{33}=\langle\phi_3|V^3|\phi_3\rangle$. The adiabatic potential differs from the Watanabe in being the average of the residual interaction V^3 over a shorter range function ($V_3\phi_3$ as compared to $|\phi_3|^2$).

A different derivation [6.103] produces the approximation

$$\bar{U} = \frac{\langle\phi_3|V^3 V_3|\phi_3\rangle}{\langle\phi_3|V_3|\phi_3\rangle} \quad . \tag{6.103}$$

This form has the advantage that systematic higher order corrections were also obtained.

The result that breakup effects can be approximately included, without leaving the DWBA structure of the matrix element, by changing the choice of distorted wave may be more general than is expected from the J-S derivation. For deuteron stripping the crucial physical point is the factorization of the energy dependence of the n-p scattering states and their coordinate space dependence within the range of their interaction. This is rigorous for a separable potential. It has been shown by BOULDIN and LEVIN [6.21] that when V_{np} is separable, a distorted wave matrix element gives the exact stripping amplitude if the Mitra spectator function [6.16] is used as the deuteron distorted wave. (See VANZANI [6.135] for the relation with Faddeev theory.)

This result can be seen within the context of our discussion in Subsection 6.1.2. The effective interaction for stripping is given by (6.28). Writing out the equation for this operator explicitly (using (6.24) rather than (6.28)) gives

$$U'_{13} = B_3 + (t_3G_0 - B_3R_3\Gamma_3L_3)U'_{33} + (t_2G_0 - B_2R_2\Gamma_2L_2)U'_{23} \quad . \tag{6.104}$$

The effect of deuteron breakup is contained in the second term on the right. If the T matrix t_3 is separable, then the proper choice of L_3 and R_3 can make this vanish exactly. If we choose the BRS form ($B_3=V_3$), and take $R_3=1$ and

$$L_3 = f_3V_3G_0 \quad , \tag{6.105}$$

where

$$f_3 = \frac{\tau_3}{\Gamma_3} \quad , \tag{6.106}$$

then the kernel term becomes

$$t_3G_0 - B_3R_3\Gamma_3L_3 = |3\rangle\tau_3\langle 3|G_0 - V_3|\phi_3\rangle\frac{\tau_3}{f_3}\langle\phi_3|f_3V_3G_0 \quad . \tag{6.107}$$

Using (1.50,55) we see that the expression vanishes.

With this off-shell transformation the first-order optical potential U'_{33} may be obtained from (6.28). Explicitly we have

$$U'_{33} = (t_1G_0 - B_1R_1\Gamma_1L_1)U'_{33} + (t_2G_0 - B_2R_2\Gamma_2L_2)U'_{23} \quad . \tag{6.108}$$

Approximating U'_{13} and U'_{23} by the Born terms B_3, and assuming that the $BR\Gamma L$ terms can be chosen to cancel the separable part of t_i (the $V_iG_iV_i$ in (1.52)), we obtain the approximation

$$U'_{33} \approx V_1G_0V_3 + V_2G_0V_3 = V^3G_0V_3 \quad , \tag{6.109}$$

so

$$\mathcal{U}'_{33} = f_3V_3G_0V^3G_0V_3 \quad . \tag{6.110}$$

This is similar to the various expressions for the J-S potential (6.102) and (6.103), since

$$\langle\phi_3|V_3|\phi_3\rangle = \langle\phi_3|G_0^{-1}|\phi_3\rangle \tag{6.111}$$

on-shell, but there are differences. The complete relation between the J-S adiabatic potential and the various possible three-body optical potentials deserves further investigation.

RAWITSCHER [6.109,110] and FARRELL et al. [6.111] and FARRELL [6.113] extend the J-S method by considering the breakup states in more detail. They both use coupled channel methods including discretizations of the continuum. Since they all truncate ill-behaved equations until they are no longer ill-behaved, there is some doubt as to the reliability of their results. They find that breakup has considerable impact on the stripping and that the high energy relative momentum components are also important.

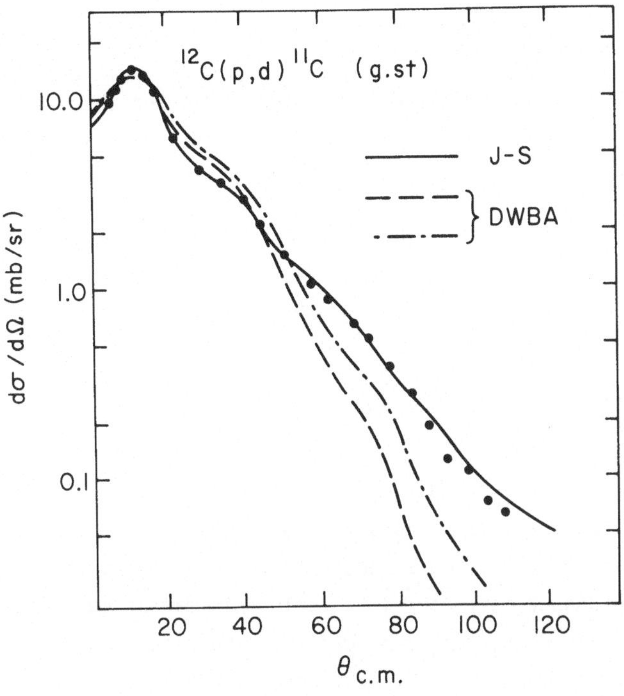

Fig.6.7. Neutron pickup on ^{12}C at 65 MeV. Comparison of Johnson-Soper (J-S) method for stripping, which includes some deuteron breakup effects, with standard DWBA. Two DWBA calculations using slightly different optical potentials are shown. From [6.112]

Experimentally, the J-S breakup correction seems to yield substantial improvements in fitting data with semi-phenomenological DWBAs. SATCHLER [6.114] has shown how phenomenological Woods-Saxon neutron and proton optical potentials can be simply modified to yield Woods-Saxon adiabatic deuteron potentials. Calculations using these potentials improve the fits to data on both light and heavy nuclei, often substantially [6.112,115,116]. The correction is particularly important above 20 MeV. One of the best J-S fits to data is shown in Fig.6.7 (from [6.112]). Two standard DWBA fits are also shown.

Three-Body Models

Both rearrangement and breakup have important effects on elastic deuteron scattering and on stripping. Since both of these effects cannot be treated simultaneously in a coupled channel model, it seems very natural that a true three-body model based on Faddeev theory should be the most natural and successful approach to deuteron reactions. To date this hope has not been realized (except, perhaps, for the special case of d+α scattering [6.117,118]).

A number of reasonably sophisticated three-body calculations have been attempted to describe deuteron elastic scattering and stripping [6.31,119,120] but the fits to data are poor in essentially all cases. Some of the difficulties are:

1) The Coulomb force must be included.

2) Spectroscopic factors must be put in.
3) Realistic nucleon-nucleus potentials are required.

Comment 3) is the most difficult and perhaps the most important. It means that Woods-Saxon type potentials are probably needed, many nucleon-nucleus partial waves must be included, and absorption effects in these interactions must be put in.

There has been promising work on each of these problems (for 1) [6.121], for 2) [6.69] and [6.122], and for 3) [6.123] and [6.124]), but these elements have never been combined in what would undoubtedly be a state of the art three-body calculation.

6.2.3 Multi-Particle Final States

Reactions with three or more particles in the final state are necessarily many-body in nature, at least in the sense of the phase space which must be considered. Often, in both nuclear and particle physics, the interactions of resonating pairs are considered in the presence of a multi-particle final state, with the deviations from phase space used to extract the resonance parameters. It is in this manner that the interactions of pairs of very short-lived particles or particles which cannot be made into targets are studied (e.g., π-π and n-n interactions). There is some evidence that the presence of additional particles in the final state can cause distortions of the apparent pairwise parameters. This is a three-body effect. We consider the model study of AMADO and NOBLE [6.125], and the experimental evidence for these effects in *Final State Interactions*, below.

A second use of multi-particle final state reactions is the study of the wave function of a particle or cluster in a many-body bound state by knocking it out. This method has recently found applications in nuclear, atomic, and molecular physics. (See the Indiana Symposium, edited by BONHAM and WALKER [6.126]). In this process, a direct reaction mechanism is assumed. Because of the three-body nature of this process, an off-shell, two-body matrix element is necessarily involved. Theoretical and experimental studies have indicated the need for including this three-body effect in order to obtain fits to data. There is also some evidence for the need for a more sophisticated three-body reaction mechanism in (p,2p) reactions at energies below 100 MeV. These aspects will be discussed below.

Final State Interactions

In many cases in nuclear and particle physics, the interaction of two (or more) particles can be most easily studied in the context of a production or breakup reaction which has many particles in the final state. An enhancement in a missing mass or inclusive spectrum is then used as evidence for a resonance or anti-bound state. Two important examples are the extraction of the neutron-neutron scattering length, and the study of meson and baryon resonances such as the ρ meson. In order to extract scattering parameters for the interacting pair, one usually makes the simplified assumption of a sequential reaction or pure final state interaction.

AMADO and NOBLE [6.125] study the validity of the simplified assumption in a three-body model. They consider the decay of a 0^+ particle (the "G") into three strongly interacting 0^+ particles (called "H"). The decay is mediated by an interaction (called "weak") which they assume can be treated perturbatively. The matrix element for decay is $\langle\chi_{3H}|K_{weak}|G\rangle$, where

$$|\chi_{3H}\rangle = [1 + G_0U_{3H}(E)]|3H\rangle \quad , \tag{6.112}$$

with $|3H\rangle$ the plane wave state of the 3H particles, and U_{3H} the full Faddeev T matrix.

The strong interaction is chosen to be a separable interaction with form factor

$$\nu(k) = k^2(k^2 + \alpha^2)^{-2} \quad . \tag{6.113}$$

A simple Yamaguchi would not suffice to produce a resonance, as monotonic potentials are well known not to produce S-wave resonances [6.127] since there is no angular momentum barrier to trap the wave function. The form above has a barrier of its own.

The weak interaction is taken to be

$$\langle \underline{p},\underline{q}|K_{weak}|G\rangle = \alpha_W^2\left[\alpha_W^2 + E(p,q)\right]^{-1} \quad . \tag{6.114}$$

The relevant parameters are the strength of the strong interaction (in dimensionless terms they write $\nu=\lambda/\lambda_0$, where λ_0 is the strength which produces a zero energy bound state), the ratio of the weak to the strong range $\rho=\alpha_W/\alpha$, and the dimensionless energy $\varepsilon=E/M_H$.

The Faddeev equations are written directly for a reduced form of the weak decay matrix element, and are solved by matrix inversion on a rotated contour.

Amado and Noble find that for narrow resonances ($\nu>0.95$) the width can be determined from the final state spectrum. As the resonance begins to broaden, three-body effects come into play, shifting the position and the widths of the decay by significant amounts. A short-range weak decay influences the width more than the position, while the opposite is true for a long-range weak decay. Another particularly interesting effect they observed was on the strength of the total decay rate and the singles spectrum. Especially when the weak force is short range, large enhancements and de-enhancements are observed. De-enhancements are seen when the two-body strong force is weakly attractive. The shape of the singles spectrum is by no means a guide to the enhancement strengths arising from three-body effects. One case ($\nu=0.55$,

$E/M_H = 0.9$) gave a singles spectrum resembling phase space very closely in shape, but having an enhancement factor of almost 1000.

These results indicate that it may be necessary to construct more sophisticated models than simple FSI for dealing with pairs interacting in the presence of other particles. Some places where these difficulties may be relevant are described in the references of Amado and Noble. Other possible examples of these effects include the modification of meson resonance widths observed in anti-proton annihilation [6.128], and the effect of deuteron breakup on the electrodisintegration of ^6Li into α+d [6.129].

The Off-Shell DWIA

As described in Subsection 6.1.1 above, knockout may be considered as a one-step direct reaction, in which the projectile interacts once strongly with the ejectile in the presence of an average nuclear potential. This leads to a DWIA matrix element for the knockout process of the form given by (6.8). If one makes a three-body model of this reaction (as in Subsect.6.1.5 above), the effective interaction producing the knockout should be a two-body T matrix imbedded in a three-body space. As a result, off-shell two-body matrix elements enter. REDISH [6.130] derives a simple form for including this off-shell effect. Specifically, if we write out (6.8) we obtain

$$\begin{aligned} M_{fi} = \int & d\underline{q}_1 d\underline{q}_2 d\underline{q}_1' d\underline{q}_2' \chi_{k_1}^{(-)*}(\underline{q}_1)\chi_{k_2}^{(-)*}(\underline{q}_2) \\ & \times \langle \underline{q}_1 - \underline{q}_2 | t\left[E - \varepsilon_\lambda - \frac{(\underline{q}_1 + \underline{q}_2)^2}{4m}\right] | \underline{q}_1' - \underline{q}_2' \rangle \delta(\underline{q}_1' + \underline{q}_2' - \underline{q}_1 - \underline{q}_2) \\ & \times \chi_{k_0}^{(+)}(\underline{q}_1')\psi_\lambda(\underline{q}_2') \quad . \end{aligned} \tag{6.115}$$

At reasonably high energies ($E_{k_0} \gtrsim 50$ MeV), the eikonal approximation is not too bad, and the distored waves are forward peaked. We expect them to be reasonable well localized in momentum space near the plane wave value. If this localization is sharp compared to the momentum shifts required to produce changes in the T matrix, then one may expand the arguments of T around the plane wave momentum values, viz.

$$\begin{aligned} & \langle \underline{q}_1 - \underline{q}_2 | t\left[E - \varepsilon_\lambda - \frac{1}{4M}(\underline{q}_1 + \underline{q}_2)^2\right] | \underline{q}_1' - \underline{q}_2' \rangle \\ & \approx \langle \underline{k}_1 - \underline{k}_2 | t\left[E - \varepsilon_\lambda - \frac{1}{4M}(\underline{k}_1 + \underline{k}_2)^2\right] | \underline{k}_0 - (\underline{k}_1 + \underline{k}_2 - \underline{k}_0) \rangle \quad . \end{aligned} \tag{6.116}$$

If the zero of energy is chosen at the threshold for breakup, this becomes a half-shell T-matrix

$$<\underline{k}_1 - \underline{k}_2|t\left[\frac{1}{2\mu}(\underline{k}_1 - \underline{k}_2)^2\right]|2\underline{k}_o - \underline{k}_1 - \underline{k}_2> \equiv <\underline{k}_f|t(\varepsilon_f)|\underline{k}_i> \quad , \qquad (6.117)$$

and the matrix element becomes

$$M_{fi} = <\underline{k}_f|t(\varepsilon_f)|\underline{k}_i>g_\lambda(\underline{k}_o,\underline{k}_1,\underline{k}_2) \quad , \qquad (6.118)$$

where

$$g_\lambda(\underline{k}_o,\underline{k}_1,\underline{k}_2) = \int d\underline{r}\chi_{k_1}^{(-)*}(\underline{r})\chi_{k_2}^{(-)*}(\underline{r})\chi_{k_o}^{(+)}(\underline{r})\Psi_\lambda(\underline{r}) \qquad (6.119)$$

is referred to as the distorted momentum distribution. The fact that the two-body T matrix is off the energy shell is specifically a three-body effect arising from the binding of the ejectile to the spectator (residual nucleus). The difference between the on- and off-shell momenta in the coplanar symmetric geometry is discussed by REDISH et al. [6.131], and arises both from the binding energy and the "recoil momentum", $\underline{k}'=\underline{k}_1+\underline{k}_2-\underline{k}_o$. (This is the momentum the struck particle would have had before the collision in the absence of distortions.)

Three-Body Effects in (p,2p) Reactions

The sensitivity of the (p,2p) reaction amplitudes to the three-body off-shell effect discussed above is studied in [6.131] and [6.132] in the energy range 100-350 MeV, for the coplanar symmetric geometry. They find that there is off-shell sensitivity for $\theta<45^\circ$ at the lower end of the energy range. As the energy goes up, the angle for which on-shell approximations suffice moves forward, until by 350 MeV eventually all of the observable region may be treated as being on shell.

For the regions of phase space in which an off-shell treatment is necessary, effects of up to a factor of two or more may be obtained as compared with on-shell treatments. When different realistic potentials are compared, however, the largest differences arise from the different quality of fits these potentials have to on-shell data. The ratio of off-shell to on-shell cross sections for different realistic and for different phase equivalent potentials varies by about 15%.

These off-shell effects have been observed at energies below 100 MeV by the Maryland group (BHOWMIK et al. [6.133]) in ^{6}Li(p,2p). Significant improvement in fitting the shapes of the angular distributions results from the use of the three-body (half-shell) prescription. Their results are shown in Fig.6.8. For further discussion of this point, see the talks of REDISH and STEPHENSON at the Indiana Conference [6.126].

Further three-body effects may be present in (p,2p) reactions on light nuclei at energies below 150 MeV. Upon analyzing the α(p,2p) experiment between 65 and 600 MeV with the DWIA of (6.118), ROOS [6.134] finds that the spectroscopic factor

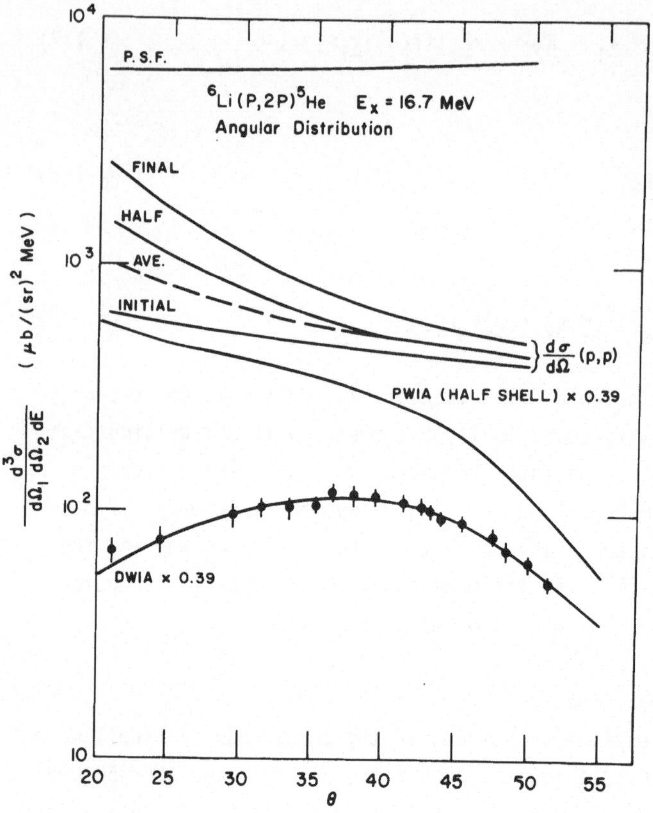

Fig.6.8. Nucleon knockout from ^{6}Li at 100 MeV. Factorized DWIA is shown using the half-shell prescription of (6.118). The PWIA, phase space factor (PSF) and various on- and off-shell p-p cross sections are also shown. The cross sections marked final, initial, and average are the on-shell p-p cross sections at ε_f, ε_i, and $(1/2)(\varepsilon_i+\varepsilon_f)$. The cross section marked half is the off-shell p-p cross section suggested by a three-body argument. From [6.133]

(ratio of theory to experiment) varies by almost a factor of two between 65 and 150 MeV. It remains reasonably constant as the energy increases above 150 MeV. A similar energy dependence was found by YOUNG and REDISH [6.56] in their model study of knockout (Subsect.6.1.5). Their DWIA/exact ratio varies with energy in the same way as does the DWIA/experiment ratio of Roos, but the size of the slope is smaller in the model than in the expermiment. This is what one would expect if many-body multiple scattering effects are creating the energy dependence. It would be extremely interesting to analyze these experiments with a more sophisticated three-body model.

6.2.4 Summary and Conclusions

Even in the study of the small number of reactions we have considered, there is considerable evidence for the presence of three-body effects.

For simple projectiles the presence of substantial reaction cross sections to three-body final states indicates the importance of three-body effects at energies above 100 MeV. Below that, optical model calculations with nucleons, pions, and kaons all indicate the need for at least including the three-body energy shift.

The presence of the sub-threshold resonance in $\bar{K}N$ scattering results in a strong modification of the kaon-nucleus optical potential. This is a particularly clear indicator that the shift occurs and is due to a three-body effect. In the area of knockout reactions from light nuclei, energy-dependent spectroscopic factors suggest the presence of more subtle three-body effects.

For deuteron projectiles, the three-body approach is very natural. Both Faddeev method model studies and non-Faddeev considerations of real data indicate that rearrangement and breakup effects play an important role in permitting a complete description of deuteron reactions.

In the study of reactions with three-body final states, the presence of three-body effects has been demonstrated both in the quasi-free and the final state interaction region. The circumstances under which these effects enter need to be understood if these reactions are to be a reliable tool for probing hadronic dynamics.

In a number of different circumstances discussed above, three-body methods have provided significant improvements in our understanding the dynamics of hadronic reactions and have led us to develop improvements of semi-phenomenological methods for describing these reactions. There is still a definite need for further investigation of these effects.

References

6.1 7th Intern.Conf. Few Body Problems in Nuclear and Particle Physics, Delhi, India, Dec.29, 1975-Jan.4, 1976, to be published

6.2 *Few Body Problems in Nuclear and Particle Physics*, ed. by R.Slobodrian et al. (University of Laval Press, Quebec 1976)

6.3 *Few Particle Problems*, ed. by I.Slaus et al. (American Elsevier, New York 1972)

6.4 *The Nuclear Many-Body Problem*, ed. by F.Cologero, C.Ciofi Degli Atti (Editrice Compositori, Bologna 1972)

6.5 J.S. Levinger: in *Nuclear Physics*, ed. by G. Höhler (Springer-Verlag, Berlin, Heidelberg, New York 1974)

6.6 E.W. Schmid, H. Ziegelmann: *The Quantum Mechanical Three-Body Problem* (Pergamon Press, Oxford 1974)

6.7 Y.E. Kim, A. Tubis: Ann. Rev. Nucl. Sci. 24, 69 (1974)

6.8 A.N. Mitra: Adv. Nucl. Phys. 3, 1 (1970)

6.9 V. Vanzani: Extended Seminar on Nuclear Physics, Sept.17-Dec.21, 1973, ICTP, Trieste

6.10 J.A. Maruhn, C.Y. Wong, T.A. Welton: Bull. Am. Phys. Soc. II, 20, 1158 (1975)

6.11 P.E. Hodgson: *Nuclear Reactions and Nuclear Structure* (Clarendon Press, Oxford 1971)

6.12 W. Tobocman: *Theory of Direct Nuclear Reactions* (Oxford University Press, Oxford 1961)

6.13 N. Austern: *Direct Nuclear Reaction Theories* (Wiley-Interscience, New York 1970)

6.14 D. Jackson: *Nuclear Reactions* (Methuen, London 1970)

6.15 F.S. Levin, H. Feshbach: *Reaction Dynamics* (Gordon and Breach, New York 1973)

6.16 A.N. Mitra: Phys. Rev. 139B, 1472 (1965)

6.17 L.R. Dodd, K.R. Greider: Phys. Rev. 146, 675 (1966)

6.18 E.O. Alt, P. Grassberger, W. Sandhas: Nucl. Phys. A139, 209 (1969)

6.19 J.V. Noble: Phys. Rev. 157, 939 (1967)

6.20 C.A. Edvi-Illes: Nucl. Phys. A164, 385 (1971)

6.21 D. Bouldin, F.S. Levin: Phys. Lett. 37B, 145 (1971)

6.22 E.F. Redish: Phys. Rev. C10, 67 (1974)

6.23 I.H. Sloan: Phys. Rev. C6, 1945 (1972)

6.24 Gy. Bencze: Nucl. Phys. A210, 568 (1973)

6.25 E.F. Redish: Nucl. Phys. A225, 16 (1974)

6.26 P. Grassberger, W. Sandhas: Z. Physik 220, 29 (1969)

6.27 A.N. Mitra, V.S. Bhasin: Phys. Rev. 131, 1265 (1963)

6.28 R.D. Amado: Phys. Rev. 122, 696 (1961)

6.29 L. Rosenberg: Phys. Rev. 135, B715 (1964)

6.30 R. Aaron, P.E. Shanley: Phys. Rev. 142, 608 (1966)

6.31 P.E. Shanley, R. Aaron: Ann. Phys. (N.Y.) 44, 363 (1967)

6.32 A.S. Reiner, A.I. Jaffe: Phys. Rev. 161, 935 (1967)

6.33 A.I. Baz, V.F. Demin, I.I. Kuz'min: Sov. J. Nucl. Phys. (U.S.A.) 4, 815 (1967)

6.34 C. Gignoux: J. Physique 30, 19 (1969)

6.35 D.P. Bouldin, F.S. Levin: Phys. Lett 42B, 167 (1972)

6.36 K. King, B.H.J. McKellar: Phys. Rev. Lett. 30, 562 (1973)

6.37 K. King, B.H.J. McKellar: Phys. Rev. C9, 1309 (1974)

6.38 B.H.J. McKellar: J. Phys. G: Nucl. Phys. 2, 180 (1975)

6.39 D.P. Bouldin: Ph.D. Thesis, Brown University 1972

6.40 M. MacFarlane: In *Proc. Intern. Conf. Properties of Nuclear States*, Montreal, Canada, August 1969, ed. by M. Harvey (U. de Montreal Press, Montreal 1969), p.385

6.41 S.T. Butler, R.G.L. Hewitt, B.H.J. McKellar, R.M. May: Ann. Phys. (N.Y.) 43, 282 (1967)

6.42 A.I. Baz, V.F. Demin, I.I. Kuz'min: Sov. J. Nucl. Phys. (U.S.A.) 4, 525 (1967)

6.43 R. Newton: J. Math. Phys. 1, 319 (1960)

6.44 I.H. Sloan: Phys. Lett. 34B, 243 (1971)

6.45 I. Lovas, J. Revai: Acta Phys. Acad. Sci. Hung. 25, 307 (1968)

6.46 P. Beregi, I. Lovas, J. Revai: Ann. Phys. (N.Y.) 61, 57 (1970)

6.47 P. Beregi, I. Lovas: Phys. Lett. 33B, 150 (1970)

6.48 I. Lovas: Magyar Fiz. Folyôirat 19, 209 (1971)

6.49 P. Beregi: Ref. [6.3], p.318

6.50 I. Lovas, E. Dénes: Phys. Rev. C7, 937 (1973)
6.51 P. Beregi, I. Lovas, Z. Physik 241, 410 (1971)
6.52 P. Beregi: Z. Physik 257, 440 (1972)
6.53 W.A. Friedman: Ann. Phys. (N.Y.) 45, 265 (1967)
6.54 H. Tyren, P. Hillman, Th.A.J. Maris: Nuovo Cim. 6, 1507 (1957)
6.55 G. Jacob, Th.A.J. Maris: Rev. Mod. Phys. 38, 121 (1966), 45, 6 (1973)
6.56 S.K. Young, E.F. Redish: Phys. Rev. C10, 498 (1974)
6.57 S.K. Young: Ph.D. Thesis, University of Maryland, 1973
6.58 R. Koshel: in Ref. [6.126]
6.59 T.K. Lim, I.E. McCarthy: Phys. Rev. 133, B1006 (1964)
6.60 P. Kazaks, R. Koshel: Phys. Rev. C1, 1906 (1970)
6.61 H.G. Pugh et al.: Phys. Lett. 46B, 192 (1973)
6.62 R.D. Amado, T. Takahashi: Phys. Rev. C12, 1134 (1975)
6.63 L.D. Rickertsen, P.D. Kunz: Phys. Lett. 47B, 11 (1973)
6.64 N.C. Francis, K.M. Watson: Phys. Rev. 92, 291 (1953)
6.65 H. Feshbach: Ann. Phys. (N.Y.) 5, 357 (1958); 19, 287 (1962)
6.66 A.K. Kerman, H. McManus, R.M. Thaler: Ann. Phys. (N.Y.) 8, 551 (1959)
6.67 J.S. Bell, E.J. Squires: Phys. Rev. Lett. 3, 96 (1959)
6.68 M. Goldberger, K.M. Watson: *Collision Theory* (Wiley, New York 1964)
6.69 E.F. Redish: Nucl. Phys. A235, 82 (1974)
6.70 F.E. Bertrand, R.W. Peele: Phys. Rev. C8, 1045 (1973)
6.71 J.R. Wu et al.: in *Clustering Phenomena in Nuclei*, U.S.E.R.D.A. publication ORO-4856-26 (1975), p.360
6.72 N.S. Wall, P.G. Roos: Phys. Rev. 150, 811 (1969)
6.73 D. Corley et al.: Nucl. Phys. A184, 437 (1972)
6.74 F.R. Kroll, N.S. Wall: Phys. Rev. C1, 138 (1970)
6.75 P.C. Tandy, E.F. Redish, D. Bollé: Phys. Rev. Lett. 35, 921 (1975)
6.76 P.C. Tandy, E.F. Redish, D. Bollé, in Ref. [6.1], and University of Maryland Technical Rpt., to be published
6.77 G.M. Lerner, E.F. Redish: Nucl. Phys. A193, 565 (1972)
6.78 G.M. Lerner: Ph.D. Thesis, University of Maryland, 1971
6.79 R.V. Reid: Ann. Phys. (N.Y.) 50, 411 (1968)
6.80 R.S. Mackintosh, A.M. Kobos: "The Real and Imaginary Proton Optical Potential: The Importance of Deuteron Channels", preprint
6.81 R.S. Mackintosh: Nucl. Phys. A230, 195 (1974)
6.82 E. Kujawski, M. Aitken: Nucl. Phys. A221, 60 (1974)
6.83 R.H. Landau, A.W. Thomas: Phys. Lett. 61B, 361 (1976)
6.84 P. Desgrolard, T.F. Hammann: Phys. Rev. C6, 482 (1972)
6.85 R. Seki, C. Wiegand: Ann. Rev. Nucl. Sci. 25, 241 (1975)
6.86 M. Alberg, E.M. Henley, L. Wilets: Ann. Phys. (N.Y.) 96, 43 (1976)
6.87 M. Krell: Phys. Rev. Lett. 26, 584 (1971)
6.88 R. Seki: Phys. Rev. C5, 1196 (1972)

6.89 M. Krell, T. Ericson: J. Comput. Phys. 3, 202 (1968)
6.90 A.R. Kunselman, R. Seki: Phys. Rev. C8, 2492 (1973)
6.91 S.D. Bloom, M.H. Johnson, E. Teller: Phys. Rev. Lett 23, 28 (1969)
6.92 J.K. Kim: Phys. Rev. Lett. 19, 1074 (1967)
6.93 W.A. Bardeen, E.W. Torigo: Phys. Lett. 38B, 135 (1972)
6.94 V. Chaloupka et al.: Phys. Lett. 50B, 1 (1974)
6.95 J. Negele: Phys. Rev. C1, 1260 (1970)
6.96 G. Baur, D. Trautmann: Nucl. Phys. A208, 261 (1973)
6.97 Gy. Bencze: Ann. Acad, Sci. Fennicae, Ser. A. 194 (1966)
6.98 Gy. Bencze, I. Szentpetery: Phys. Lett. 30B, 446 (1969)
6.99 J. Testoni, L.C. Gomes: Nucl. Phys. 89, 288 (1966)
6.100 K. Ueta, K. Hara: Z. Physik 248, 311 (1971)
6.101 R.C. Johnson, P.J.R. Soper: Phys. Rev. C1, 976 (1970)
6.102 J.D. Harvey, R.C. Johnson: J. Phys. A. 7, 2017, (1974)
6.103 R.C. Johnson, P.C. Tandy: Nucl. Phys. A235, 56 (1974)
6.104 G.H. Rawitscher: Phys. Rev. 163, 1223 (1967)
6.105 G.H. Rawitscher: Phys. Rev. Lett. 20, 673 (1968)
6.106 G.H. Rawitscher, S.N. Mukherjee: Phys. Rev. 181, 1518 (1969)
6.107 G.H. Rawitscher, S.N. Mukherjee: Ann. Phys. (N.Y.) 68, 57 (1971)
6.108 S. Watanabe: Nucl. Phys. 8, 484 (1958)
6.109 G.H. Rawitscher: Phys. Rev. C9, 2210 (1974)
6.110 G.H. Rawitscher: Phys. Rev. C11, 1152 (1975)
6.111 J.P. Farrell Jr., C.M. Vincent, N. Austern: Ann. Phys. (N.Y.) 96, 333 (1976)
6.112 G.L. Wales, R.C. Johnson: University of Maryland Technical Rpt. 76-127, to be published
6.113 J.P. Farrell, Jr.: Ph.D. Thesis, University of Pittsburgh, 1974
6.114 G.R. Satchler: Phys. Rev. C4, 1485 (1971); also see correction in Ref. [6.112]
6.115 M.D. Cooper, W.F. Hornyak, P.G. Roos: Nucl. Phys. A218, 249 (1974)
6.116 P.G. Roos et al.: Nucl. Phys. A255, 187 (1975)
6.117 J.L. Gammel, B.J. Hill, R.M. Thaler: Phys. Rev. 119, 267 (1960)
6.118 P.E. Shanley: Phys. Rev. 187, 1328 (1969)
6.119 K.A.-A. Hamza, S. Edwards: Phys. Rev. 181, 1494 (1969)
6.120 Gy. Bencze: "Three-Body Model Calculations for the Deuteron Nucleus Collision", unpublished preprint, Finnish Summer School, Liperi, 1973
6.121 Gy. Bencze: Nucl. Phys. A196, 135 (1972)
6.122 L. Rosenberg: Phys. Rev. C13, 1406 (1976)
6.123 K.H. Yang, W.J. Gerace, J.F. Walker: Nucl. Phys. A240, 189 (1975); A242, 365 (1975)
6.124 K.H. Yang: Ph.D. Thesis, University of Massachusetts, Amherst, 1975
6.125 R.D. Amado, J.V. Noble: Phys. Rev. 185, 1993 (1969)
6.126 Symp. Momentum Space Wave Functions in Atoms, Molecules, and Nuclei, Bloomington, Indiana, ed. by R. Bonham et al., May 31-June 4, 1976, to be published

6.127 R.G. Newton: *Scattering Theory of Waves and Particles* (McGraw Hill, New York 1966)

6.128 A.M. Gleeson, W.J. Meggs, M. Parkinson: Phys. Rev. Lett. 25, 74 (1970)

6.129 Z. Gromadzki, J.V. Noble: Phys. Lett. 51B, 9 (1974)

6.130 E.F. Redish: Phys. Rev. Lett 31, 617 (1973)

6.131. E.F. Redish, G.J. Stephenson, Jr., G.M. Lerner: Phys. Rev. C2, 1665 (1970)

6.132 G.J. Stephenson, Jr., E.F. Redish, G.M. Lerner, M.I. Haftel: Phys. Rev. C6, 1559 (1972)

6.133 R.J. Bhowmik, C.C. Chang, P.G. Roos, H.D. Holmgren: Nucl. Phys. A226, 365 (1974)

6.134 P.G. Roos: Phys. Rev. C9, 2437 (1974)

6.135 V. Vanzani: Nuovo Cim. 16A, 449 (1973)

6.136 N.S. Wall: Univ. of Maryland Technical Rpt. 76-060, 1975, unpublished

6.137 T. Revai: Nucl. Phys. A205, 20 (1973)

6.138 J.D. Harvey, R.C. Johnson: Phys. Rev. C3, 636 (1971)

Subject Index

SPRINGER TRACTS IN MODERN PHYSICS

Strong Interaction Physics

Heidelberg-Karlsruhe International Summer Institute in Theoretical Physics (1970)

Vol.57 (1971), Pp.270

With contributions by: *D. Atkinson, R.A. Brandt, A.P. Contogouris, J. Hamilton, R. Oehme, W. Rühl, H.R. Rubinstein, H. Satz, E.J. Squires, K. Symanzik, G. Wanders, J. Zinn-Justin*

Symposium on Meson-, Photo-, and Electroproduction at Low and Intermediate Energies

Bonn, September 21-26, 1970

Vol.59 (1971), Pp.222

With contributions by: *M. Ademollo, H. Fischer, L. Foà, G. von Gehlen, G. von Holtey, D. Lüke, D. Schwela, P. Söding, C. Verzegnassi, G. Wolf*

Photon-Hadron Interactions I

International Summer Institute in Theoretical Physics DESY, July 12-24, 1971

Vol.62 (1972), Pp.147

R. Jackiw: Canonical Light-Cone Commutators and Their Applications

H.D. Dahmen: Local Saturation of Commutator Matrix Elements

P.V. Landshoff: Duality in Deep Inelastic Electroproduction

C.H. Llewellyn Smith: Parton Models of Inelastic Lepton Scattering

H.R. Rubinstein: Duality for Real and Virtual Photons

V. Rittenberg: Scaling in Deep Inelastic Scattering with Fixed Final States

K. Huang: Duality and the Pion Electromagnetic Form Factor

K. Huang: Deep Inelastic Hadronic Scattering in Dual-Resonance Model

G. Furlan, N. Paver, C. Verzegnassi: Low Energy Theorems and Photo- and Electroproduction Near Threshold by Current Algebra

Photon-Hadron Interactions II

International Summer Institute in Theoretical Physics DESY, June 12-24, 1971

Vol.63 (1972), Pp.189

J. Frøyland: High Energy Photoproduction of Pseudo-scalar Mesons

K. Schilling: Some Aspects of Vector Meson Photoproduction on Protons

D. Schildknecht: Vector Meson Dominance, Photo- and Electroproduction from Nucleons

F.M. Renard: ρ-ω Mixing

A. Donnachie: Exotic Electromagnetic Currents

A.P. Contogouris: Regge Analysis and Dual Absorptive Model

P.D.B. Collins, F.D. Gault: The Eikonal Model for Regge Cuts in Pion-Nucleon Scattering

Vol.65 (1972), Pp.145

H. Theissen: Spectroscopy of Light Nuclei by Low Energy (<70 MeV) Inelastic Electron Scattering

H. Arenhövel, H.J. Weber: Nuclear Isobar Configurations

K. Heinloth: Experiments on Electroproduction in High Energy Physics